METEOROLOGY IN NINETEENTH-CENTURY SOCIETY

Nineteenth-Century Science, Technology and Medicine: Sources and Documents

Forthcoming:

Evolution in Victorian Culture
4 volume set
Edited by Piers Hale

Nineteenth Century Euclidean Geometry
4 volume set
Edited by Alice Jenkins

Meteorology in Nineteenth-Century Society
4 volume set
Edited by Aitor Anduaga

Nineteenth-Century British Microscopy
4 volume set
Edited by Meegan Kennedy

Nineteenth-Century Natural Theology
4 volume set
Edited by Adam R. Shapiro

The Science of Mind in the Nineteenth Century
4 volume set
Edited by Stephen T. Casper & Jon D. Goss

Science and Sound in Nineteenth-Century Britain
4 volume set
Edited by Edward J. Gillin

Scientific Methodology in Nineteenth Century Britain
4 volume set
Edited by Charles H. Pence

Women in Medicine in the Long Nineteenth Century
4 volume set
Edited by Claire Brock

Insects in Victorian Culture
4 volume set
Edited by John Clark

Electric Power in Victorian Britain
4 volume set
Edited by Nathan Kapoor

Women, Gender and Science in Victorian Culture
4 volume set
Leila McNeill & Anna Reser

METEOROLOGY IN NINETEENTH-CENTURY SOCIETY

Edited by
Aitor Anduaga

Volume I
Weather Forecasting and Nation-State Building

Routledge
Taylor & Francis Group
LONDON AND NEW YORK

First published 2026
by Routledge
4 Park Square, Milton Park, Abingdon, Oxon OX14 4RN

and by Routledge
605 Third Avenue, New York, NY 10158

Routledge is an imprint of the Taylor & Francis Group, an informa business

British Library Cataloguing-in-Publication Data
A catalogue record for this book is available from the British Library

ISBN: 978-1-032-54825-8 (set)
ISBN: 978-1-032-54827-2 hbk (vol I)
ISBN: 978-1-003-65029-4 ebk (vol I)

DOI: 10.4324/9781003650294

Typeset in Times New Roman
by Apex CoVantage, LLC

CONTENTS

ACKNOWLEDGEMENTS

I would like to thank Amaia Guerrero and Usua Arrizabalaga at the Main Library of the University of the Basque Country and Jon Euba at the Library of the University of Deusto for their help in locating and providing copies of some of the hard-to-find primary sources included in this four-volume collection. I am also grateful to Dr. Anton Erkoreka and Dr. Begoña Madarieta, director and research fellow, respectively, of the Basque Museum of the History of Medicine and Science at the University of the Basque Country, which enabled me to devote time to conceptualising the sections and preparing the lengthy editing work. Finally, I would also like to express my most sincere gratitude to Rachel Douglas and Simon Alexander at Routledge, to Marie Roberts for her excellent copy-editing work and her kindness, and to Meegan Kennedy and Piers Hale as General Editors for the new series of Routledge Historical Resources: Nineteenth-Century Science, Technology and Medicine, for their confidence and valuable help from the very beginning to the final completion.

Research was supported with funding from the Basque Government (IT1441–2022) to the Basque University System's Biography & Parliament Research Group and from the project "Epistemic Network: Between locality and globality" (PID2023–147611NB-I00), funded by the Ministry of Science, Education and Universities (Spain) (MICIU/AEI/10.13039/501100011033) and by FEDER/UE.

SERIES PREFACE

Nineteenth-Century Science, Technology and Medicine: Sources and Documents

GENERAL EDITORS' INTRODUCTION

This volume is part of the Routledge Historical Resources primary source collection, "Nineteenth-Century Science, Technology and Medicine: Sources and Documents." This series offers a fresh look at the historical development of science, technology, and medicine through the eyes of subject experts and specialists from around the world. The series provides curated access to significant primary-source documents on these subjects, collating and contextualizing these valuable resources so that they may ground not just scholarly research but undergraduate and graduate research and teaching as well. As well as being published in print and ebook formats, the contents of this and all other volumes in this series will be available through the Routledge Historical Resources database, where they will be presented in a fully integrated and searchable format, together with supplemental secondary sources and images and newly commissioned scholarly written and video essays.

Our editors have framed their volumes around themes they consider most pertinent to understanding nineteenth-century developments in their chosen field and its significance to broader changes in nineteenth-century British culture and society. Often, the editors have chosen to provide short extracts to illustrate particular points or themes; in other cases, they have provided entire essays, book chapters, or periodical articles as appropriate. Each volume is prefaced by an introductory essay in which the editors review their topic and discuss the significance of their sources. They also frame each entry or collection of related entries with explanatory headnotes. Thus, each volume or set of volumes will provide a useful resource of easy-to-navigate, well-contextualized sources curated by an expert eye, which will serve as a useful introduction for those unfamiliar with the field. Since the editors build each volume to represent a unique, deeply considered scholarly perspective on the foundational primary sources for a particular subject, the series should also prove valuable to established scholars, even those with deep experience in the field. Indeed, given the range of topics covered and sources collated in this series, we anticipate that this project will foster an exciting and insightful interdisciplinary perspective on the histories and cultures of nineteenth-century

British science, technology, and medicine. We as general editors have learned a lot through our engagement with the editors who have contributed to this project, and we believe that wider audiences will, too.

Throughout the project we have asked editors to comment on the state of the existing historiographical debate, and to specify where their own work fits in. Some editors have chosen to do this through the provision of recommendations for further reading; others do so through explicit commentary on this topic in their introductory essays, and in the headnotes and footnotes to each of the sources they have selected.

The histories of science, technology, and medicine are wide and often disparate fields, even though they are often encountered by students in the context of a single university department. Although they have much in common, scholars of science, technology, and medicine also propose field-specific questions and approaches; these are represented in the project through the disparate interests of the range of scholars contributing to the series. We have also worked to bring the insights of literary scholars as well as historians to the project.

We have made every effort to cover major nineteenth-century themes and topics in the scientific and technological disciplines as well as to give space to specialist topics. Of course, as is inevitably the case in a project of this nature, which must rely on the availability of expert scholars in particular fields, there will remain gaps. We have attempted to address these where we can in the form of introductory essays and video essays in the database. We do believe, though, that the reader and researcher will find that this project presents the major themes, as well as many specialist topics, in detail, and with scholarly rigour.

We are very grateful to Routledge for their invitation to take on this project, and we would especially like to thank the commissioning editor Rachel Douglas, her successor, Laura Pilsworth, and Simon Alexander. Most of all, however, we are grateful to the extensive list of scholars from around the world who have agreed to contribute, and who make this series what it is.

Meegan Kennedy and Piers J. Hale

GENERAL INTRODUCTION

Meteorology in Nineteenth-Century Society

The last twenty-five years, and especially the last decade, have seen a "paradigm turn" in the study of the history of meteorology. Whereas in the twentieth century meteorology tended either not to be a subject worthy of historiographical study or was often seen – from a presentist perspective – as a field evolving in a linear and teleological progression towards current knowledge, it is now taken as a subject of study *in extenso* and from a multidimensional perspective.[1] The turn is *paradigmatic* insofar as it entails a broad rethinking and restructuring of analytical frameworks, which are now aligned with interdisciplinary integration, societal considerations and spatial analysis. This can be attributed, in part, to the increasing cross-fertilisation between Science and Technology Studies (STS) and environmental history following its emergence as a distinct discipline, and also in part to growing concern about anthropogenic climate change.[2] Since the fin-de-siècle evidence for global warming, the standard histories of atmospheric science no longer appear plausible. As predictions of future climate have led to calls for bolder societal action to mitigate the resulting environmental damage, scholars have seen the need to understand the deep intertwining of atmospheric science with climate change in its political, social and cultural dimensions. This has had particular significance for the study of nineteenth-century meteorology since it has traditionally been portrayed as a Baconian fact-gathering activity on *both* weather and climate. As Martin Mahony and Angelo Matteo Caglioti have argued, this has contributed to the 'interdisciplinary, social and spatial turn' of new scholarship in the history of meteorology.[3] Reflecting this turn, against a backdrop of growing public interest and awareness of climate change, 'the focus has shifted from scientific theories about the weather to meteorological practices, spaces, material and instrumental cultures'.[4] However, this turn towards what these two authors define as 'relocating meteorology' needs to be nuanced.

Applied to the shifting focus on practices, spaces, cultures, experiences and the shaping of the social world, the choice of the term 'relocation' may seem logical. However, as James R. Fleming and Vladimir Janković have pointed out, the history of the atmospheric sciences has been and 'is always about studying society, vicariously or not'.[5] The first systematic attempts to elucidate the historical development of meteorology took place in the twentieth century and came from scientists (not historians) who tended to focus on theory-builders, notorious controversies and the anatomical dissection of emblematic institutions, almost always in a decontextualised way. Meteorology did not seem to interest historians. As Stefano Casati and Marco Ciardi noted in their entry on the disciplinary history of meteorology in Gregory Good's 1998 magnum opus, *Sciences of the Earth: An Encyclopedia of Events, People, and Phenomena*, 'meteorology is generally neglected by historians of science' and its literature is 'really meager'.[6] However, the few historians of science who problematised and historicised meteorology sought, with varying degrees of depth and success, to understand, if not completely, then at least in some respects, its imbrication with the politics, culture and society of the time.

Let us consider several examples. Although Robert M. Friedman's 1989 work on Vilhelm Bjerknes and the construction of modern meteorology is often exemplified as a study of theory-building, it is indeed much more than a biography of a school builder: it shows how the polar front was constructed not only from the theorisation but also the instrumental practice of the Bergen meteorologists.[7] In his influential 1990 book, *Meteorology in America, 1814–1874*, James R. Fleming argues convincingly that meteorology is not just about the correct rational understanding of weather phenomena but about the formulation of a science with theoretical, observational and institutional dimensions, in which theorists, observers and administrators interacted to address a variety of problems embedded in a specific American social context.[8] Frederik Nebeker's no less influential *Calculating the Weather* (1995) is often cited for demonstrating an analytical framework according to which twentieth-century numerical methods unified disparate strands from three meteorological traditions: empirical (climatology), theoretical (atmospheric physics) and practical (weather forecasting). Less well known, however, is that the work provides interesting insights into the ways in which military and institutional factors, such as operational application in the Second World War, drove the science towards computational approaches.[9] Although Mark Monmonier's 1999 work on how meteorologists learned to map, predict and dramatise the weather is likely to disappoint those interested in a social history of weather mapping, since it dispenses with – among others – issues such as astrometeorology, prophecy and weather rules, it does meritoriously analyse the practices of visual representation by exploring the heuristics and symbolism of such central tools for the meteorologist as isobar, isotherm or wind-direction arrow.[10] When it comes to demonstrating how truth-making claims about the weather were entangled with the cultural and political world of the proponents, these works probably have little to offer. For the present discussion, the importance of their contribution lies not in the cultural conditioning to the achievements under consideration, but in their formulation of a set of links of material practices to specific historical and political contexts that turned them into expressions that were by no means timeless or suprahistorical.

When expanded by new multidimensional perspectives – Mahony and Caglioti, we should recall, have in mind to incorporate emerging historiographical trends, geographical spaces and innovative concepts – the epistemic turn can in many respects be seen as a delocalisation – rather than a relocation – of meteorology. That they considered it a relocation is because, as both authors acknowledge, they were inspired by Kapil Raj's seminal work *Relocating Modern Science,* in which he argues for a change of focus – hence the title – in the way we understand the diffusion of the so-called Western sciences;[11] thus, Raj challenges the traditional diffusionist model of linear transfer of knowledge from the metropolis to the periphery, proposing instead spaces of circulation and processes of intercultural encounter in the production and transformation of knowledge.[12] After all, "relocating" implies moving from one specific focus to another. However, emerging scholarship in the history of meteorology does not simply move to a new specific focus but also broadens its epistemic scope through two main actions: embracing a plurality of perspectives and decentralising the context. First, the shift from national narratives and careers of knowledge-leaders and institution-builders to focusing on environmental, cultural, sociological and geographical histories means moving away from a geographically or ideologically fixed centre. This was clearly so among the historians, sociologists and anthropologists of environmental knowledge who gathered at the conference 'Weather, Local Knowledge and Everyday Life' in Rio de Janeiro in 2008. Its organisers, Janković and Christina Barboza, sought to release weather 'from disciplinary shackles' and urged participants to 'see weather as lived rather than theorized'. Their hope, as they state, was nothing less than to launch 'a longitudinal and global approach to weather experiences' as they are produced or masked in 'routines, rituals and everyday practices'.[13] In the context of nineteenth-century meteorology, the examination of "secondary" voices, such as seamen, military personnel, engineers and insurance clerks, is of special importance given the practical application of knowledge that characterises the experience of weather. Secondly, "de-localising" suggests breaking down the boundaries of dominant theories and national narratives and adopting a decentralised and networked understanding of meteorology as a discipline shaped by social, environmental and global dynamics. This is evident, for example, among the indigenous employees who worked in the meteorological and astronomical observatories of European colonies. As Fiona Williamson points out, the indigenous contribution was particularly important at the Hong Kong Observatory, since the effectiveness of meteorology lay in the ability to collect and process large amounts of instrumental data.[14] By delocalising, the history of meteorology shifts from a narrative of centralised achievements to a cross-cultural and networked history.

A second reason for qualifying (though by no means rejecting) the proposed "relocation of meteorology" is the appeal made by its proponents to place it in a broader, global and imperial context. In contrast to the prevailing twentieth-century view of meteorology as an acontextual and placeless enterprise, the imperial and global contexts of science are increasingly under scrutiny. It is worth remembering, however, that diffusionist historians – notably Lewis Pyenson, perhaps the leading exponent of this theory in the exact sciences, including the observatory sciences – already dealt in the 1980s with what might be called the empires of science.[15] To begin with, he reminds us that the peripheral science developed in the observatories of the European colonies was merely an extension of the metropolitan science and argues that this transmission of knowledge took place through two types of vectors: the 'functionary' and the 'seeker'. While the functionary sought to integrate his original European institution into the social fabric of his destination, drawing on science as a mere stimulus, the seeker aimed to use his host institution to produce and publish original research.[16] According to Pyenson, the Jesuits, and many other functionaries, were not "genuine" seekers, so to speak, or authentic researchers engaged in "pure" cognitive enquiry.[17] Rather, they were committed to spreading Catholic doctrine in the missions, and it was only with great difficulty and internal conflict that they managed to reconcile their civil service role with the task of research.[18] As Pyenson's analysis reveals, the missionary and colonialist function of functionaries and seekers was of crucial epistemological significance. The role of both the functionary and the seeker in the empire counted in the conception and perception of science as a means or an end. The Jesuits in the mission observatories regarded science as a stimulus, an incentive, a means to higher ends, and often cultivated it in a quasi-self-taught or circumstantial way. On the contrary, seekers in colonial observatories considered it as an end in itself and per se, that which, *stricto sensu,* was their *raison d'être.* According to Pyenson, the function assigned to a mission or a colony qualified the author of meteorological or astronomical observations at a given observatory as a 'pure' researcher or not. Scientific knowledge was thus located in both missionary and imperial spaces.

Recent scholarship on the history of meteorology and the geophysical sciences, however, has refuted Pyenson's diffusionist model of knowledge in the observatory sciences by highlighting the similarities between the goals of physical inquiry in the scientific communities involved in colonies and missions, and by emphasising the role of epistemic networks in the processes of knowledge circulation and globalisation.[19] Thus, these recent studies have shown how the Spanish and French Jesuits in the observatories of these missions, stimulated by local social and commercial elites in Havana, Manila and Shanghai, made scientific research an end in itself, a duty of conscience and social justice.[20] At the same time, these studies have demonstrated how profoundly missionary science was embedded in a wider culture of scientific knowledge adaptation. In the field of tropical cyclone forecasting, for example, they have shown how Jesuits in mission observatories proceeded in two ways, first by projecting shared expertise to the missions and then by adapting it to specific local contingencies and needs within a cycle of exchanges. The evidence gathered reveals that the Jesuits' predictive probabilism, through the search for correlations between different physical phenomena (both atmospheric and terrestrial), was indeed the heuristic reflection of their adaptive capacity. Nothing could be further from the idea of a centre of calculation through which the Western dominance of knowledge over the periphery was produced, as advocated by diffusionist theories, than the Jesuits' ability to adapt, modify and produce methods of cyclonic prediction in local and missionary contexts.[21] While the perspectives of these studies may open up new avenues in the study of missionary sciences and colonial and imperial sciences, they also provide a point of reflection to question – or at least refine and sharpen – certain reasons adduced for the call for a "relocation of meteorology". The important practical aspect of these studies is that what characterises Pyenson's work is not the lack of attention to imperial and international contexts, as evidenced by his works on French and German overseas expansion, but rather the lack of consistency and congruence of his diffusionist perspective focused on the centrifugal spread of Western science throughout the world.

Until the late eighteenth century, meteorology was preoccupied with extraordinary meteorological events, agricultural seasons, and qualitative predictions.[22] Whether it was the prognostics of storms, the augury of earthquakes, the expectation of waterspouts or the announcement of the aurora borealis, in short, the various "meteors" – understood in the Aristotelian sense – and their fleeting and unstable nature, the concern was to report on the phenomena (meteoric reportage), to discern the trace of divine providence, and to vindicate the predictive – and therefore social – power of the traditional rules of the weather.[23] As Janković holds, 'meteors, weather, and seasons were among the most powerful forces informing the physical, moral, financial, and political landscapes of the eighteenth century'.[24] Things changed substantially in the nineteenth century. By the 1810s, observations of unusual meteorological events had been replaced by the more tedious tables of means of temperature and barometric pressure in Britain's most widely read scientific periodicals. The attraction for the extraordinary and the infrequent gave way to a concern for the systematic and the measured. Meteorology now focused on topics such as massive data collection and systematic analysis; the combination of thermodynamics, fluid dynamics and optics to understand atmospheric phenomena; and the synoptic forecasting of weather and major cyclones.[25] Interest shifted, as it were, from the discontinuity of meteors to the continuum of weather; from particularity and eventuality to generality and periodicity; from local pursuit to global inquiry; from powerful reportage to the power of isobars and storm laws; from prognostics and portents to forecasts and warnings.[26] In this century, increasing attention was paid to a whole suite of new issues rotating, in various ways and to a greater or lesser extent, around such topics as the need for standardisation, the uniformity of observing systems, the comparison of data and records, the simultaneity of observations, and the transformation of local observations into more general terms, such as the reduction of sea level adjusted values.[27] Just as the science of weather ousted the meteoric tradition and scientific expertise displaced amateur experience, so too did quantitative and synoptic meteorology replace qualitative and speculative meteorology.

In analysing what meteorology meant in the nineteenth century, we are confronted with the difficulty of the deep imbrication of what weather science and climate science meant to society in this period. Scientific meteorology, as a science devoted to the study of the condition of the atmosphere and its variability in time and place, was cultivated primarily, though not exclusively or omnipresently, in observatories, educational institutions and scientific associations. Its cultivators, usually statistically oriented scientists and empirically minded natural philosophers, tended to present climate as the average weather. By this, they meant the statistical description of meteorological quantities in terms of mean and variability, independent of its human dimension. However, in nineteenth-century Western climate discourses, climate was seen not only as a statistical index of average parameters, but also as an agent or force with a powerful influence on all domains of social life. Indeed, this second conception has been the more dominant over the centuries, and the historical concern has been with 'what it *does* rather than what it *is*'.[28] As Fleming and Janković have argued, the definition of climate 'as a statistical index is an anomaly' in time and is only comprehensible 'in connection to an instrumental, quantitative, weather-biased understanding of the atmosphere'.[29] Those who conceived of climate as an agent were united by their concern with health, human physiology, geography and everyday life, and ranged from naturalists who sought to relate climate to social welfare through the study of medical topographies, and physicians interested in the influence of weather on epidemics and health in general, to geographers with agricultural and demographic concerns, and colonists who argued about the climatological reasons for racial, moral and political differences between regions.[30] Although rich in tradition and worthy of study, this vast climate-related collective is not an integral part of modern scientific meteorology and, as such, has not been included among the sources extracted in this compilation of primary sources.

The historical relationship, that is, in the nineteenth century, between the science of weather and the science of climate has important implications for how the category of meteorology was understood at the time. For the cultivators of scientific meteorology, the weather and climate sciences were in many respects interdependent and mutually constitutive until the middle of the century. Attempts by statistically oriented scientists in the first half of the century to analyse the variability of the weather and its deviations from mean states as part of the study of climate anticipate some of the arguments used in the second half of the century in the search for laws and rules governing storm phenomena. The transition from the search for the patterns and laws behind periodic regularities to the quest for the laws governing non-periodic phenomena, which took place in the middle decades of the century, was not without tension. While statistically oriented scientists were mostly devoted to the former – analysing historical weather records – for their part, empirically minded natural philosophers and seamen focused on the latter – studying short-term, more localised and immediate phenomena. In the second half of the century, with the study of storms and cyclones, the paths diverged (although in many ways they remained interdependent): meteorology set eyes on weather forecasting and climatology expanded into global studies.

That nineteenth-century scholarship has found Victorian meteorology so fascinating is, at least in part, because it magnificently incarnates the flow of influence and power from science to society – and vice versa. The nineteenth century, and the Victorian period in particular, inherited a deep sense of national climate that had emerged in the eighteenth century from the systematic recording of weather.[31] This science asserted its authority and predictive power in maritime economics, nation-state building, and the expansion of the imperial project.[32] It found its cultural battle in the debates that followed the establishment in 1854 of a government office originally founded to collect meteorological statistics, and which, when the office subsequently turned to forecasting, revealed its vulnerability to the scientific elite and to public expectations.[33] At the same time, it operated as a knowledge-creating force that acted intrinsically in colonial enterprises, both shaping and being shaped by inter-colonial and imperial politics. Ruth Morgan has noted 'the importance of predicting the future for the realisation of the colonial project' in the context of forecasting drought and famine in British India and the Australian colonies.[34] Beyond the national and imperial framework, however, Victorian meteorology must be placed in a wider, continental, international and global context of the quest for order, security and centralisation in processes of nation-state building articulated not only in Britain but also in other European countries. Meteorological practices, material and instrumental cultures and the articulation of scientific values in society also need to be examined in international and global contexts, since they often attest to Victorians' impressions of and responses to weather science, rather than weather science per se.[35]

WEATHER FORECASTING AND NATION-STATE BUILDING

In meteorology, studying weather – understood as 'the condition of the atmosphere at any given time and place' – means studying processes and flows that are highly dynamic and changing, rather than phenomena that are localised to particular sites.[36] 'Weather' thus becomes something that is intrinsically three-dimensional and changing: the defining elements of weather are its spatiality in the Earth's atmosphere and its variability in time and space. Historical studies of weather and climate knowledge tend to be presented in static, horizontal scenarios of observation and surveillance: observation systems, networks of observatories and stations, hubs, fixed coordinates, centres of calculation in accordance with actor-network theory, etc., often in conjunction with developments of cartographic techniques through two-dimensional isobaric weather maps, or with historiographies of national and imperial progress through the installation and expansion of weather stations. Both time and the vertical dimension, as spaces of action and objects of inquiry, have been largely overlooked in these discursive traditions. However, the history of weather forecasting is an ideal subject for investigating not only the horizontal expansion and two-dimensional mapping of weather and climate sciences but also its imbrications, both conceptual and social, with questions of verticality, territoriality, autonomy, security and order. Recent studies on nineteenth-century meteorology and weather forecasting shed light on the social aspects behind observing systems and organisation and point to the role of certain social and political factors in shaping the practices and methods of weather forecasting in this period.

In the science of meteorology, weather forecasting engages with the study of air motion or fluid dynamics. Theorists distinguish between two ways of describing fluid flows. In the so-called Eulerian approach, the properties of fluid flow are expressed as 'fields' and each field is expressed as a function of space and time. The object of study is the pressure, velocity, etc. of a fluid at a given location, not the position or velocity of a particular particle. This approach produces isobars and wind streamlines, which are usually plotted on a map. In contrast, the Lagrangian approach tracks individual fluid particles, not the fluids themselves, and expresses their velocities, positions, etc. only as a function of time. Instead of maps, this approach produces pathways and trajectory analysis.[37]

The reasons why weather forecasting was particularly susceptible to being shaped by specific social factors lie in the very intrinsic nature of the Eulerian and Lagrangian approaches. Throughout the 1850s and 1860s, when the methodological foundations of weather forecasting practice were being laid in Europe, meteorological observations were mostly surface-based, direct high-altitude measurements were rare, and there were hardly any permanent high-altitude weather stations. With its spatially fixed coordinate system and its focus on routine surface observations, the Eulerian approach had characteristics that made it particularly suited to observing systems that were centralised and standardised. This was in stark contrast to the characteristics of the Lagrangian approach, which was typically associated with non-routine asymptotic observations and familiar with observing systems that were not easily controllable and manageable. Furthermore, the Lagrangian approach seems to denote freedom, imagination and autonomous observation systems. What was at stake, therefore, was not a mere choice of approach but the very imbrications of science, knowledge and power: while the use of the Lagrangian approach implies unmanageability and verticality, the use of the Eulerian approach brings with itself territoriality and horizontality.[38] This fact, which until recently had gone unnoticed by environmental historians, geographers, historians, and sociologists of science, allows us to study interesting transitions between the scientific and social worlds.

Weather forecasting is a knot at the intersection of the relationship between science, politics, and society. Its origins date back to the 1840s, when a number of opposing theories of storms based on mechanical and thermal principles were elaborated in Europe and the United States. The so-called 'storm controversy' between the whirlwind theory and the convective theory later gave rise to the proposal of two wave-like solutions: John Herschel and William Birt's atmospheric waves in Britain and Elias Loomis's cold waves in America.[39] Statistical methods were applied to meteorology in the pre-synoptic era – i.e., before the first maps of weather patterns in the mid-1850s. However, one key event changed the history of predictive meteorology.[40] In 1854, the Black Sea storm led Urbain Le Verrier of the Paris Observatory to adopt an Eulerian approach to weather forecasting.[41] State interests rather than scientific reasons pushed him to take this approach: the new weather maps of isobars (on which the Eulerian method was based) were not only able to reveal weather patterns but also served as a tool to achieve order, security and centralisation – especially in France and Italy. Interestingly, in Britain, a new approach – the Lagrangian one – acted as a counterweight to the Eulerian approach. It was characterised by a quest for autonomy and individual judgment in weather forecasting.[42] Thus, Robert FitzRoy, George Jinman and Francis Galton proposed storm flow-models and iconic – instead of isoline-based – weather maps. FitzRoy's attempts to forecast the weather by combining empirical rules and instrumental readings are worth mentioning.[43] He championed a Lagrangian approach that was diametrically opposed in forms and purpose to the methods used in continental Europe. Although not exclusively British or representative of British meteorological science – the Royal Society came to doubt his methods to the point of suspending forecasts after his death – the Lagrangian approach was the one that initially set the pace for weather prediction in Britain.

In the long run, however, his efforts were in vain. The convergence between statistical science and the synoptic method of isobaric weather charts led to the hegemony of the Eulerian approach from the 1870s to the 1910s in Britain and elsewhere. This convergence was partly driven by the analytical method of weather maps devised by Alexander Buchan in Scotland. The search for "weather types" and isobaric forms characterised synoptic meteorology in the 1880s and 1890s. In Britain, the head of the Meteorological Office, Robert H. Scott, welcomed Buchan's analytical method and synoptic weather maps. Later, the meteorologist Ralph Abercromby proposed seven forms of isobars and attempted to associate them with particular types of weather. Abercromby's isobar geometry became the epitome of the use of the Eulerian method in weather forecasting.[44]

The nineteenth century witnessed the triumph of the Eulerian over the Lagrangian approach in the practice of weather forecasting. But it was not only that. Meteorological science took on certain statistical forms that affected the standardisation and coordination of observation systems. A new – official and state-sponsored – synoptic meteorology was projected in international meetings and national governments. The allocation of public funds to weather forecasting in Britain and the United States had an unequivocal effect: state patronage led to a public service-oriented meteorology rather than a research-oriented meteorology. As a result, meteorological research remained subordinate to public service throughout the long nineteenth century.[45]

CYCLONES, TRADE AND NAVIGATION

Much recent work by geographers and historians of meteorology has been concerned with the diverse and complex ways in which this science was embedded in trade, navigation and maritime economics. The general atmospheric circulation and dominant wind flows (which in turn determined the flows of ocean currents) influenced both the pace and place of maritime trade and the nature and contours of European maritime empires. As the historian Greg Bankoff has noted, the expansion of these powers towards global dominance and imperial hegemony 'is usually characterised as maritime but should more fittingly be described as Aeolian or "wind-driven"'.[46] This reorientation of scholarship towards the role of winds, currents and spaces requires a shift in focus from imperial maritime narratives to an examination of how particular winds and currents, in particular places and times, benefited or harmed the social, economic and geostrategic interests that empires pursued. The creation of ordered spaces in imperial worlds through a scientific understanding of the forces of nature was essential if such spaces were to be navigated, managed, controlled and exploited. As Michael Reidy has suggested, in the nineteenth century 'the British Admiralty, maritime community, and scientific elite collaborated to bring order to the world's seas, estuaries and rivers'.[47] The transformation of the vast oceans into 'an ordered and bounded grid, inscribed with isolines of all kinds – tidal, magnetic, thermal, and barometric – in areas uncharted and on coasts unseen', was vital to imperial aspirations.[48] As Simon Naylor has shown, these institutions and communities fostered a 'culture of instrumental meteorology onboard voyages of exploration, and on Royal Naval and Hydrographic Office survey ships'; it was an 'informal, even idiosyncratic, culture of meteorological inquiry' that became increasingly formalised.[49] As early as 1845, in his address to the British Association for the Advancement of Science, the astronomer John F.W. Herschel described ships equipped with scientific instruments and manned by disciplined observers as 'itinerant observatories'.[50] However, Naylor argues that the ships of the British Navy and other countries cannot be considered as such because, despite explicit support for science, the Admiralty was 'slow and conservative' in implementing the new observing regimes and instrumental practices.[51] Be that as it may, for many of the officers and sailors involved in nautical meteorology, 'the motivation for collecting data lay not in improving our understanding of the atmosphere but rather our ability to use it', as Duncan Agnew has rightly observed.[52]

Meteorology emerged as a key ally of imperial maritime trade, providing captains with the knowledge to take full advantage of seasonal winds and to avoid, or at least reduce the risk of, tropical storms. Naval officers, engineers and colonial officials such as Henry Piddington (who coined the term "cyclone"), William Reid and Charles Meldrum studied the development of storms in the Caribbean and Southeast Asia, collecting observations from ships' logbooks and from observers on land.[53] At the same time, the British Admiralty, under the leadership of Francis Beaufort, was transforming the navy's logbooks into standardised meteorological records that would be useful to both scientific research and the state.[54] This was by no means fortuitous. Behind this interest lay the opportunity to produce practical rules for the safe navigation of ships in storm-prone areas. It was at this time that the US Navy, under the command of Lieutenant Matthew F. Maury, established a depot to map the winds over the oceans.[55] By establishing standardised systems of uniform observations at sea, the naval and merchant marine authorities sought both to optimise shipping routes by taking advantage of seasonal winds and to ensure the safety of voyages, minimising economic losses to merchant companies and chambers of commerce.

Given the growing concern for maritime trade and navigation – following the replacement of sails by steam engines in maritime transport – and in particular for storm warnings to coastal towns in the 1850s and 1860s, the question of the storm warning system was posed in terms of a 'search for rules of thumb'. While in the Netherlands Christophorus Buys Ballot formulated his famous wind rule and lobbied internationally for his law to be adopted as the basis for a warning system, in Britain this system met with opposition from Admiral FitzRoy, head of the Board of Trade's Meteorological Department, who established his own system based on the distribution of barometers to poor fishing communities.[56] A controversial aspect of this choice was that poor fishermen could become amateur observers, contributing to meteorological science while remaining independent of the elitist forms of the Royal Society.[57] A somewhat similar initiative had been promoted by Herschel in the 1830s, when he organised a scheme of hourly meteorological observations at the Admiralty; Herschel's scheme was not controversial, however, because he chose naval officers (not illiterate fishermen) as the 'ideal observers' who provided reliable and disciplined records, and his purpose was to gather information about storms – rather than weather forecasting, a thorny subject at the time.[58] Ironically, after FitzRoy's death, the Royal Society introduced the term "Buys Ballot law" into its official reports, thus reinstating elitist ways of knowing.

ASTROMETEOROLOGY AND POPULAR METEOROLOGY

The prediction of meteorological phenomena had not only commercial and naval implications but also astrological and cultural ones. For centuries, the ancient practice of predicting these phenomena based on the position of celestial bodies had been closely associated with astrology, and, in particular, with astrometeorology.[59] Victorian astro-meteorologists proposed various theories about the lunar influence on the weather and attempted to relate these ideas to natural philosophy. It is, therefore, unsurprising that astrometeorologists and weather prophets interpreted the meaning and practice of predicting natural phenomena and cycles very differently from how observers attached to meteorological observatories usually did so.[60] In fact, astro-meteorologists often criticised conventional institutional meteorology, arguing that it was a limited and overly empirical form of knowledge, and that the accumulation of data per se was useless without a broader theoretical framework. For their part, popular weather prophets played a very active role in debates on methodological issues that marked the development of meteorology as a science.[61] In this context, the boundaries between the two worlds of knowledge were often blurred. Distinguished natural philosophers such as Herschel and figures in weather science such as FitzRoy himself discussed the nature and extent of lunar and planetary influences on the weather, and sought the connections of such influences with electrical and magnetic phenomena. This is why all these astrometeorological and lunar ideas were widely disseminated through various media and formats, not only through almanacs and magazines, but also through scientific works and popular writings.[62] The almanac, in particular, became the medium par excellence through which all these ideas and theories habitually circulated. As historian Katherine Anderson has pointed out, 'the almanac helped define the interactions between elite and popular scientific knowledge'.[63] The sources in this collection illustrate the multidimensional nature of meteorological experience in which synoptic weather forecasts and maps could coexist with astrometeorological ideas and popular superstitions and beliefs.

METEOROLOGY AND RELIGION

The involvement of religious orders in the observation of atmospheric phenomena has always been evident in the public domain, even before the establishment of meteorological societies and national weather services. One order in particular stood out: the Jesuits. The Society of Jesus focused its activities on two apostolates: education and overseas missions. In both, they sought to reconcile religion and science.[64] To understand how the Jesuits produced, disseminated and structured knowledge about weather, it is necessary to understand their global (transnational) and networked approach. Rather than a set of unconnected and isolated observatories, the Jesuits built an expansive, decentralised network of knowledge that culminated in the establishment of the world's first tropical cyclone warning systems.[65] Jesuit observatories and their methods of cyclone forecasting were unrivalled in geopolitically and commercially competitive cyclone-prone regions such as the West Indies and the Far East, and elsewhere in the Americas and Asia.[66] The directors and staff of these observatories received scientific training and instruction at the Collegio Romano and Stonyhurst College in England, the main Jesuit platforms for science. Underlying this success is the fact that the Jesuits fused prediction and correlation in the language of their missionary culture. By understanding Jesuit scientific knowledge as a result of this fusion, we gain a more accurate picture of the culture of prediction and its accommodation across religious frameworks in the nineteenth century.[67]

Manyof the sources concerning religious observers involved in the science of meteorology focus on the diverse and complex ways in which religious beliefs are culturally, socially and materially embedded in their scientific practice and work. This framework has more to do with the everyday reality of religion and lived religious experience than with the defence of firm theological doctrines and grand, hypostatised convictions. Influential theologians and natural philosophers such as William Whewell and William Prout were not alone in publicly advocating mutual understanding between Christian religion and science when considering certain aspects of meteorology in relation to natural theology.[68] Many others have welcomed the empirical inquiry of the atmosphere as a science that reveals relationships and laws that imply the existence of a creative and governing intelligence.[69] Quakers and Anglican clergy publicly embraced this reconciliatory position. Quakers, such as Luke Howard and John Fletcher Miller, as a result of their worldview in which faith, science and morality were intertwined.[70] Anglican clergymen, for whom science was a noble enterprise that reflected the beauty and magnificence of God's handiwork while fulfilling a moral and social commitment to protect the community it served from natural hazards.[71] Clerics such as Joseph Bancroft Reade and Charles Lowndes strove to distance the science of weather from the ideas and practices of astrometeorology by founding the British Meteorological Society (later renamed the Royal Meteorological Society) in 1850, at a time when the two sciences were perceived to be confused in many respects.[72] Others, fewer in number, chose to be staunch defenders of sacred scripture, advocating the alignment of scientific truths with biblical truths. Taking into account the religious experiences of all these weather observers, as well as their institutional roles, the sources in this collection demonstrate the conciliatory nature of meteorological activity in which the empirical investigation of atmospheric phenomena coexisted with orthodox and heterodox elements of Christianity.

METEOROLOGY AND EMPIRE

The relevance of imperial and colonial enterprises to the production of meteorological knowledge is manifold, and the sources in this collection reveal some of the settings in which scientists explored weather and climate, and some of the places in which they displayed the results of their enquiries. The precise identification and study of these sites (geopolitically, physically and socially) offer insights into the nature and practice of meteorology as a *colonial science*.[73] However, this concept does not only refer to how, when and where the politics of empire-building or colonial systems shaped the weather and climate sciences, nor to how these sciences benefited from the interplay between empire, knowledge and power. It also has a lot to do with the legitimation of colonialism through these sciences. It is about the contribution of these sciences to colonial rule and the occupation and control of other peoples' territories. Knowledge was not usually created in the colonies as a neutral entity. It was habitually created for contextual reasons. Whether one considers Piddington's *Sailor's Horn-book* or Meldrum's practical rules, Reid's storm laws or Doberck's typhoon laws, the production and circulation of knowledge were intimately bound up with the defence and legitimation of colonial and imperial enterprises. This is the terrain that the reader of this collection will traverse here, observing the fundamental importance of the imperial framework in which meteorological knowledge was produced and circulated, as throughout the nineteenth century and well into the twentieth, 'science and empire grew together'.[74]

The implications of all this are not inconsiderable for understanding the reasons that led all of these weather scientists to study weather and climate in different kinds of natural and humanised environments. Indeed, meteorology mattered to the Empire in several critical respects. First, it showed how powerful empirical science could be in protecting and stimulating maritime and colonial trade when an atmospheric phenomenon seriously threatened the local environment. In the context of the southern Indian Ocean, for example, Meldrum's rules for ship handling were highly particularistic because of their umbilical ties to the immediate physical environment.[75] It is also worth remembering that it was the meteorological instruments that proved to be so liberating to ship captains, who were normally threatened by tropical storms, in making the most of the seasonal winds. A second issue with implications on the history of meteorology revolves around what might be called the strategies of colonialism. Whether in predicting monsoon patterns in India, issuing warnings to protect sugar cane crops in Mauritius, identifying periods of drought in Australi, or recommending safe shipping routes in Hong Kong, the significance of the colonial utility of meteorology was brought to the surface. Meteorology helped to make intelligible the unknown, often inhospitable environments in which settlers sought to settle. In its authoritative capacity as a tamer of the weather and a prognosticator of the future, meteorology helped to reinforce the discourse of scientific and technical superiority of the colonial powers, justifying their dominance under the discourse of civilising progress. But while it had a direct impact in facilitating economic exploitation, territorial expansion and social control, the objective of being a key cog in the colonial governance machine was only partially fulfilled, depending on when and where. Finally, one theme has, in my view, been underdeveloped in the analysis of the relationship between meteorology and empire: the space of education. The interest here lies in analysing how particular educational environments facilitated or constrained meteorological observations. For meteorological observatories sometimes shaped and were shaped by education. The establishment of observatories in educational institutions such as colleges and universities is inextricably linked to the value of education as an instrument of imperial control. In the Spanish colony of Cuba, between the 1850s and the 1870s, a *plate shift* in the educational system remarkably shaped the practice of meteorology on the island, privileging the Jesuit Observatory of the Colegio de Belén, which represented the interests of the Society of Jesus and the metropolis, over the Havana Physical-Meteoric Observatory of Andrés Poey, which represented the interests of the Creole bourgeoisie.[76] All these aspects were implicated in the production and circulation of meteorological knowledge.

Charting the geographies of weather and climate science has fostered a similar interest in another type of physical, social and architectural entity: observatories. Tracing the stories of their functions, such as keeping time in ports, describing local and regional climates, mapping regional magnetic variations, recording seismic movements, large-scale statistical endeavours – not to mention maps, charts, weather forecasts, surveys, etc. – brings to view the extent to which knowledge of observatory sciences 'served imperial systems of government'.[77] Paying attention to what David Aubin, Charlotte Bigg and H. Otto Sibum call 'observatory techniques' and their association with a 'culture of precision' opens up interesting possibilities for reflecting on the history of nineteenth-century meteorology through the various techniques – observation, measurement, mapping, data reduction, tabulation and preservation, etc. – that were employed in colonial observatories. The observatory functioned as an experimental space where a diverse array of technological tools were systematically used, tested, calibrated and sometimes even designed. As these three historians have noted, 'the observatory helped to further the culture of precision that transformed scientific practices in the nineteenth century'.[78] Important too, of course, is the setting and infrastructure on which the observatories were established according to the plans of their promoters in the imperial metropolises. The account of the conception of the "magnetic crusade" as a coordinated plan for the global study of terrestrial magnetism and related phenomena provides a nice example: the plan, forged by eminent scientific and imperial authorities such as John Herschel, Francis Beaufort and Edward Sabine, not only attracted the scholarly elite of the Royal Society and the newly formed British Association for the Advancement of Science, but also financially and operationally involved the British government through the Admiralty and the Army. Observatories set up for this purpose in places such as the Cape Colony, Hobart Town and Toronto were joined by stations operated by the British East Company in Madras, Simla, Bombay and Singapore.[79] As Herschel pointed out in a lecture on magnetism held at Cambridge in 1845, 'the gigantic problems of meteorology, magnetism, and oceanic movements can only be resolved by a far more extensive geographical distribution of observing stations, and by a steady, persevering, systematic attack'.[80] The observatory thus emerged as the altar where the globalising ambitions of the Empire were married to the needs of global science.

In light of these contextual cartographies, meteorological publications displayed distinctive colonial logics of production.[81] Instructions for making meteorological observations, for example, took different forms in different places, although they all pursued the common goal of helping to standardise observations, unify protocols and use instruments appropriately to avoid measurement errors. In South Africa, for example, Herschel drew up his *Instructions* (1835) to implement a 'concerted plan of contemporaneous observations' at the imperial level.[82] In Hong Kong, Doberck's *Instructions* (1883) responded to the strong rivalry with the Jesuits' meteorological activities at the Zikawei Observatory in Shanghai. Both contrasted with the works published in London by Robert H. Scott and William Marriott, whose instructions followed metropolitan guidelines and formulations that did not take into account the specificities of colonial meteorology. Elsewhere, contexts conditioned storm theories. In Barbados, for example, Reid argued that the study of storm tracks from the tropics to higher latitudes allowed the elucidation of the mechanisms governing wind behaviour, an approach that was at odds with studies in Europe that focused only on high-latitude cyclones. Meanwhile, in Mauritius, Alexander Thom proposed that cyclones in the Indian Ocean were caused by the meeting between the northwest monsoon and the southeast trade wind. Contextual spaces, therefore, turned out to be of critical significance in delineating cartographies of production. The place and context in which observations were made – colonial, imperial, metropolitan – were of great importance in the production of meteorological knowledge.

THE VOLUMES

This collection provides new tools and perspectives for investigating the relationship between meteorology, culture, society and empire in Britain and its colonial dominions in the nineteenth century. Aimed at scholars and undergraduate and postgraduate students, it offers a rich selection of thematically focused and curated primary sources. It aims to move beyond the traditional Victorian periodisation and the confinement of meteorology to national and imperial preoccupations and interests in order to revisualise the meteorological space of the period. The collection does not claim to provide a comprehensive overview of nineteenth-century meteorology. Rather, the purpose is to present various ways of conceptualising nineteenth-century meteorology by emphasising both the most committed voices and often overlooked perspectives. A key challenge of the collection is to integrate transnational, continental and inter-imperial currents into our interpretations, understanding and representations of the world of nineteenth-century meteorology in Britain. This would fulfil the aim of projecting emerging historiographical trends and broadening the epistemic and geographical scope of a field in which the study of institutions, practices and material cultures has dominated historical and cultural studies.

The collection consists of four volumes. Each is structured around a different conception of meteorology. Volume I is concerned with 'Weather Forecasting and Nation-State Building'. It examines the key role played by nation-state building and statistics in shaping weather forecasting in Britain and the Western world. It shows the confrontation between two rival approaches (Eulerian and Lagrangian) to weather forecasting and how states legitimised the former and contributed to its hegemony in practice. Volume II on 'Cyclones, Trade, Navigation and Society' deals with other sides of the analytical polyhedron of meteorology. It considers the importance of weather science for safe navigation and maritime commerce, expressed through storm laws, practical rules for avoiding cyclones and the introduction of standardised observation systems. It also brings alternative popular meteorological currents and convictions together with forms of astrometeorology and prophecy that abounded at the time. Volume III, 'Meteorology and Religion', focuses on the "religious component" as a distinct and analytical category for understanding the ways in which religion is expressed and embodied in the science of meteorology. With particular attention to the contributions of the Jesuits, the volume brings together sources from various Christian orders and traditions to show how different religious convictions and perspectives were engaged in fields as diverse as correlation, prediction, nephology, cyclonology and meteorological spectroscopy. Volume IV, on 'Weather, Climate and Empire', extends the scope of the previous volumes beyond Britain to its colonial dominions and also includes climate. It sheds light on how British empire-building shaped the sciences of weather and climate, and how these sciences in turn helped to legitimise the imperial project. Each volume can be consulted separately, but together they help to elucidate the complex and polyhedral nature of nineteenth-century meteorology.

The choice of source material has sparked important debates about the influence of digitisation on the way scholars and students alike access and explore primary documents. While platforms such as the Haithi Trust Digital Library, archive.org and Project Gutenberg offer a wealth of publicly available nineteenth-century texts, a growing number of educational organisations and cultural institutions are also digitising archival documents and other primary sources for research and teaching purposes. In various forums, scholars working in universities, libraries, and archives have expressed their belief that the use of primary and first-hand sources can better clarify and illustrate issues such as weather forecasting, climate change and the impact of atmospheric phenomena.[83] These scholars argue that students' use of historical documents about scientific data, methods, theories and results not only helps develop their analytical skills and critical thinking, but also promotes deeper learning about severe weather and climate, as well as society's preparedness and response to these challenges.[84] As experts in these fields have recognised, knowledge of the history of meteorology and climatology since their inception as physical sciences in the nineteenth century is fundamental to understanding the causes and historical patterns of severe weather and climate change that so concern society today.[85] So, rather than merely a compendium of primary sources, this collection uses primary sources as a way of thinking about history, while also looking at history as a way of building effective strategies for dealing with future weather and climate challenges.

Each primary source is preceded by a headnote which begins with an overview of the main issue addressed, condensing what is considered to be the current consensus. Next, the source and, where feasible, the author, are placed in their social and cultural context, highlighting the historical moment, debate or stage of development that is essential to understanding them. A brief analysis of the text is provided, identifying the author's aims, basic ideas, proposals and solutions. It also examines the impact and legacy of the source on society and, as far as possible, the context of its publication. Where several related sources are included, as pairs or small groups of texts that complement each other, their connection and the meaning they acquire when examined together are explained. These headnotes, as well as the excerpted sources, are intended to illustrate the procedural richness, critical reflection and farsightedness with which nineteenth-century authors engaged in the production and circulation of meteorological knowledge.

The collection serves as a tool for researchers from different disciplinary fields interested in nineteenth-century meteorology by structuring and presenting the texts under specific thematic categories. In a number of cases, translations of extracts from texts originally published in German, French, Spanish and Italian have been included, giving researchers and students access to difficult materials that have so far remained unpublished in English. The reproduction of the sources has been carefully adapted to preserve their integrity while seeking accessibility. The different spellings used in the extracts have not been changed or standardised, as priority has been given to preserving the particular character and essence of the original language. However, long dashes, the style of quotation marks and the formatting of subtitles in the selected texts have been standardised to ensure a consistent presentation. Since covering the wide range of perspectives in nineteenth-century meteorology requires coverage of scientific and broader sociocultural manifestations, we have drawn on the varied discourses through which practice and knowledge were articulated. In practical terms, this means using a wide variety of materials, ranging from scientific books and articles to newspaper notes, periodical bulletins, logbooks, correspondence, popularisations of scientific knowledge, almanacs, maps and atlases, among others. Each volume contains detailed bibliographies of the specific editions used to compile the extracts and create the subject indexes. In addition, an extensive bibliography of historical and scientific works on meteorology in the nineteenth century is presented at the end of Volume IV. The diversity and breadth of literature reflect the vitality and scope of the "epistemic turn" in scholarship in the weather and climate sciences that has taken place over the last two decades. We conclude with the hope that the material presented in this collection will inspire new lines of inquiry.

Notes

1 Martin Mahony and Angelo Matteo Caglioti, 'Relocating Meteorology', *History of Meteorology*, 8 (2017), pp. 1–14; Martin Mahony and Georgina H. Enfield, 'Climate and Colonialism', *Wiley Interdisciplinary Reviews: Climate Change*, 9(2) (2018), e510 https://doi.org/10.1002/wcc.510; Harriett Mercer and Thomas Simpson, 'Imperialism, Colonialism, and Climate Change Science', *WIREs Climate Change*, 14(2023), e851. https://doi.org/10.1002/wcc.851

2 Mark D. Hersey and Jeremy Vetter, 'Shared ground: Between environmental history and the history of science', *History of Science,* 57(4) (2019), pp. 403–440 (pp. 404–405).

3 Mahony and Caglioti, 'Relocating Meteorology', p. 4.

4 Mahony and Caglioti, 'Relocating Meteorology', p. 3.

5 James R. Fleming and Vladimir Janković, 'Revisiting Klima', *Osiris*, 26 (2011), pp. 1–15 (p. 10).

6 Stefano Casati and Marco Ciardi, 'Meteorology: Disciplinary History', in Gregory Good (ed.), *Sciences of the Earth: An Encyclopedia of Events, People, and Phenomena.* Volume 2 (New York and London: Garland, 1998), pp. 564–571 (p. 571).

7 Robert Marc Friedman, *Appropriating the Weather: Vilhelm Bjerknes and the Construction of a Modern Meteorology* (Ithaca, NY: Cornell University Press, 1989).

8 James Rodger Fleming, *Meteorology in America, 1800–1870* (Baltimore: Johns Hopkins University Press, 1990).

9 Frederik Nebeker, *Calculating the Weather. Meteorology in the 20th-Century* (San Diego: Academic Press, 1995).

10 Mark Monmonier, *Air Apparent: How Meteorologists Learned to Map, Predict, and Dramatize Weather* (Chicago: The University of Chicago Press, 1999).

11 Mahony and Caglioti, 'Relocating Meteorology', p. 4.

12 Kapil Raj, *Relocating Modern Science: Circulation and the Construction of Knowledge in South Asia and Europe, 1650–1900* (Basingstoke: Palgrave Macmillan, 2007).

13 Christina Barboza and Vladimir Janković, 'The Many Lives of Weather', in Vladimir Janković and Christina Barboza (eds), *Weather, Local Knowledge and Everyday Life: Integrated Climate Studies* (Rio de Janeiro: MAST, 2009), pp. 11–17 (pp. 6 and 12).

14 Fiona Williamson, 'Just Doing their Job: The Hidden Meteorologists of Colonial Hong Kong c.1883–1914', *British Journal for the History of Science*, 54(3) (2021), pp. 341–359.

15 See Pyenson's trilogy: Lewis Pyenson, *Cultural Imperialism and Exact Sciences: German Expansion Overseas, 1900–1930* (New York: Lang, 1985); Lewis Pyenson, *Empire of Reason: Exact Sciences in Indonesia, 1840–1940* (Leiden: Brill, 1989); Lewis Pyenson, *Civilizing Mission. Exact Sciences and French Overseas Expansion, 1830–1940* (Baltimore: The Johns Hopkins University Press, 1993). See also: Michael Adas, *Machines as the Measure of Men: Science, Technology, and Ideologies of Western Dominance* (Ithaca: Cornell University Press, 1989).

16 Lewis Pyenson, 'Functionaries and Seekers in Latin America: Missionary Diffusion of the Exact Sciences, 1850–1930', *Quipu*, 2 (3) (1985), pp 387–420 (p. 389).

17 For a critique of Pyenson's categorisation and terminology, see: Thomas F. Glick, 'Crítica a N. Stepan y L. Pyenson', *Quipu*, 2(3) (1985), pp. 437–442 (pp. 438–440).

18 Lewis Pyenson, 'Pure Research, Jesuit Institution, and Metropolitan Ambitions: The Evolution of French Policies Overseas, 1880–1940', in *Chine et Europe: Évolution et particularités des rapports est-ouest du XVIe au XXe siècle. Actes du IVe Colloque International de Sinologie de Chantilly* (París: Institut Ricci, 1991), pp. 249–271 (pp. 266–267).

19 Aitor Anduaga, *Cyclones and Earthquakes. The Jesuits, Prediction, Trade, and Spanish Dominion in Cuba and the Philippines, 1850–1898* (Quezon City, Ateneo de Manila University Press, 2017); Aitor Anduaga, 'Los jesuitas y el estudio de los tifones en Filipinas, 1865–1899', *Illes i Imperis*, 22 (2020), pp. 63–83; Aitor Anduaga, 'The Jesuits and Meteorological Observatories'. *Quaderni di Storia della Fisica*, 27 (2023), 85–100.

20 Anduaga, *Cyclones and Earthquakes*. See also: Agustín Udías, *Searching the Heavens and the Earth: The History of Jesuit Observatories* (Boston: Kluwer, 2003); Gregory Cushman, 'The Imperial Politics of Hurricane Prediction: From Calcutta and Havana to Manila and Galveston, 1839–1900', in Erika Marie Bsumek, David Kinkela, and Mark Atwood Lawrence (eds.), *New Approaches to International Environmental History* (Nueva York: Oxford University Press, 2013), pp. 137–162; Anduaga, 'Los jesuitas'.

21 Aitor Anduaga, 'Epistemic Network: The Jesuits and Tropical Cyclone Prediction, 1860–1900', *Isis*, 113(3) (2022), pp. 513–536, pp. 529–531.

22 Jan Golinski, *British Weather and the Climate of Enlightenment* (Chicago: University of Chicago Press, 2007).

23 Anne Lawrence-Mathers, *Medieval Meteorology. Forecasting the Weather from Aristotle to the Almanac* (Cambridge: Cambridge University Press, 2020).

24 Vladimir Janković, *Reading the Skies: A Cultural History of English Weather, 1650–1820* (Chicago: University of Chicago Press, 2001), p. 10.

25 Gisela Kutzbach, *The Thermal Theory of Cyclones: A History of Meteorological Thought in the Nineteenth Century* (Boston, MA: American Meteorological Society, 1979).

26 Aitor Anduaga, *Politics, Statistics, and Weather Forecasting, 1840–1910. Taming the Weather* (London and New York: Routledge, 2020).

27 Katharine Anderson, *Predicting the Weather: Victorians and the Science of Meteorology* (Chicago: The University of Chicago Press, 2005).

28 Fleming and Janković, 'Revisiting Klima', p. 2.

29 Fleming and Janković, 'Revisiting Klima', p. 2.

30 Vladimir Janković, *Confronting the Climate: British Airs and the Making of Environmental Medicine* (New York: Palgrave Macmillan Ltd., 2010).

31 Jan Golinski, *British Weather and the Climate of Enlightenment* (Chicago: University of Chicago Press, 2007).

32 James Burton, 'The History of the British Meteorological Office to 1905' (The Open University, PhD diss., 1988); J. Malcolm Walker, *History of the Meteorological Office* (Cambridge: Cambridge University Press, 2012).

33 Anderson, *Predicting the Weather*; Pauline Halford, *Storm Warning: The Origins of the Weather Forecast* (Stroud: Sutton Publishing, 2004).
34 Ruth Morgan, 'Prophecy and Prediction: Forecasting Drought and Famine in British India and the Australian Colonies', *Global Environment,* 13(1) (2020), pp. 95–132 (p. 130).
35 Anduaga, *Politics, Statistics, and Weather Forecasting.*
36 C. Donald Ahrens, *Meteorology Today: An Introduction to Weather, Climate, and the Environment*, 6th edition (Pacific Grove: Brooks and Cole, 2000), p. 15.
37 For an overview of the Eulerian and Lagrangian fluid descriptions, and the differences between the two methods, see: Donald F. Young, *A Brief Introduction to Fluid Mechanics*, 5th ed. (Hoboken, NJ: Wiley, 2011), pp. 103–107.
38 Anduaga, *Politics, Statistics, and Weather Forecasting*, p. 10.
39 Fleming, *Meteorology in America*; Vladimir Janković, 'Ideological Crests versus Empirical Troughs: John Herschel's and William Radcliffe Birt's Research on Atmospheric Waves, 1843–50'. *British Journal for the History of Science*, 31 (1998), pp. 21–40.
40 John D. Cox, *Storm Watchers: The Turbulent History of Weather Prediction from Franklin's Kite to El Niño* (Hoboken, NJ: John Wiley & Sons, Inc., 2002), pp. 85–90; James Lequeux, *Le Verrier–Magnificent and Detestable Astronomer* (New York: Springer, 2013), pp. 280–284.
41 For the Paris Observatory and its contribution to forming an 'intellectual and material culture' in the field of dynamic meteorology, see: Fabien Locher, *Le savant et la tempête. Étudier l'atmosphère et prévoir le temps au XIX siècle* (Rennes: Presses Universitaires de Rennes, 2008), pp. 139–145. See also: Christina Helena da Motta Barboza, *As viagens do tempo. Uma história da meteorologia em meados do século XIX* (Rio de Janeiro: E-papers Serviços Editoriais Ltda, 2012).
42 Sarah Dry, 'Safety Networks: Fishery Barometers and the Outsourcing of Judgement at the Early Meteorological Department', *The British Journal for the History of Science*, 42 (1) (2009), pp. 35–56.
43 James Burton, 'Robert FitzRoy and the Early History of the Meteorological Office', *British Journal for the History of Science*, 19 (1986), pp. 147–176; Halford, *Storm Warning*; Anderson, *Predicting the Weather*; Walker, *History of the Meteorological Office.*
44 Burton, 'Robert FitzRoy', p. 146; Walker, *History of the Meteorological Office*, p. 124.
45 Anduaga, *Politics, Statistics, and Weather Forecasting*, pp. 298–341.
46 Greg Bankoff, 'Aeolian Empires: The Influence of Winds and Currents on European Maritime Expansion in the Days of Sail', *Environment and History,* 23 (2017), pp. 163–196 (pp. 164–165).
47 Michael S. Reidy, 'From the Oceans to the Mountains: Spatial Science in an Age of Empire', in Jeremy Vetter (ed.), *Knowing Global Environments* (New Brunswick, N.J.: Rutgers Univ. Press, 2011), pp. 17–38 (p. 34).
48 Michael S. Reidy, *Tides of History: Ocean Science and Her Majesty's Navy* (Chicago: Univ. Chicago Press, 2008), p. 6.
49 Simon Naylor, 'Weather Instruments all at Sea: Meteorology and the Royal Navy in the Nineteenth Century', in Fraser MacDonald, Charles W.J. Withers (eds.), *Geography, Technology and Instruments of Exploration* (London: Routledge, 2015), pp. 77–96 (p. 78). See also: Simon Naylor, *The Observatory Experiment. Meteorology in Britain and Its Empire* (Cambridge: Cambridge University Press, 2024), pp. 22–65.
50 Alison Winter, 'Compasses All Awry': The Iron Ship and the Ambiguities of Cultural Authority in Victorian Britain', *Victorian Studies*, 38 (1994), pp. 69–98, (p. 75). See also Simon Naylor, 'Log Books and the Law of Storms: Maritime Meteorology and the British Admiralty in the Nineteenth Century', *Isis*, 106 (4) (2015), pp. 771–797.
51 Naylor, 'Log Books and the Law of Storms', p. 78.

52 Duncan Carr Agnew, 'Robert Fitzroy and the Myth of the "Marsden Square": Transatlantic Rivalries in Early Marine Meteorology', *Notes and Records of the Royal Society of London [Great Britain]* 58(1) (2004), pp. 21–46 (p. 24).
53 Tirthankar Roy, 'The Law of Storms': European and Indigenous Responses to Natural Disasters in Colonial India, *c.* 1800–1850', *Australian Economic History Review*, 50(1) (2010), pp. 6–22; Martin Mahony, 'The 'Genie of the Storm': Cyclonic Reasoning and the Spaces of Weather Observation in the Southern Indian Ocean, 1851–1925', *British Journal for the History of Science*, 51(4) (2018), pp. 607–633.
54 Dennis Wheeler and Clive Wilkinson, 'The Determination of Logbook Wind Force and Weather Terms: The English Case', *Climatic Change*, 73 (2005), pp. 57–77; Naylor, 'Log Books and the Law of Storms'; Naylor, *The Observatory Experiment,* pp. 22–65.
55 Chester G. Hearn, *Tracks in the Sea: Matthew Fontaine Maury and the Mapping of the Oceans* (Camden, ME: International Marine/McGraw-Hill, 2002).
56 Azadeh Achbari, Frans van Lunteren, 'Dutch Skies, Global Laws: The British Creation of "Buys Ballot's Law»', *Historical Studies in the Natural Sciences*, 46 (1) (2016), pp. 1–43.
57 Sarah Dry, *Fishermen and Forecasts: How Barometers Helped Make the Meteorological Department Safer in Victorian Britain* (London: London School of Economics and Political Science, 2007), discussion paper nº 46.
58 Gregory A. Good, 2006. 'A Shift of View: Meteorology in John Herschel's Terrestrial Physics', in James Rodger Fleming, Vladimir Janković, and Deborah R. Coen (eds.), *Intimate Universality. Local and Global Themes in the History of Weather and Climate* (Sagamore Beach, MA: Watson Publishing International, 2006), pp. 35–67.
59 Patrick Curry, *A Confusion of Prophets: Victorian and Edwardian Astrology* (London: Collins & Brown, 1992).
60 Anderson, *Predicting the Weather*, pp. 52–55.
61 Katharine Anderson, 'The Weather Prophets: Science and Reputation in Victorian Meteorology', *History of Science*, 37 (1999), pp. 179–216.
62 Carey Gibbons, 'Grasping the Elusive: Victorian Weather Forecasting and Arthur Hughes's Illustrations for George Macdonald's *At the Back of the North Wind'*, in Nancy Rose Marshall (ed.), *Victorian Science and Imagery: Representation and Knowledge in Nineteenth-Century Visual Culture* (Pittsburgh, Pa.: University of Pittsburgh Press, 2021), pp. 79–110.
63 Anderson, *Predicting the Weather*, p. 54.
64 Agustín Udías, *Searching the Heavens and the Earth: The History of Jesuit Observatories* (Boston: Kluwer, 2003); Agustín Udías, *Jesuit Contribution to Science: A History* (New York: Springer, 2015).
65 Anduaga, 'Epistemic Network'.
66 Gregory Cushman, 'The Imperial Politics of Hurricane Prediction: From Calcutta and Havana to Manila and Galveston, 1839–1900', in Erika Marie Bsumek, David Kinkela, and Mark Atwood Lawrence (eds.), *New Approaches to International Environmental History* (New York: Oxford University Press, 2013), pp. 137–162; Anduaga, *Cyclones and Earthquakes.*
67 Aitor Anduaga, 'The Jesuit Culture of Correlation in Observatory Sciences', *Isis*, 116:1 (2025), pp. 3–22.
68 Jonathan Topham, 'Science and Popular Education in the 1830s: The Role of the Bridgewater Treatises', *British Journal for the History of Science,* 15 (1992), pp. 397–430.
69 Frank Turner, *Between Science and Religion: The Reaction to Scientific Naturalisn in Later Victorian England* (New Haven, CT: Yale University Press, 1974); Frank Turner, 'The Victorian Conflict between Science and Religion: A Professional Development', *Isis,* 69 (1978), pp. 356–376; B. M. G. Reardon, *Religious Thought in the Victorian Age*. 2nd ed. (London: Longman, 1995).
70 Geoffrey Cantor, 'Quakers and Science', in Stephen W. Angell and Pink Dandelion (eds.), *The Oxford Handbook of Quaker Studies* (Oxford: Oxford University Press,

2013), pp. 520–534; Geoffrey Cantor, *Quakers, Jews and Science* (Oxford: Oxford University Press, 2013)

71 John H. Brooke, *Science and Religion* (Cambridge: Cambridge University Press, 1991); Jim A. Bennett, *Church, State and Astronomy in Ireland – 200 years of Armagh Observatory* (Armagh: Armagh Observatory, 1990).

72 George J. Symons, 'History of the English Meteorological Societies', *Quarterly Journal of the Royal Meteorological Society,* 7 (1881), pp. 66–107.

73 Martin Mahony and Georgina Endfield, 'Climate and Colonialism'. *Wiley Interdisciplinary Reviews: Climate Change*, 9(2) (2018), e510; Martin Mahony, 'Meteorology and Empire', in Andrew Goss (ed.), *The Routledge Handbook of Science and Empire* (London and New York: Routledge, 2021), pp. 47–58; James Beattie, Emily O'Gorman, and Matthew Henry, *Climate, Science, and Colonization: Histories from Australia and New Zealand* (Basingstoke: Palgrave Macmillan, 2014); Harriett Mercer and Thomas Simpson, 'Imperialism, Colonialism, and Climate Change Science', *WIREs Climate Change*, 14(2023), e851. https://doi.org/10.1002/wcc.851; Deborah R. Coen, *Climate in Motion: Science, Empire, and the Problem of Scale* (Chicago: University of Chicago Press, 2018); Angelo Matteo Caglioti, *Meteorological Imperialism. Climate Science, Environment, and Empire in Liberal and Fascist Italy (1870–1940)*, PhD diss. (University of California, Berkeley, 2017).

74 Andrew Goss, 'Introduction: An Imperial Turn in the History of Science', in Andrew Goss (Ed.), *The Routledge Handbook of Science and Empire* (London and New York: Routledge, 2021), pp. 1–9 (p. 1).

75 Martin Mahony, 'The 'Genie of the Storm': Cyclonic Reasoning and the Spaces of Weather Observation in the Southern Indian Ocean, 1851–1925', *British Journal for the History of Science* 51(4) (2018), pp. 607–633.

76 Anduaga, *Cyclones and Earthquakes*, pp. 168–214.

77 Simon Naylor and Matthew Goodman, 'Atmospheric Empire: Historical Geographies of Meteorology at the Colonial Observatories', in Martin Mahony and Samuel Randalls (eds.), *Weather, Climate, and the Geographical Imagination* (Pittsburgh, PA: University of Pittsburgh Press, 2020), pp. 25–42 (p. 27). See also: Naylor, *The Observatory Experiment.*

78 David Aubin, Charlotte Bigg and H. Otto Sibum, 'Introduction: Observatory Techniques in Nineteenth-Century Science and Society', in David Aubin, Charlotte Bigg and H. Otto Sibum (eds.), *The Heavens on Earth: Observatories and Astronomy in Nineteenth-Century Science* (Durham, NC: Duke University Press, 2010), pp. 1–32 (p. 11).

79 John Cawood, 'The Magnetic Crusade: Science and Politics in Early Victorian Britain', *Isis*, 70 (1979), pp. 492–518; Christopher Carter, 'Magnetic Fever: Global Imperialism and Empiricism in the Nineteenth Century', *Transactions of the American Philosophical Society*, New Series, 99(4) (2009), pp. 1–163; Matthew Goodman, 'Proving Instruments Credible in the Early Nineteenth Century: The British Magnetic Survey and Site-Specific Experimentation', *Notes and Records: The Royal Society Journal of the History of Science*, 70 (2016), pp. 251–268.

80 John Frederick William Herschel, 'An Address to the British Association for the Advancement of Science at the Opening of their Meeting at Cambridge, June 19th, 1845', in John Herschel, *Essays from the Edinburgh and Quarterly Reviews* (London: Longman, Brown, Green, Longmans & Roberts, 1857), pp. 634–682 (p. 652).

81 Sarah Wright and Matalena Tofa, 'Weather geographies: Talking about the weather, considering diverse sovereignties', *Progress in Human Geography,* 45(5) (2021), https://doi.org/10.1177/0309132520982

82 John Frederick William Herschel, *Instructions for Making and Registering Meteorological Observations in Southern Africa and other Countries in the South Seas, as also at Sea* (London: Bradbury and Evans, 1835), p. 1.

83 The Library of Congress supports educators with access to its vast digital archives, offering classroom tools and professional development to facilitate the use of primary sources in teaching. See: www.loc.gov/classroom-materials/weather-forecasting/?loclr=blogtea
84 Danna Bell, 'Primary Sources in Science Classrooms: Severe Weather and Community Resilience', May 24, 2016. Library of Congress blogs, https://blogs.loc.gov/teachers/2016/05/primary-sources-in-science-classrooms-severe-weather-and-community-resilience/
85 Joshua P. Howe, Paul S. Sutter, *Making Climate Change History: Documents from Global Warming's Past* (Seattle: University of Washington Press, 2017), pp. 3–4.

INTRODUCTION TO VOLUME I

Weather Forecasting and Nation-State Building

Nineteenth-century meteorological science is a child of statistics and nation-state building. Not surprisingly, it was during the processes of nation-state building and their concomitant centralising impulse in the second half of the nineteenth century – arising in part after the revolutions of 1848 that triggered the republican revolts against the European monarchies – that the practice of weather mapping was officially recognised and aroused the interest of both governmental authorities and weather scientists. The close relationship between weather maps and state building has its *raison d'être*. In addition to illustrating what the weather will be like in nearby places, maps contribute to creating an imaginary national space and accentuate our perceptions of common belonging. As geographer David N. Livingstone states, maps are 'repositories of trust' regarded by society as 'an accurate representation of the world under scrutiny'. Furthermore, insofar as they convey an impression of objectivity, precision and accuracy, maps have become 'an analogy for scientific theory itself'.[1] Weather maps are able to simplify the complexity of atmospheric phenomena for the sake of public information and understanding. Hence, as Mark Monmonier suggests, the reader must have a 'healthy degree of scepticism' about the reality that maps purport to represent.[2] Last but not least, maps are a symbol of power, a symbol that is usually associated with the reinforcement of order, security and the status quo. In his work on maps, knowledge and power, J. Brian Harley argues that 'maps are preeminently a language of power, not of protest'.[3] For all these reasons, any history of meteorology that ignores or forgets the social and political significance of weather mapping is doomed to be ahistorical.

Nineteenth-century meteorological science is intertwined not only with state building but also with the rise of statistics and its related historical processes, such as industrialisation, metropolisation, globalisation and the foundations of the information society. Drawing upon mathematical tools such as mean values and probability, statistical science was central to weather scientists in their attempt to order, interpret and decipher a diversity they deemed as seemingly inapprehensible. In fact, the reason we envisage weather forecasts today partly as the meteorologists of the time did – that is, by synoptic weather maps – has much to do with the rise of statistics to a position of influence and hegemony. In other words,

the fact is strongly linked to the increasing use of statistics not only as a mode of graphical representation but also as a politico-administrative tool in nation-state building.[4] The texts in this volume attest to a desire to promote the practice and production of weather maps through the two senses of the concept of statistics, that is, the organisation of data collecting and processing (whether by the state or private agencies) on the one hand, and the science for the mathematical treatment of quantitative data on the other.[5]

Attention should also be paid to instances where tensions arose between those who championed statistical and synoptic-cartographic methods and those who questioned them. These tensions played out in a variety of ways and on different issues, from official inquiries against empiricists such as Robert FitzRoy for their "obscure" methods of forecasting (based on local rules) to elite debates conceived by the Royal Society on the merits of precision, determinism and statistical knowledge.[6] Francis Galton's influential report, laid before both Houses of Parliament in 1866, led to the cancellation of FitzRoy's storm-warning service by stressing the unreliability of forecasts.[7] Later, in the 1870s and 1880s, the rise of statistical techniques and synoptic cartography helped to endow weather maps with scientific legitimacy. This volume shows the importance of issues such as security, order and control in the science of meteorology, highlighting how governments geared national weather systems towards forecasting and storm warnings rather than research.

TWO APPROACHES (EULERIAN VS LAGRANGIAN); TWO ORGANISATIONAL MODELS

The idea that the Black Sea storm (also known as the Great Storm of Balaklava of 14–15 November 1854 in Crimea) was the starting point of modern meteorology has become commonplace in the annals of meteorology.[8] As a result of this event, Urbain Le Verrier and the Paris Observatory are said to have turned their attention to storm forecasting and led the creation of an international meteorological network in Europe.[9] Historians have profusely discussed its impact and the nature of synoptic meteorology in its early stage; Christina Barboza has emphasised the role of Le Verrier's telegraph network in his attempt to control politically the Paris Observatory's activities;[10] and Fabien Locher has linked the production of synoptic weather maps to the formation of an 'intellectual and material culture' in the field of dynamic meteorology.[11] However, a fundamental aspect of this historical episode has gone unnoticed until recently by historians of science: the confrontation between two approaches – the Eulerian versus the Lagrangian – that defined the conceptual framework of nineteenth-century weather forecasting.[12]

The origins of this confrontation date back to the 1840s, when a series of opposing storm theories, based on mechanical and thermal principles, were formulated in Europe and the United States. The "storm controversy", as these debates between the (mechanical) whirlwind theory and the (thermal) convective

theory were called, will be discussed in Volume II.[13] Two wave solutions to this controversy were proposed on different continents: Elias Loomis's cold waves in America and John Herschel and William Birt's atmospheric waves in Britain. Their seminal texts are collected in this volume.[14] These theories form the pre-1850 conceptualisations of storms. Soon after, statistical methods began to be applied to meteorology; Adolphe Quetelet devised a graphical method for representing atmospheric waves and thus ascertaining the regularities that lie behind non-periodic barometric changes. Insofar as it laid the foundations for a dynamic conception of meteorological space, Quetelet's method became crucial in bridging pre-synoptic techniques and synoptic weather mapping.

In this framework, a key episode – the Black Sea storm of 1854 – marked the course of predictive meteorology. However, it did so not so much because it stimulated storm warning services, but because it drove Le Verrier to adopt a Eulerian approach to weather forecasting in the interests of national security and order. This approach focused on the movement of air passing fixed observation points (usually ground-based weather stations). Thus, the recorded air movement patterns were projected onto a synoptic map. It is interesting to note that the reason that led Le Verrier to embrace this approach had much to do with the accommodation of his scientific thought to the state's interests. State interests, rather than scientific reasons, pushed him to take this approach: the new weather maps of isobars (on which the Eulerian method was based) were not only able to reveal weather patterns but also served as an instrument to achieve national order and security, as well as the control and centralisation of scientific activities. The primary sources collected in this volume show how the Eulerian approach was first adopted in France and Italy and then spread to other countries. The establishment of a storm warning service by Jesuit Father Angelo Secchi in Pontifical Rome in the late 1850s was conceived on the basis of this approach. Secchi's cartographic method prompted Hippolyte Marié-Davy from the Paris Observatory to develop a synoptic method of weather forecasting in 1863, in which 'low-pressure areas' or depressions became the objects of study.[15]

Interestingly, in Britain, a new approach – the Lagrangian one – acted as a counterweight to the Eulerian approach. It was characterized by the quest for autonomy and individual judgement in weather forecasting. Unlike the Eulerian one, the Lagrangian approach focused on following an individual air parcel as it moved through space and time. Thus, its object of enquiry was the air-trajectory or the air-mass concept – not the synoptic map. Thus, in the 1860s Robert FitzRoy, George Jinman and Francis Galton proposed storm-flow models and iconic – instead of isoline-based – maps of weather.[16]

FitzRoy's attempts to forecast the weather by combining empirical rules and instrumental readings are worth mentioning. As the head of the Board of Trade's Meteorological Department, he championed a Lagrangian approach diametrically opposed in forms and aims to the methods used in continental Europe. For it was through his so-called 'fishery barometer programme' that FitzRoy conferred

authority and individual autonomy on local seamen – whom he trusted more than observatory weather scientists.[17] He issued storm warnings – which he called 'cautionary signals' – after the violent storm of October 1859 that caused the *Royal Charter* disaster. His was also the first general 'weather forecast' (as he coined it) ever published, on the first of August 1861, issued by the UK's Met Department. Historian James Burton noted that the accuracy of FitzRoy's storm warnings varied between 76 and 90 percent.[18] Even though the Lagrangian approach was neither British in origin nor entirely representative of British meteorological science – its credibility indeed came under scrutiny by the Royal Society, leading to the suspension of forecasts posthumously – it was the approach that first guided weather prediction practices in Britain. In this volume, we will move away from the historiography focused on FitzRoy's discordances with the values of Victorian culture towards thinking in more nuanced and panoramic ways about his championing of the Lagrangian approach and its associated observational system that confronted Europe's mainstream approach.

The confrontation between the Eulerian and Lagrangian approaches set the conceptual framework for the practice of weather forecasting in the 1850s and early 1860s. However, it should not be thought that the adoption of these approaches implied the application of Euler's equations (or their Lagrangian equivalent) to the patterns of air motion, as illustrated in fluid mechanics treatises and textbooks.[19] These approaches also had much to do with organisational aspects of science and its observational frames of reference. It is for this reason that both approaches incorporated with them a social perspective regarding the way scientists conceptualised weather prediction and envisaged observational systems and their respective worldviews. At this point, their worldviews were almost antagonistic: while the proponents of the Lagrangian approach defended autonomy, local power and freedom of individual judgement, those of the Eulerian approach advocated centralism, authority and state control of meteorological activities.[20]

In the long run, however, FitzRoy's efforts were wasted. The convergence between statistical science and the synoptic method of weather isobaric maps led to the hegemony of the Eulerian approach from the 1870s to the 1910s in Britain and elsewhere. The beginnings of this movement date back to the 1850s, when the science of meteorology began to take certain statistical forms that concerned not only graphical procedures and methods, but also organisational aspects related to standardisation and coordination in the modes and systems of observation.[21] One of the main contributions in this field was Christophorus Buys Ballot's calculation method based on deviations from mean values; according to him, it was deviations – rather than direct measurements, which in his view were not trustworthy values – that informed us about the atmospheric state.[22] This method allowed him to derive a relationship between wind direction and atmospheric pressure, a local rule of thumb that would be transformed into a widely recognised law (the so-called "Buys Ballot's baric wind law") by the favourable reports of the Royal Society of London and would serve as the basis for a system of storm warnings.

A few years earlier, the practice of storm warnings in Britain had been abolished by the Royal Society as "unscientific" following FitzRoy's death in 1865.[23]

The convergence between statistical science and the synoptic method of weather isobaric maps in Britain took place not only by the influence of Buys Ballot's work but also – and above all – by the achievements of a Scottish group in the field of vital statistics. Alexander Buchan's and James Stark's ideas and methods reflect how the intersection between meteorology and vital statistics provided an impetus for this convergence. They criticised the record compilation and interpretation methods of James Glaisher from the Royal Observatory at Greenwich, and demanded strict observing regulations and rigour in vital statistics. They then applied this statistical rigour to the mapping of average atmospheric variables in Britain.[24] For Buchan, and contrary to what Buys Ballot held, deviations from means did not adequately reflect the barometric pressure distribution. The pressure reduced to sea level, however, did. From this perspective, he devised an analytical method of synoptic weather mapping that would share essential features of Marié-Davy's Eulerian approach: it was the pressure field that determined the patterns of temperature, wind and precipitation associated with a storm.[25]

The search for "weather types" and isobaric forms characterised synoptic meteorology in the 1880s and 1890s. In Britain, the Meteorological Office welcomed Buchan's analytical method and synoptic weather maps, as shown in the texts of the Office's head, Robert H. Scott. The search for statistical regularities in weather patterns, as well as the use of rules of thumb for methodical forecasting, became widespread at this time. Examples of these efforts were Edward B. Garriott's "weather types" in the United States or Gabriel Guilbert's rules for weather forecasting in France, which soon became popular in Britain and Europe in general. None, however, achieved in the islands the glory of Ralph Abercromby, who attributed a definite weather type to a given form of isobars.[26] In fact, Abercromby proposed seven forms of isobars and attempted to identify them with certain weather types. By assigning a geometric form to the abstract notion of weather type, he aimed to inspire forecasters to think that analogous synoptic situations in the past could be repeated in the near future. Abercromby's isobaric geometry became the epitome of the use of the Eulerian method in weather forecasting.

The science of meteorology took on certain statistical forms that affected the standardisation and coordination of observational systems. A new – official and state-sponsored – synoptic meteorology was projected in international meetings and national governments. Protocols, standardisation of methods and definitions, uniform and coordinated systems of observations and other issues were discussed at the first international meetings in Leipzig (1872) and Vienna (1873), where the barometric gradient was postulated as the indicator of the degree of approach of storms.[27] The language of synoptic maps, in symbiosis with that of statistical quantification, paved the way for the legitimisation and entrenchment of national weather forecasting services.

PUBLIC SERVICE AND WEATHER MAP PRODUCTION

Both the commitment and the subordination of research to public service, and the subsequent bureaucratisation and stagnation of predictive practice, characterised national weather agencies between 1870 and the First World War. These agencies were primarily concerned with providing public service in their quest for national surveillance and control by using increasingly advanced mapping technologies.[28] The technical issue was not a trivial one. In fact, no institution or observatory other than a national agency had the necessary technical resources to daily produce weather maps in massive quantities.[29] In Britain, as elsewhere, the allocation of public funds for weather forecasting had an unequivocal impact: state patronage led to a public service-oriented meteorology rather than a research-oriented meteorology. As a result, meteorological research always remained subordinate to public service.[30] As Tor Bergeron from the Bergen School of Meteorology in Norway acknowledged in retrospect, 'the meteorological institutes of the world – some of them with the main task of forecasting weather – were rapidly transformed into administrative–statistical bureaus, mainly loaded with the burden of ruling administratively and digesting statistically the observational data'.[31]

Weather maps embodied the practice of scientific prediction for public service. As visual objects, they were inscribed in a language that transcended the cultural boundaries derived from languages and verbal expression, and that proved to be very accessible and comprehensible to the general public. Their visuality and intelligibility seemed to make them ideal tools for simplifying the complexity of atmospheric phenomena and written verbal descriptions.[32] As stated in a book on weather forecasting, 'once the symbolism is known, the [weather] map allows us to assimilate a vast amount of information at a single glance. To transmit the same information in writing, a long, hard-to-read, and tough-to-memorise text would be required'.[33] The mass production of daily weather maps became a priority for the main national meteorological institutes: emulating the path initiated by the Paris Observatory, the U.S. Weather Service was the first to become involved in map production in 1871, followed by the British Met Office in 1872, and so on up to almost twenty by 1900. As the articles reproduced in this volume by Robert H. Scott, Frederic Gaster, H.H.C. Dunwoody and other employees of these institutions show, the common feature behind the imprint of most weather and climatological maps was the configuration of a national space: the creation of the official picture of a national meteorological space that theretofore did not exist. It was meant to provide a tangible and national form to the intangible and universal idea of the weather.

The primary sources in the last section of this first volume evince the extensive development of more than six decades of meteorological cartography.[34] They show how, in a first transition, around the 1850s, maps changed their function from being tools for weather *diagnosis* to tools for weather *prognosis*. Or, put in another way, how they went from being a means of elucidating the general natural

laws governing atmospheric changes to being, thanks to the advent of telegraphy, a means of forecasting the progress of atmospheric disturbances.[35] They also show how, in the 1850s and 1860s, synoptic charts reflected the coexistence of two opposing approaches – the Eulerian and the Lagrangian – to weather mapping and the study of forecasting in general. While the maps of Marié-Davy and Buchan, among others, emphasised the visual power of isolines and the determining role of the baric field in weather conditions, the maps of FitzRoy and Francis Galton resorted to symbols and trajectories to show the motion of air masses. The concluding article of the volume, Mark Harrington's 'History of the Weather Map', shows how support from government agencies reinforced the isolinear character of the weather maps.[36]

VOLUME CONTENTS

This volume illustrates the intertwining of meteorological science and nation-state building in Britain and the influence received from surrounding countries in the nineteenth century. It covers a wide range of factors and currents that link synoptic weather science to state interests. Its central theme is the confrontation of two approaches (Eulerian and Lagrangian) and the subsequent hegemony of the former over the latter that defined the conceptual framework of weather prediction in the Western world. This is a novel departure from conventional narratives for describing the science of meteorology in this period conceived from cultural and national/imperial approaches. The imperial, instead, is addressed in Volume II on 'Cyclones, Trade, Navigation and Society', and especially in Volume IV on 'Weather, Climate and Empire', as a constituent element of the transformation of the science of meteorology. Instead, this volume opens by proposing the confrontation of two approaches incorporating two different observation systems and social organisational models: one, the Eulerian, in favour of centralism, authority and state control; the other, the Lagrangian, in favour of autonomy, local power and freedom of individual judgement.

Section One explores the dissemination of Herschel and Birt's atmospheric wave theory in Europe and Quetelet's graphical method of representing these waves. It traces the adoption of this method in France and Italy as a result of accommodation of scientific thought to state's interests in their pursuit of order, security and centralism. Section Two demonstrates how there was a confrontation between a Eulerian approach (developed in France and Italy) and a Lagrangian approach (adopted mainly in Britain), which acted as a counterweight from 1855 to 1865 and which advocated the quest for autonomy and individual judgement in weather prediction. The texts in this section show how the latter championed an approach diametrically opposed in form and goals to the methods used in continental Europe.

Section Three illustrates the hegemony of the Eulerian approach from the 1860s onwards, largely as a result of the convergence between statistical science and the synoptic method of weather representation. The texts in this section show how

meteorologists embarked on the search for objective criteria to establish methodical forecasting, especially the search for empirical rules of thumb and statistical regularities in weather patterns. Several texts highlight the importance of international meteorological meetings in promoting an official, state-funded science that combined the languages of statistical quantification and cartography.

Section Four turns to national weather services and the mass production of synoptic maps of isobars in the final decades of the nineteenth century, with a special focus on the British Met Office and its allocation of public funds. The texts in this section show how weather forecasts and storm warnings were prepared and disseminated. They also evince how state support led to a public service-oriented meteorology, in which research was always subordinated to public duty. Finally, Section Five offers an overview of nineteenth-century meteorological cartography, from the coexistence of two opposing approaches in the 1850s and 1860s to the primacy of isolines characteristic of the Eulerian method from the 1870s onwards. The emphasis on local weather forecasting shifts focus from observation points to isolines (in particular, isobars), providing insights into how the expansion of statistical cartography and public funding helped make isobars the dominant feature of synoptic weather maps.

The primary sources in this volume cover a wide range of genres and forms and encompass articles, essays, reports, letters, treatises, conference papers and scientific works. A considerable number of sources are taken from periodical publications. While some were published in reputable scientific journals, such as *Nature* and *Philosophical Magazine*, most appeared in periodicals attached to a specific institution, such as the Meteorological Societies of Britain, Scotland, France and America, as well as in annals and bulletins of national weather services. In nineteenth-century meteorology, it became almost the rule for every national service, amateur association and national observatory to have one periodical organ of diffusion. Periodicals facilitated the creation of communities whose members shared, to a greater or lesser extent, values and cultural aspects. In sociologist Gary Alan Fine's terms, they helped to create a 'culture of prediction' by sharing knowledge, methods, values and rules around the practice of weather forecasting.[37] This opportunity to share a culture was especially important for those who embraced a specific approach and method, as it contributed to a recognition of belonging and, in some cases, creating a cohesive group identity. However, national narratives and many cultural narratives often tend, intuitively rather than deliberately, to pigeonhole the knowledge produced by these actors into the categories of "national" and "foreign", and, on a more general level, "local" and "global". In doing so, they often draw clear lines of demarcation between monolithic blocks. Our sources, however, dismantle these arguments. For example, the empiricist Christophorus Buys Ballot proposed a local rule of thumb as the basis for his system of storm warnings which, although relatively successful within Dutch borders, was met with opposition and scepticism among his overseas rivals – including FitzRoy, who devised his own system of warnings. Ironically, it was the Royal Society – which after FitzRoy's death took the British Met Office under

its control – that eventually turned Buys Ballot's local rule into a publicly distinguished meteorological law in Britain.[38]

It is important to highlight that, while it is difficult to reflect so in a work of this nature, the positions and actions of the actors at stake can be interpreted as a struggle of forces expressed at three levels: the local, the national and the global. The merits of precision, validity and statistical knowledge in the field of forecasting were the subject of intense rivalry and controversy even within the same country.[39] British seamen (including FitzRoy) and the Royal Society interpreted the value and nature of weather forecasts in diametrically opposed terms, as did officers and scientists from the French Navy and the Paris Observatory. For example, FitzRoy's forecasts found a warm reception among the French Navy, to the extent that the commander-in-chief of the English Channel and North Sea fishery districts sought to establish an autonomous weather service dependent on FitzRoy's warnings in the French Navy.[40] Not only that; FitzRoy's gale warnings were telegraphed to weather departments in Sweden, Denmark and some Germanic states. These examples illustrate the extent to which tensions between advocates of the two approaches at stake crossed national boundaries and were expressed at the local, national and global levels.

Nation-state building and statistics shaped weather forecasting across the long nineteenth century. The pieces chosen for this volume testify to a multitude of aspects that characterised the confrontation of two approaches and the subsequent hegemony of one of them, and each section serves as a basis for understanding the various stages of this long and gradual procedural movement. These texts offer insight into the intertwining of cognitive and organisational aspects (and observational frames of reference) that were inherent in the two approaches. But each section cannot do justice to the diversity of voices that contributed and challenged each discussed method. Nonetheless, the volume demonstrates the variety and multiplicity of state interests related to order, security and surveillance in the practice of weather forecasting, and shows how the latter became a public service in the pursuit of these interests through increasingly advanced mapping technologies.

Notes

1 David N. Livingstone, *Putting Science in its Place: Geographies of Scientific Knowledge* (Chicago: University of Chicago Press, 2003), p. 154.
2 Mark Monmonier, *Air Apparent. How Meteorologists Learned to Map, Predict, and Dramatize Weather* (Chicago: The University of Chicago Press, 1999), p. 1.
3 John Brian Harley, 'Maps, Knowledge and Power', in Denis Cosgrove and Stephen Daniels (eds), *The Iconography of Landscape. Essays on the Symbolic Representation, Design and Use of Past Environments* (Cambridge: Cambridge University Press, 1988), pp. 277–312, on p. 301.
4 Theodore M. Porter, *The Rise of Statistical Thinking 1820–1900* (Princeton: Princeton University Press, 1986); Stephen M. Stigler, *The History of Statistics. The Measurement of Uncertainty before 1900* (Cambridge, Mass.: The Belknap Press of Harvard University Press, 1986).

5 Oscar Boris Sheynin, 'On the History of the Statistical Method in Meteorology', *Archive for History of Exact Sciences*, 31:1 (1984), pp. 53–95.

6 James Burton, 'Robert FitzRoy and the Early History of the Meteorological Office', *British Journal for the History of Science*, 19 (1986), pp. 147–76; Pauline Halford, *Storm Warning. The Origins of the Weather Forecast* (Stroud: Sutton Publishing, 2004); Naylor, *The Observatory Experiment*; J. Malcolm Walker, *History of the Meteorological Office* (Cambridge: Cambridge University Press, 2012).

7 Katharine Anderson, *Predicting the Weather: Victorians and the Science of Meteorology* (Chicago: The University of Chicago Press, 2005), pp. 123–25, 157–59.

8 See, e.g.: John D. Cox, *Storm Watchers: The Turbulent History of Weather Prediction from Franklin's Kite to El Niño* (Hoboken, New Jersey: John Wiley & Sons, Inc., 2002), pp. 85–90; Fabien Locher, *Le savant et la tempête. Étudier l'atmosphère et prévoir le temps au XIX siècle* (Rennes: Presses Universitaires de Rennes, 2008); Christina Helena da Motta Barboza, *As viagens do tempo. Uma história da meteorologia em meados do século XIX* (Rio de Janeiro: E-papers Serviços Editoriais Ltda., 2012); James Lequeux, *Le Verrier–Magnificent and Detestable Astronomer* (New York: Springer, 2013), pp. 280–84.

9 Lequeux, *Le Verrier.*

10 Barboza, *As viagens*; Christina Helena da Motta Barboza, 'Nice Weather, Meteors at the End of the Day', in Stefan Emeis and Cornelia Lüdecke (eds), *From Beaufort to Bjerknes and Beyond: Critical Perspectives on Observing, Analyzing, and Predicting Weather and Climate* (Augsburg: Rauner, 2005), pp. 157–68.

11 Locher, *Le savant*, p. 139–45.

12 Aitor Anduaga, *Politics, Statistics and Weather Forecasting, 1840–1910. Taming the Weather* (London & New York: Routledge, 2020).

13 James Rodger Fleming, *Meteorology in America, 1800–1870* (Baltimore: Johns Hopkins University Press, 1990).

14 Vladimir Jankovic, 'Ideological Crests versus Empirical Troughs: John Herschel's and William Radcliffe Birt's Research on Atmospheric Waves, 1843–50', *British Journal for the History of Science*, 31 (1998), pp. 21–40.

15 Locher, *Le savant.*

16 Gisela Kutzbach, *The Thermal Theory of Cyclones: A History of Meteorological Thought in the Nineteenth Century* (Boston: American Meteorological Society, 1979), pp. 82–3; Halford, *Storm Warning*, pp. 177–88; Anduaga, *Politics*, pp. 165–67, 181–93.

17 Sarah Dry, 'Safety Networks: Fishery Barometers and the Outsourcing of Judgement at the Early Meteorological Department', *The British Journal for the History of Science*, 42:1 (2009), pp. 35–56; Anduaga, *Politics*, pp. 171–75.

18 James Burton, *The History of the British Meteorological Office to 1905* (The Open University, PhD diss., 1988), pp. 56–9.

19 Olivier Darrigol, *Worlds of Flow. A History of Hydrodynamics from the Bernoullis to Prandtl* (Oxford: Oxford University Press, 2008), pp. 23–30; Andrew Bennett, *Lagrangian Fluid Dynamics* (Cambridge: Cambridge University Press, 2006), pp. 1–24.

20 Anduaga, *Politics*, p. 7.

21 Sheynin, 'On the History'.

22 Marcel Boumans, *Science Outside the Laboratory: Measurement in Field Science and Economics* (Oxford: Oxford University Press, 2015), pp. 68–76.

23 Azadeh Achbari and Frans van Lunteren, 'Dutch Skies, Global Laws: The British Creation of "Buys Ballot's Law"', *Historical Studies in the Natural Sciences*, 46:1 (2016), pp. 1–43.

24 Walker, *History*, pp. 32–4; Anduaga, *Politics*, pp. 213–24.

25 Kutzbach, *The Thermal Theory*, pp. 71–5.

26 Burton, *The History*, p. 146; Walker, *History*, p. 124; Anderson, *Predicting*, pp. 205–07.

27 Aleksandr Khristoforovich Khrgian, *Meteorology: A Historical Survey* (Jerusalem: Israel Program for Scientific Translations, 2nd ed, 1970), p. 178.
28 Donald R. Whitnah, *A History of the United States Weather Bureau* (Urbana: University of the Illinois Press, 1961); Walker, *History.*
29 Katharine Anderson, 'Mapping Meteorology', in James Rodger Fleming, Vladimir Jankovic, and Deborah R. Coen (eds), *Intimate Universality. Local and Global Themes in the History of Weather and Climate* (Sagamore Beach: Watson Publishing International, 2006), pp. 69–91, on p. 77.
30 Anduaga, *Politics*, pp. 298–341.
31 Tor Bergeron, 'Synoptic Meteorology: An Historical Review', *Pure and Applied Geophysics*, 119 (1981), pp. 443–73, on p. 448.
32 Monmonier, *Air Apparent*; Anderson, 'Mapping Meteorology'.
33 Translation mine from French original. Jean-Pierre Javelle, Michel Rochas, Claude Pastre, Michel Beaurepaire and Bruno Jacomy, *La météorologie, du baromètre au satellite. Mesurer l'atmosphère et prévoir le temps* (Lausanne: Éditions Delachaux et Niestlé, 2000), p. 86.
34 Arthur H. Robinson and Barbara Bartz Pentchenik, *The Nature of Maps: Essays towards Understanding Maps and Mapping* (Chicago: Chicago University Press, 1976); Monmonier, *Air Apparent*; Anderson, 'Mapping Meteorology'; Deborah R. Coen, 'Scaling Down: Mapping the Austrian Climate between Empire and Republic', in Fleming, Jankovic and Coen, *Intimate Universality*, pp. 115–40.
35 Arthur H. Robinson and Helen Wallis, 'Humboldt's Map of Isothermal Lines: A Milestone in the History of Thematic Cartography', *Cartographic Journal*, 2 (1967), pp. 119–23; *Jane R. Camerini, 'The Physical Atlas of Heinrich Berghaus: Distribution Maps as Scientific Knowledge', in Renato G. Mazzolini (ed), Non-Verbal Communication in Science Prior to 1900* (Florence: Leo S. Olschki, 1993), pp. 479–512.
36 Mark W. Harrington, 'History of the Weather Map', in Oliver L. Fassig (ed), *Report of the International Meteorological Congress, Chicago, Ill., August 21–24, 1893. Part II* (Washington D.C.: Weather Bureau, 1895), pp. 327–35.
37 Gary Alen Fine, *Authors of the Storm: Meteorologists and the Culture of Prediction* (Chicago: University of Chicago Press, 2007), p. 2.
38 Azadeh Achbari, Frans Van Lunkeren, 'Dutch Skies, Global Laws: The British Creation of "Buys Ballot's Law"', *Historical Studies in the Natural Sciences*, 46:1 (2016), pp. 1–43, on pp. 32–4.
39 Fiona Williamson, 'Uncertain Skies: 'Forecasting' Typhoons in Hong Kong c. 1874–1906', *Quaderni Storici*, 52:3 (2017), 777–802.
40 Locher, *Le savant*, pp. 49–51.

Part 1

SYNOPTIC WEATHER SCIENCE AND STATE INTERESTS

1.1

PRECURSORS

Editorial Headnote

1.1 Precursors

The compilation of extracts from mid-nineteenth-century meteorological studies offered here provides some evidence of the concerns and graphical representations that gave rise to synoptic meteorology. In some respects, the 1850s represent a distinctive period in the development of new synoptic techniques for the study of storms but, at the same time, it is not possible to trace a complete break with their precursors. Although the first five extracts in Part 1 differ in many ways from the rest of the section, there are also significant continuities: both the "cold waves" proposed by Elias Loomis (1811–1889) and the "atmospheric waves" by John Herschel and William Birt constituted the first wavelike solutions to what was then known as the "storm controversy". These solutions helped to lay the foundations for a dynamical-predictive conception of atmospheric weather.

Although it would have been possible to begin Part 1 with this controversy of opposing (mechanical and thermal) storm theories (see Vol. II), our aim to show the close relationship between weather forecasting and nation-state building is most conclusively achieved by tracing the early wavelike formulations of storms. We have therefore chosen these extracts of Part 1 to demonstrate a) a growing interest in conceptualising storms in terms of wave-like motions of the body of the atmosphere; b) the rise of statistical and graphical thinking; and c) the development of the synoptic chart, or, in other words, a graphical representation technique that showed the successive simultaneous states of the atmosphere at a given time over a large area, based on measurements of various meteorological variables taken at the surface at specific times.[1]

Tracing the early roots of synoptic meteorology is useful because the relationship between weather forecasting and state gradually became a key point of interest in European and American national observatories. Loomis is included here as the most important of the precursors, but most of the authors in this section share many of the fundamentals of the wave approach.

We begin with two seminal papers by Loomis, the first and last in a series he published from 1841 to 1845. Although the studies of the exponents of mechanical and convective storm theories – William C. Redfield and James P. Espy, respectively – were influential (see vol. II), Loomis' work is of greater significance for present purposes.[2] 'On the Storm Which Was Experienced throughout the United States about the 20th of December, 1836' (1841) and 'On two Storms Which Were Experienced throughout the United States, in the Month of February, 1842' (1845) deserve to be considered pioneers of the wave approach, even if the reality is more interesting and multifaceted.[3] In his 1841 paper, Loomis was certainly novel in introducing the concept of a *wave of cold air* moving faster than the storm centre, in defining what would later be called a *mid-latitude cyclone*, and in approaching it through a case study of a single storm (rather than averaging the results of many storms), but he also illustrated his idea on a chart and, by tracing its shape and progress, came to represent 'the first cross section of what today is called a cold front'.[4]

Loomis' 1841 study opens with a description of his participation – at the invitation of the British astronomer John Herschel – in the 1835 campaign of hourly meteorological observations during the equinoxes and solstices. His participation meant an acknowledgement of the importance of the first-ever international campaign of synchronised and uniform observations, which was promoted by the British Association for the Advancement of Science at selected stations around the world. Loomis drew on this campaign to study a large-scale storm that occurred during the winter solstice of 1836. By this time, Loomis was a well-known professor of mathematics and natural philosophy at Western Reserve College in Hudson, Ohio, having been a tutor at Yale College, where he graduated in 1830. In 1836 he went to Paris for a period of study, attending the classes of François Arago and Jean-Baptiste Biot, among others. He chose this storm because he had at hand an unusually large number of observations covering the whole of the USA and beyond to northern Canada and southern Mexico: he had barometric readings from twenty-seven stations and hourly readings from eight eastern stations from Quebec to Baltimore. No one had ever provided such a comprehensive overview and a study of a single storm in such detail. Loomis chose to present his study to the American Philosophical Society of Philadelphia, a prestigious association founded in 1743 by Benjamin Franklin and others, which fostered intellectual exchange and ongoing scientific debate; publishing the study in its journal (*Transactions*) brought him not only prestige and credibility but also an impact and was well received. Given the then-recent controversy between Redfield and Espy over the pattern of surface winds in storms (the mechanical whirlwind theory versus the thermal convective theory controversy), Loomis' paper is a timely attempt to settle this debate through a thorough case study of a single storm with highly distinctive characteristics, following its progress as extensively and minutely as possible. It is also true, however, that Loomis was unable to confirm which of the two theories was correct.

Loomis' 1845 paper "On two storms", the first example of a synoptic weather map, is refreshingly suggestive and describes the course of two winter storms in 1842. Our extract focuses on Loomis' proposals to extend the undulant form to include large-scale meteorological elements such as barometric oscillations, and to represent these elements on a synoptic weather map drawn long after the storms had occurred. With regard to the first proposal, his main empirical innovation was that the velocity and form of the barometric wave could be calculated by plotting and analysing synoptic maps. The determination of this wave by drawing barometric curves from a number of stations and plotting the lines of greatest depression for every six hours on a map of the USA is the major proof of what he pursued. The plotting of lines of maximum depression, and, in particular, the construction of lines of equal barometric deviation, should be placed in the context of graphical representation within the so-called 'storm controversy'. In fact, while Redfield represented the wind field by lines and arrows, and Espy plotted the minimum pressure lines, Loomis charted deviations from normal pressure and lines of equal deviation from temperature. The effect is unquestionable: Loomis definitely achieved more complete weather charts. As historian Mark Monmonier points out, the Loomis maps of 1845 are the 'first printed examples of the *synoptic weather map*', combining 'information for pressure, temperature, wind, sky conditions, and precipitation'.[5] The use of maps was intended to rule out those theories that were wrong. As Loomis goes on to point out, the function of synoptic mapping techniques was to elucidate the correct theory on the progress of storms and their physical laws:

> It appears to me, that if the course of investigation adopted with respect to the two storms of February, 1842, were systematically pursued, we should soon have some settled principles in meteorology. If we could be furnished with two meteorological charts of the United States, daily, for one year, charts showing the state of the barometer, thermometer, winds, sky, etc., for every part of the country, it would settle for ever the laws of storms. No false theory could stand against such an array of testimony. Such a set of maps would be worth more than all which has been hitherto done in meteorology.
>
> (p. 183)

Notes

1 Aitor Anduaga, *Politics, Statistics, and Weather Forecasting, 1840–1910. Taming the Weather* (London and New York: Routledge, 2020), pp. 42–7.

2 Hubert Anson Newton, 'Biographical Memoir of Elias Loomis, 1811–1889', *Biographical Memoirs, National Academy of Sciences*, 3 (1895), pp. 213–52; Gisela Kutzbach, 'Loomis, Elias', in Charles Coulston Gillispie (ed.), *Dictionary of Scientific Biography* (New York: Charles Scribner's Sons, 1973), vol. 8, p. 487.

3 John D. Cox, *Storm Watchers: The Turbulent History of Weather Prediction from Franklin's Kite to El Niño* (Hoboken, NJ: John Wiley & Sons, Inc., 2002), pp. 41–9.

4 Gisela Kutzbach, *The Thermal Theory of Cyclones: A History of Meteorological Thought in the Nineteenth Century* (Boston, MA: American Meteorological Society, 1979), p. 30.

5 Mark Monmonier, *Air Apparent. How Meteorologists Learned to Map, Predict, and Dramatize Weather* (Chicago, IL: The University of Chicago Press, 1999), p. 36.

1

ON THE STORM WHICH WAS EXPERIENCED THROUGHOUT THE UNITED STATES ABOUT THE 20TH OF DECEMBER, 1836

Elias Loomis

Source: Elias Loomis, 'On the Storm Which Was Experienced throughout the United States about the 20th of December, 1836'. *Transactions of the American Philosophical Society,* 7 (7), 1841, pp. 125, 128–130, 156–163.

On the Storm which was experienced throughout the United States about the 20*th of December,* 1836. *By Elias Loomis, Professor of Mathematics and Natural Philosophy in Western Reserve College. Read March* 20, 1840.

BEING well convinced that meteorology is to be promoted, not so much by taking the mean of long-continued observations, as by studying the phenomena of particular storms developed over a widely extended country, I resolved to select some single storm of strongly marked characteristics, and trace its progress as extensively and minutely as possible. For this investigation I made choice of the storm which occurred in the United States about the 20th of December, 1836; not only because it seems well suited to my purpose, but because I found ready furnished for my use a considerable number of most valuable observations. In the eastern states this storm occurred within the period recommended by Sir John Herschel for hourly meteorological observations, and all the phenomena of the storm were most carefully and minutely recorded at eight different stations, namely, at Baltimore, New York, Albany, Flushing, New Haven and Gardiner, in the United States; as, also, at Montreal and Quebec, in Lower Canada. These observations, with the exception of those at Baltimore, are published in the Report of the New York University Register for 1837. I addressed a letter to each individual who, so far as I could ascertain, kept a meteorological register, requesting an extract from the same for the period in question. The result is that I have obtained barometric observations from twenty-seven different stations within the United States and the neighbouring British possessions.

DOI: 10.4324/9781003650294-3

[. . .]

I have also collected a great variety of observations from stations beyond the probable limits of the storm, to which reference will be made in the sequel. I think it unnecessary to transcribe all the above mentioned observations at large, because they would occupy a great space, and the substance of them will presently be presented in a condensed form.

In discussing these observations, four distinct subjects of inquiry present themselves.

I. A remarkable oscillation of the barometer.
II. A sudden depression of the thermometer.
III. Rain – amount – with time of beginning and end.
IV. Wind – its duration and velocity.

I. THE OSCILLATION OF THE BAROMETER

A bare inspection of the meteorological registers is sufficient to show that, during the period in question there was every where throughout the United States a sudden depression of the barometer, immediately succeeded by an equally sudden rise; that the barometric minimum occurred first in the western states, and later in the eastern, passing like a wave over the entire country, from west to east. In order to present this fact in the most striking light, I have projected on Plate I., Fig. 1, curves representing this oscillation of the barometer at each station. The abscissas represent the times of observation, and the ordinates the barometric heights on a scale reduced one-third; that is, a difference of two-thirds of an inch in the ordinates of a curve is intended to show an absolute oscillation of the barometer to the amount of an inch. It will be observed that at most of the stations there is a striking resemblance in the curve described, so that we need not hesitate to call it the *wave;* just as an elevation or depression of the surface of the ocean which arrives successively at distant stations, we call one wave, understanding, of course, that the same wave, at successive instants, may be formed by entirely different particles of water. The same atmospheric wave, then, passed over the continent from west to east, from the valley of the Mississippi to the Bermudas and Newfoundland. We wish now to be able to assign the *rate* of progress of this wave, and in order to institute the comparison, we must fix upon some particular *phase* of the wave. We might make the comparison for the times of either the barometric maximum or minimum. The latter, however, is much the best suited to our purpose, as the barometer remained much longer near the high than the low extreme. There is a difficulty, however, in ascertaining the precise instant of even the barometric minimum. When observations are taken but two or three times a day, it is highly improbable that the minimum should occur at the precise instant of either observation. There may, therefore, be an uncertainty in the time to the amount of twelve hours, or more. Fortunately, however, this uncertainty is, in a measure, obviated in the present instance. The hourly observations made at Baltimore, Montreal, Albany, Flushing, New Haven and Gardiner, exhibit the precise form of the barometric curve described, and the instant of greatest depression. Moreover, as will appear in the sequel, this minimum was contemporaneous with a remarkable change of wind, and the time of this change was noted at several stations; as, for example, at Springfield, Syracuse, Philadelphia, New York, and Hanover. At nearly half of the stations, then, the time of minimum is certainly known; and the form of the curve at these stations will guide us in completing the curve at the remaining stations. This conjectural completion of the curves is represented on the chart by dotted lines. We are then enabled to trace the progress of this wave with very considerable precision. I accordingly drew upon a map of the United States lines connecting all those places where the barometer attained its minimum at the same instant. The result is shown on the accompanying chart, Which exhibits the lines of greatest depression for every six hours from the morning of December 20th to the noon of the 23d. The velocity of the wave appears not to have been uniform. Thus, on the southern border of the United States its velocity varied from seventeen to twenty-nine statute miles per hour; and on the northern borders from seventeen to thirty-seven miles per hour. No allowance is here made for difference of longitude, which would increase this velocity by about one twenty fifth part. The form of these curves for the north-west part of the United States rests mainly upon the thermometric observations to be described presently. The front of this wave appears, in many places, decidedly convex, to an amount much beyond the possible errors of observation.

Having thus discovered the form and velocity of the wave, let us attend to the *amount* of the depression of the barometer. The following table exhibits the difference between the barometric minimum and the subsequent maximum, the observations being arranged in the order of latitude.

	Latitude.	Bar. Range.		Latitude.	Bar. Range.
Indian Key,	24° 48′ N.	.26 nich.	Twinsburgh,	41 18	.68 nich.
Pensacola,	30 28	.46	New Haven,	41 18	.98
Natches,	31 34	.79	New Bedford,	41 38	1.02
Savannah,	32 5	.65	Do.	" "	.97
Bermuda,	32 34	.41	Boston,	42 21	.92
Lexington,	38 6	.96	Albany,	42 39	1.173
Washington,	38 53	.98	Syracuse,	43 1	.95
Baltimore,	39 17	1 02	Rochester,	43 8	1.03
Marietta,	39 25	.98	Hanover,	43 41	1.14
Springfield,	39 53	1.00	Gardiner,	44 10	1.00
Philadelphia,	39 57	.97	Halifax,	44 39	.52
New York,	40 43	.97	Montreal,	45 31	1.266
Flushing,	40 45	1.042	Quebec,	46 49	1.57
Sunbury,	40 53	1.00	St. Johns,	47 34	.85

It will be observed that the range of the barometer increases generally with the latitude. A more particular consideration of the subject, however, is deferred for the present. I come, therefore, to consider,

II. THE MOVEMENT OF THE THERMOMETER

The movement of the thermometer during the period in question was quite as remarkable as that of the barometer, and generally in the opposite direction. Thus, while the barometer was falling under the influence of the storm, the thermometer was every where rising; and as the barometer rose the thermometer fell with extraordinary rapidity. It is not thought necessary to insert here all the observations which have been collected. The following, being about half the whole number, are selected as being, from their distribution, a fair representative of the whole.

[. . .]

I come now to inquire for the cause of the rain. Rain is always owing to one cause, namely, a sudden reduction of the temperature of the atmosphere below the dew point. We run no risk in always assigning this as the immediate cause of rain. But how is this reduction of temperature effected? It may occur in a great variety of ways, among which are the following: –

I. Radiation.
II. Warm air coming in contact with cold earth or water.
III. A warm current coming in contact and mingling with a cold one.
IV. Air suddenly transported into elevated regions.

It is doubtful whether the first of these causes ever produces rain, for the reduction of temperature is too gradual. It is very common for the thermometer to sink during the night below the dew point of the preceding evening, and without any change of wind; yet not only no rain follows, but the atmosphere may remain perfectly transparent the whole time. Such circumstances produce a most copious dew, but seldom if ever rain.

The second case is one which is often observed, but ordinarily produces merely a fog. Thus the warm air from the gulf stream rolling over the cold banks of Newfoundland produces the densest fogs; and in winter, the air from the sea flowing in upon the frozen earth has its moisture abundantly condensed. Yet the reduction of temperature effected in this way is also so gradual, it is believed never to produce any considerable rain.

The third cause, though very similar in principle to the second, is believed to be more efficient than either of the preceding, because it will produce a more sudden change of temperature. The phenomenon, moreover, of atmospheric currents flowing one over the other in different directions may be observed almost every day of the year. But the fourth cause named, admitting its existence, must be allowed to be by far the most efficient of all. For the reduction of temperature may be conceived to be effected with almost any degree of suddenness, and to any extent. Thus, in midsummer, air with a dew point of 80°, being suddenly elevated two or three miles above the earth's surface, would, from change of temperature, lose almost the totality of its vapour. By supposing some cause which should continually renew the operation over a limited locality, we could account for the most violent and abundant rains on record. This case, therefore, demands particular consideration. This effect might be produced, 1. By a horizontal current impinging upon the side of a mountain. The current might thus be forced up to an elevation, where it would experience a very great and sudden reduction of temperature, and a portion of its vapour be instantly precipitated. 2. A large body of air might be suddenly and mechanically elevated by a volcanic eruption. The direct effects of these two causes must, however, be quite local. If they ever influence distant stations, it must be indirectly, by means of currents here set in motion. 3. Air may be elevated by a whirlwind; for even in horizontal whirlwinds there is ordinarily, if not always, an upward motion in the centre of the vortex. 4. When a hot and cold current, moving in opposite directions, meet, the colder, having the greatest specific gravity, will displace the warmer, which is thus suddenly lifted from the surface of the earth, is cooled, and a part of its vapour precipitated. This is a cause which may operate in any locality, and with almost any degree of energy. It is believed, therefore, to be, at least in this latitude, the most common cause of rain. Let us now review the circumstances of the rain which fell during the period under examination. This occurred during the winter, and with a southerly wind. We have here, then, the second cause for the reduction of atmospheric temperature, operating under the most favourable circumstances. Yet the precipitation arising from this cause must begin at the surface of the earth, and proceed very gradually. But in the present case the rain came from a considerable elevation, and poured down in torrents. Although, then, this be admitted as a cause of rain, I cannot regard it as the main cause, nor any thing more than a very feeble auxiliary. Can the third cause be the true one? We have, in the present case, a warm current from the south, and, without doubt, a little above it, a cold current from the west. The consequence would be, a certain amount of precipitation from the partial mingling of these two currents where they came in contact. The whole amount of air which could be cooled in this way is very small, and the consequent precipitation would be only a small fraction of that which was observed. Let us, then, inquire if the fourth cause could be supposed to operate. We have already shown from the observations of the wind, without the aid of any hypothesis, that the southerly current must have disappeared by being elevated into the upper regions of the air. This I regard as a necessary deduction from the observations. We find, then, a warm current suddenly cooled, and its moisture must, of course, be in part precipitated. I find by computation, that if the entire atmosphere, saturated with moisture at the temperature of 70°, be cooled 5°

	70°		5°	one inch of water will be precipitated.
"	60	"	7	
"	50	"	10	
"	40	"	16	
"	30	"	25	

But the average amount of water which fell throughout the United States was somewhat less than one inch. The cause we have assigned, then, seems adequate to account for the phenomena. We have simply determined, as yet, that the south-east wind must have disappeared by being lifted from the earth's surface. But in what direction did it continue to move? It might be supposed to continue on its course towards the north-west, above the other current, or to return towards the south-east. Being a current of more than a thousand miles in breadth, we can hardly suppose it to escape either by the northeast or the south-west. Did it, then, continue its direction towards the northwest? We should then have a warm and moist current flowing at a great elevation over an exceedingly cold one. The consequence would be, that its moisture would not only be precipitated, but frozen, and would descend to the earth in the form of snow or hail. This is conformable to observation. Snow and hail did fall at nearly all of the northern stations, after the north-west wind set in. But the amount was small; much less than must necessarily result if this entire southerly wind had flowed over the northerly, and had its moisture precipitated by it. Still, it seems probable that a part of this southerly wind did continue on its course, and produce the snow which was observed to fall. I infer that the current was mainly turned back upon itself, so that the moisture, as fast as precipitated, fell through the lower current, still blowing from the south-east. My idea may, perhaps, be best illustrated by a diagram.

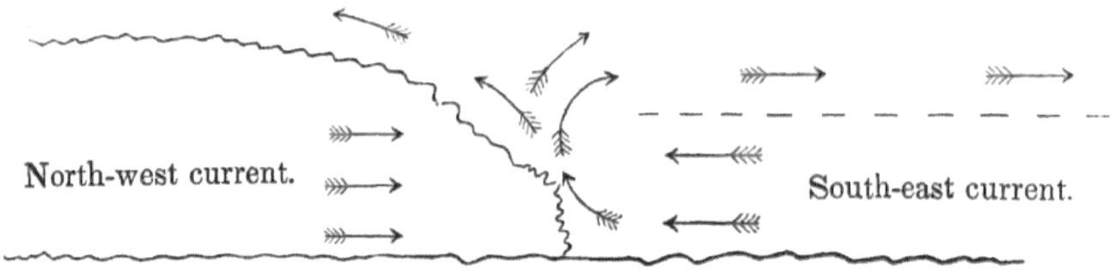

A part of the moisture of this south-east current might be frozen from the cold of elevation, even if it were driven back in the direction from which it came. But, falling through a warm stratum, before it reached the earth's surface it would be entirely melted. At several stations hail fell in small quantities very soon after the wind changed. This might be owing to the moisture of the warm current suddenly condensed by contact with the cold one, and sustained by this upward motion, until the frozen drops had acquired the size actually observed, which in no case was very great.

Why more rain fell at some stations than at others only moderately distant I cannot certainly say. The causes, doubtless, were local, and it would be unsafe to assign particular reasons without a good knowledge of the localities. Such an effect, however, might be produced by a hill of moderate elevation, which would raise the lower current, and, consequently, the returning upper current, to an unusual elevation, producing a corresponding depression of temperature.

The observations of the thermometer on pages 131–133 are believed to present nothing very difficult of explanation. That the temperature should rise under a southerly wind was to be anticipated. The wind blew from a southern quarter more than twelve hours, in which time it probably travelled nearly five hundred miles, or, making allowance for the obliquity of its course, between four and five degrees of latitude. The difference of temperature corresponding to this difference of latitude, for winter, in the United States, is about ten degrees. But the temperature of the air generally rose more than this number of degrees above the mean. Moreover, there prevailed in the northern states a vast body of snow and ice, which rapidly absorbed the caloric of this southern current. The explanation, then, is insufficient. We shall find an additional cause for this elevated temperature in the abundant precipitation of aqueous vapour. When vapour becomes reduced to the liquid state, it gives up a large amount of latent heat, which will, of course, be taken up by the surrounding air. The temperature of the eastern states was found to be a little greater than that of the western. At the former, the southerly wind came from the ocean, which, in winter, has a higher temperature than the land. It is not clear, however, that more rain fell in one section than the other.

That the north-west wind was cold is not remarkable. It came from a high latitude, where the prevalent temperature is, probably, far below zero. Its severity, however, became somewhat tempered in its progress, the mean of the minima at the eastern stations being fourteen degrees higher than at the western.

I come now to the phenomenon probably the most difficult of all to be explained, namely, the oscillation of the barometer. For the movements of this instrument various causes have been assigned, some of them so unphilosophical that it is really astonishing that they could ever have been seriously proposed by intelligent men.

1. The oscillations of the barometer have been ascribed to the destruction of large masses of air in the higher regions by electricity. The supposition is too gratuitous to deserve serious consideration.
2. They have been ascribed to the diminished pressure resulting from the loss of rain. But the amount of rain which fell in the case under consideration would be balanced by a column of mercury about one-fifteenth of an inch in height.
3. Heat, by expanding a column of air, causes it to ascend to a greater height, and thus changes its centrifugal force arising from the earth's rotation. This cause is too insignificant to produce the effect in question.
4. They have been ascribed to the attractions of the sun and moon. Laplace estimates the greatest oscillation of the barometer due to this cause to be, at the equator, 0.025 inch.

5. Leslie ascribes them to the centrifugal force arising from violent winds. But, in the case of a hurricane, this would not produce an oscillation of the barometer amounting to the thousandth part of an inch.
6. The opposition of winds. This might produce a small movement of the barometer. We shall presently inquire whether this cause operated in the case under consideration.
7. The barometer has frequently been observed to fall under the influence of a whirlwind. But in the present case there was no whirlwind.
8. These oscillations have been ascribed to sudden changes in temperature and in the amount of aqueous vapour. An elevation of temperature of the entire atmosphere could not directly affect its pressure, for, in proportion as its density is diminished, its height will be increased. But if, by any means, a portion of hot air can be made to displace an equal bulk of cold air, the weight of the column must be diminished. It is obvious that this cannot be a state of permanent equilibrium; yet it is worthy of inquiry whether it may not temporarily exist under the influence of winds. On the 20th of December, 1836, the air over nearly the whole of the United States became unusually heated, and its specific gravity was, of course, diminished. If, then, the height of the atmosphere remains invariable, a diminution of pressure ought to be the consequence. But, although a fall of the barometer is usually accompanied by an elevation of temperature, the reverse is sometimes the case. Thus the fall of the barometer in Europe, which I have represented on Plate 2, and which, at most places, amounted to more than an inch, was accompanied by a steady fall of the thermometer. The barometer, in this case, fell in spite of the increased specific gravity of the air. We may naturally presume, then, that a change in the specific gravity of the air produces only a secondary effect on the oscillations of the barometer.
9. A wind blowing upward or downward would affect the pressure of the air. This is a cause whose existence we have proved in the case in question. Its effect upon the mean pressure of the air in the equatorial regions is unequivocally maintained in the Instructions for the British Scientific Expedition to the Antarctic Regions, recently prepared by the President and Council of the Royal Society, causing the barometer at the equator to stand permanently lower than in latitude 30°, by about a quarter of an inch. The ascending current of December 20th could not, however, exert any direct influence upon the barometer, except near the centre of the storm. To account for the entire oscillation, I think, we must admit another principle quite distinct in its operation.
10. Let a wind blow ever so violently over the earth's surface, and the diminution of gravity arising from the centrifugal force must be inconsiderable. But, imagine the different parts of the current to travel with unequal velocity, and there will arise a mechanical condensation, or rarefaction. When air is at rest, or in motion, with a uniform velocity, its particles are maintained at a constant distance from each other. But let the velocity of one section be increased

beyond that of the succeeding, and the same particles of air are forced to fill a greater space. Such is the principle of the undulations which produce the sensation of sound. It appears to me that a similar effect must have been produced in the storm of December 20th. The southeast wind which accompanied the rain moved with an accelerated velocity. The particles, therefore, of air at one extremity of the current must have left those of the other extremity at an increased distance. Hence a mechanical rarefaction, and, of course, diminished pressure. The reverse effect must have taken place after the storm had passed. A north-west wind sets in with great violence. A vast body of air is precipitated towards the southeast. The partial vacuum which at first existed is very soon supplied. Yet, though the first impelling cause has ceased to act, the momentum of the excited current still urges it onward. The front of the wave is impelled by the momentum of the mass in the rear, and a mechanical condensation results, bringing, of course, increased barometric pressure. The cause, however, which produces this extraordinary rise, being temporary in its nature, soon ceases, and the barometer falls. The causes which I have here assigned for the oscillation of the barometer appear to me to be such as are known to be true, and that they are sufficient to account for the phenomena.

I have thus analyzed the phenomena of this somewhat remarkable storm, and given such explanations of them as have appeared to me most satisfactory. It is a most interesting and important inquiry, how far the conclusions at which I have here arrived may be safely generalized. I have already made some progress in this investigation; but as the present paper has now grown to an unexpected length, I reserve my remarks upon that point to some future opportunity. I trust it will not be inferred from my silence with respect to the labours of others in this important field, that I am wholly ignorant of them, or am insensible of their value. I have availed myself of the labours of others as far as was in my power. To have credited every suggestion to its original author would have been inconvenient, and generally superfluous, being found in most treatises on meteorology. I am happy, however, to express my particular obligations to the labours of Messrs. Redfield, Espy, and Col. Reid, and shall esteem myself well repaid if the present communication shall contribute something to the progress of that science which they have done so much to promote.

2

ON TWO STORMS WHICH WERE EXPERIENCED THROUGHOUT THE UNITED STATES, IN THE MONTH OF FEBRUARY, 1842

Elias Loomis

Source: Elias Loomis, 'On two Storms Which Were Experienced throughout the United States, in the Month of February, 1842'. *Transactions of the American Philosophical Society*, 9 (2), 1846, pp. 161, 164, 179–184.

> *On two Storms which were experienced throughout the United States, in the month of February,* 1842. *By Elias Loomis, Professor of Mathematics and Natural Philosophy Western Reserve College. Read May* 26, 1843.

THE favour with which my account of the storm of December 20, 1836, has been received, has induced me again to undertake a similar investigation. A remarkable tornado was experienced in the north-east part of Ohio, February 4, 1842, an account of which may be found in the American Journal of Science, Volume xliii., pages 278–300. I was induced to hope that a careful investigation of the phenomena of the general storm which accompanied the tornado, might throw some light upon the origin of the tornado itself. Hence this storm was selected for investigation. Another very remarkable storm having been experienced about the middle of the same month, it occurred to me that I might obtain the materials for an investigation of both storms with about the same ease as for one. In collecting these materials, I have to acknowledge my obligations to my correspondents. To Professor Bache, I am indebted for a copy of the observations made in the State of Pennsylvania. To Professor Espy, I am indebted for a large mass of information he has, with great pains, collected, and which he has generously placed at my disposal. The remaining observations were chiefly obtained by direct application to the observers. I have thus obtained sixty-eight registers, containing barometric observations.

[. . .]

 DOI: 10.4324/9781003650294-4

Storm of February 16, 1842.

I have represented the principal phenomena of this storm on the accompanying charts, one to five. Those regions, where the sky was unclouded, or where the cloudiness was less than one half, are coloured blue; those where the sky was entirely overcast, or the cloudiness exceeded one half, but without rain or snow, are coloured brown; the fall of snow is indicated by the green colour, and rain by the yellow. The direction of the wind is represented by the arrows, and its force is, to a certain extent, indicated by their length. When, however, the winds were faint, the arrows, if drawn of a strictly proportionate length, would have been too short to attract notice, and are therefore somewhat magnified. A calm is represented by a cipher, (0). I have also represented the barometric, and thermometric observations in the following manner: Having determined, as well as I was able, the mean height of the barometer at each station, I compared each observation with the mean. I then drew a line passing through all the places, where the barometer stands at its mean height. This line is marked – – 0, and may be called the line of mean pressure. I then drew a line through all the places where the barometer stands two inches above the mean. This line is marked – – – +.2, and so of the others. In like manner, a line joining all those places where the thermometer stands at its mean height for that hour and month, is marked . . . 0°, and may be called the line of mean temperature. Another line joins all those places where the thermometer is 10° above the mean, and is marked . . . 10°, and so of the others. Nearly every circumstance essential to a correct understanding of the phenomena of the storm is thus presented to the eye at a single glance.

Chart 1. for the morning of February 15, represents the storm near its commencement, and before it had acquired much violence, or its features had become very distinctly marked. In the extreme north-eastern part of the United States, is seen a corner of a storm rapidly passing off. In the vicinity of Buffalo, a few particles of snow are falling, and the sky in its vicinity is overcast. Perhaps this may be ascribed to the vapour arising from the lakes, aided by the rise of ground as the current travels eastward, by which means the moist air at the surface is forced up to a region of greater elevation and cold, and its vapour condensed. With these exceptions, nearly the whole eastern and central portions of the United States are unclouded. But throughout the valley of the Mississippi, the heavens are obscured, and in the northern part some snow is falling. This is the commencement of the storm which I propose particularly to consider. Throughout the entire United States the barometer stands a little above the mean. The line of – – – +.2, is marked upon the chart. The thermometer no where stands much above its average for February. The line of mean temperature passes nearly through the middle of the United States; and that of 10° below the mean, is an undulating line in the eastern part of the United States. The direction of the winds is exceedingly various. In the eastern part of the United States, the winds are generally north-west, the result of the preceding storm. The weight and density of the atmosphere were here both greater than farther west. As you proceed westward, the wind moderates, becomes calm at many places, and in the central parts of the United States begins to set towards the west. Throughout the Mississippi valley the winds are neither very strong nor uniform in their directions; – a result to be expected, as there was no where any very considerable disturbance of mean temperature or pressure.

[...]

GENERALIZATIONS

I. DIRECTION OF THE WIND.

The question which has, within a few years, excited most interest in Meteorology, is, whether storms are rotary or centripetal, and it has been supposed that either diagram, No. 1 or No. 2, must represent the motion of the wind.

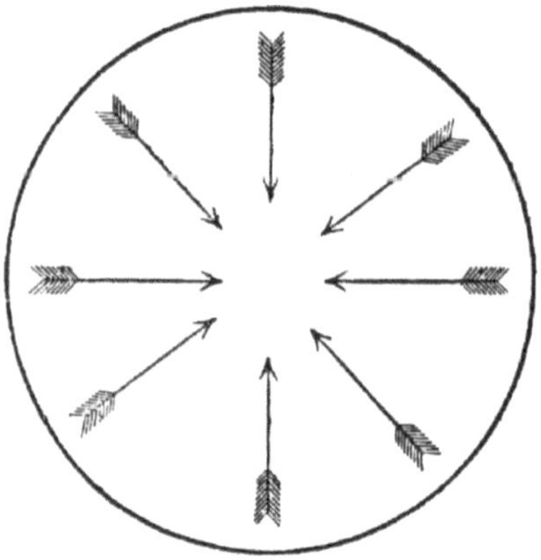

No. 1.

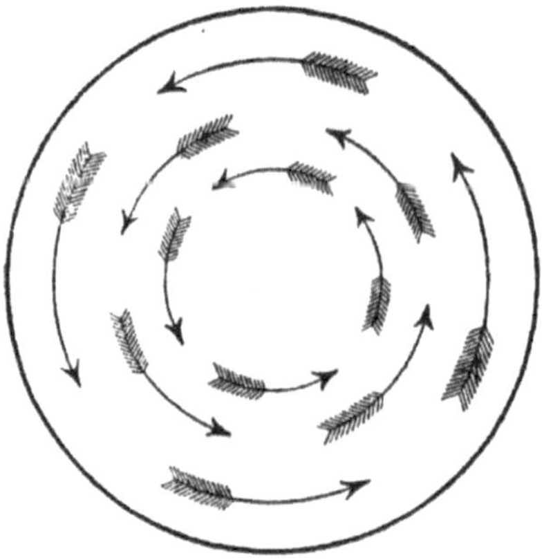

No. 2.

The accompanying charts will show that neither of these diagrams faithfully represents the storms here investigated, and it is doubtful whether either of them ever accurately represents the motion of the wind over any large portion of the earth's surface. The storm of December 15, 1839, has been quoted as a strong case of the revolving kind, and if we take only a semicircle of two hundred miles radius on the north-west side of the centre, the correspondence is very good, (see Mr. Redfield's diagram, Philosophical Transactions Volume viii., page 81,) though, even here, most of the arrows are inclined to the circumferences drawn, and, as Mr. Redfield has remarked, the direction is generally *inward* towards the centre of the storm. But if we take the other half of the semicircle, which Mr. Redfield has omitted, the circles are not completed. There have been several cases, of limited extent, in which the winds might be represented by figure No. 1.; but I have seen no account of a storm, of a hundred or more miles in diameter, in which the winds could be faithfully represented by either of the above diagrams. A combination of the two

is, however, frequently seen. No. 3 is intended to represent the motions of the wind February 16,1842, at sunset, near the centre of the storm. Substantially the same diagram is applicable for the morning of the seventeenth, so far as appears from the observation, and tolerably well for the morning of the sixteenth, although there is here more irregularity, the storm being of larger dimensions and less violent. The lower half of this diagram represents, very faithfully, the observations for the morning of December 21, 1836; and it may be inferred that if observations could be procured from more northern stations, they would show the other half of the diagram. The same figure represents, pretty accurately, the storm of January 26, 1839, also the great British storm of August 17, 1840, and that of January 6, 1839, except that in the south-west quarter of the latter, the winds were, perhaps, more south-westerly.

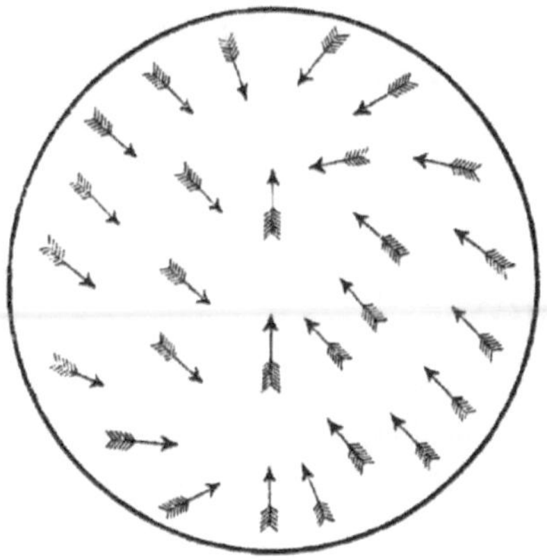

No. 3.

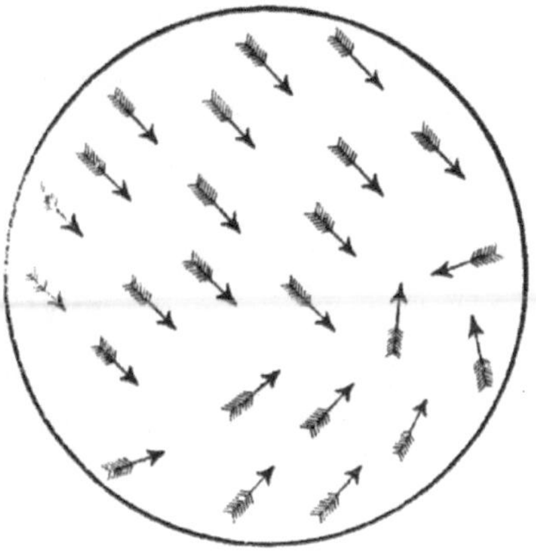

No. 4.

Diagram No. 4, represents the storm of March 17, 1838, the north-east wind covering more than half the region of rain and snow. The storm of December 15, 1839, was quite similar to this last, though the north-east wind covered a smaller portion of the storm. In some storms there is a predominance of north-east winds; in others of the south-east; in some north-west, and others south-west winds; but in all we find certain common characteristics, namely, an inward motion, with a tendency to circulate against the sun. These are the ordinary features of those storms which are accompanied by a gust and sudden oscillation of the barometer. The storm of February 3, 1842, can hardly be comprehended under this class. It was so extensive, the motion of the barometer so slow, and the winds moderate on the second and third, that it seems more properly to belong to a distinct class; although on the fourth and fifth, when the winds sprung up with some force, we find the characteristics mentioned above.

II. PROGRESS OF THE STORM.

A second question of some importance is, in what direction and with what velocity does the storm advance? It is not difficult to perceive that the general progress of storms is eastward, but, when we undertake to assign the direction with minute precision, we feel the necessity of a precise definition of the terms "storm," and the "centre of the storm." I define the centre of a storm to be the point of greatest barometric depression. The storm of February 16 travelled, then, from sunrise of the sixteenth to sunrise of the seventeenth, five hundred and sixty statute miles in a direction of about north 53° 18′ east, although to sunset of the sixteenth, its direction had been about north 74° 46′ east. It seems probable that its path was not a straight line, although it is possible that its centre, on the morning of the seventeenth, was farther south than I have here assigned, there being a deficiency of observations on the southern quarter. Velocity twenty-three miles per hour. The storm of February 2 and 3, remained nearly stationary. On the fourth and fifth its progress can be pretty accurately traced. It travelled north 62° 17′ east, eight hundred and sixty miles in twenty-four hours, or thirty-six miles per hour. The direction of the storm of December 20,1836, cannot be precisely assigned. It was clearly eastward, but having only observations from the southern half of the storm, it is impossible to assign, accurately, the place of greatest barometric depression.

CONCLUSION.

The following, then, is my view of the origin of such storms as I have been investigating. This generalization will probably include the greater part of winter storms, but will require some modification when applied to summer storms. Imagine a time perfectly clear, when the wind is from the west, or a few degrees south of west, with the barometer and thermometer at their mean heights. This may be regarded as the normal state of the atmosphere, and the whole body of air from its upper limit to the surface of the earth is moving on harmoniously in one direction. How is rain produced in such an atmosphere? The first requisite appears to be a change of direction of the lower stratum of air. This appears, in winter, to be frequently the effect of a preceding storm. The prevalent westerly current being temporarily checked in its progress by a violent storm, soon acquires force sufficient to break down all opposition. It supplants the rarefied air of the storm, and not only restores the barometer to its mean height, but the momentum of the excited mass carries it considerably above the mean. This excess of pressure causes a reverse current a little to the westward of a violent storm; and hence we sometimes have a long series of violent storms succeeding each other, at nearly equal intervals; and hence also a violent storm succeeded by an unusually high barometer affords ground for expecting a second storm within one or two days. But this explanation will not apply to all cases, for then, if the barometer should ever settle down to its mean height all over the globe, we never could have another storm. The case here supposed is not likely ever to happen, but even if it should, we cannot admit the consequences attributed to it. Admit such a case to occur, and the sun's heat would be competent to generate a new storm. Different portions of the earth's surface absorb the sun's rays in unequal degrees, and afford unequal quantities of moisture for evaporation. The result is, that we find bodies of air in close proximity of unequal density, arising from unequal temperature or humidity. Either case would be sufficient to cause a deflection of the lower stratum of air from its normal direction. Suppose, then, we have the mass of the atmosphere pursuing its wonted course from west to east, while a stratum of a mile or so in height, next to the earth's surface, blows in some different direction. If this direction be from the south to the north, then this current must be cooled in its progress by change of latitude. This effect may be aided by the inequalities of the earth's surface, and by friction upon the upper stratum of colder air. At the surface of the earth, where the temperature is probably five or ten degrees above the dew point, no remarkable effect may follow. But at a certain elevation, the air is always saturated with vapour. A very slight reduction of temperature causes cloud, and its density and extent will be proportioned to the energy of the causes in operation. If the wind should blow from the north, it might happen that no cloud would be formed; but if the direction should be easterly, being partly opposed to the normal current, some portion of this mass would almost necessarily be elevated from the earth's surface, and being cooled, its vapour be condensed. The first stage of the process, then, is an abnormal current at the earth's surface; the second is the production of cloud. At this stage the sky is covered with a veil which checks radiation; the thermometer rises above the mean from this cause, and also from the heat liberated in the condensation. This only adds to the energy of the first abnormal current. More cloud is thus formed, and presently the particles of water having acquired sufficient size, fall rapidly to the earth. The wind being southerly, the thermometer rises. A portion of the atmosphere being thus unusually heated and loaded with vapour, while the upper limit of the atmosphere remains nearly invariable, the barometer necessarily falls. Thus, these causes might continue to operate a long time, acquiring energy by their own action. A limit, however, is soon attained. The rarefaction thus produced creates a tendency in the surrounding colder and heavier air, to rush in and occupy its place. Moreover, if the wind be at all easterly, as is usually the case, it partially obstructs the progress of the normal current. This temporary retardation only gives it accumulated energy, and it is soon reinstated with unwonted violence. When the rarefaction is considerable, this rush of air upon the last half of a storm is not generally in the precise direction of the upper current, but more northerly; this air being the denser. Our southerly wind is thus supplanted by a violent north-wester. We have thus a great rarefaction and elevation of the temperature under a south or south-east wind with rain, extending over a large territory. This may be called the third phase of the storm, although it differs from the second only in intensity. There is now a general rush of heavier air to fill this void. This rush is chiefly from the north; but an independent cause, that which imparts direction to the upper current, would give us a west wind. Under these two forces the resulting current is chiefly north-west, but every where upon the borders, the tendency will be inward. The air thus flowing inward towards a central area, forces upwards the warmer air which rises in the middle, and being cooled by elevation, discharges a greater quantity of rain. The currents moving centrally from every point of the compass, interfere with each other, and pursue their routes spirally inward. We have thus a species of rotation, which, in the centre of the storm may have a destructive violence, as at Mayfield, February 4, 1842. This is the fourth phase, and is the case of a violent storm fully organized. This west or northwest wind carries the storm off from a fixed locality, and it is thus transferred successively to points farther and farther east. But this action cannot continue indefinitely. There is a cause in operation which will soon terminate its violence. This westerly wind travels more rapidly than the easterly. The rarefaction at the centre of the storm is a cause which acts equally upon both winds; but the one is opposed to the upper current, and the other nearly coincides with it: hence the one is accelerated, and the other retarded. The result is, that at successive points, farther and farther east, the same storm (after the north-west wind has begun to blow with great violence) has a less duration; the thermometer rises to a less height, the barometer has a smaller oscillation, and thus at a point far eastward, the oscillation becomes nearly extinct, and the only peculiarity observed in the wind is a stronger westerly current succeeding a calm. This is the fifth and final phase of the storm.

It appears to me, that if the course of investigation adopted with respect to the two storms of February, 1842, were systematically pursued, we should soon have some settled principles in meteorology. If we could be furnished with two meteorological charts of the United States, daily, for one year, charts showing the state of the barometer, thermometer, winds, sky, &c., for every part of the country, it would settle for ever the laws of storms. No false theory could stand against such an array of testimony. Such a set of maps would be worth more than all which has been hitherto done in meteorology. Moreover, the subject would be well nigh exhausted. But one year's observation would be needed; the storms of one year are probably but a repetition of those of the preceding. Instead, then, of the guerilla warfare which has been maintained for centuries with indifferent success, although at the expense of great self-devotion on the part of individual chiefs, is it not time to embark in a general meteorological crusade? A well arranged system of observations spread over the country, would accomplish more in one year, than observations at a few insulated posts, however accurate and complete, continued to the end of time. The United States are favourably situated for such an enterprise. Observations spread over a smaller territory would be inadequate, as they would not show the extent of any large storm. If we take a survey of the entire globe, we shall search in vain for more than one equal area, which could be occupied by the same number of trusty observers. In Europe there is opportunity for a like organization, but with this incumbrance, that it must needs embrace several nations of different languages and governments. The United States, then, afford decidedly the most hopeful field for such an enterprise. Shall we hesitate to embark in it; or shall we grope timidly along, as in former years? There are but few questions of science which can be prosecuted in this country to the same advantage as in Europe. Here is one where the advantage is in our favour. Would it not be wise to devote our main strength to the reduction of this fortress? We need

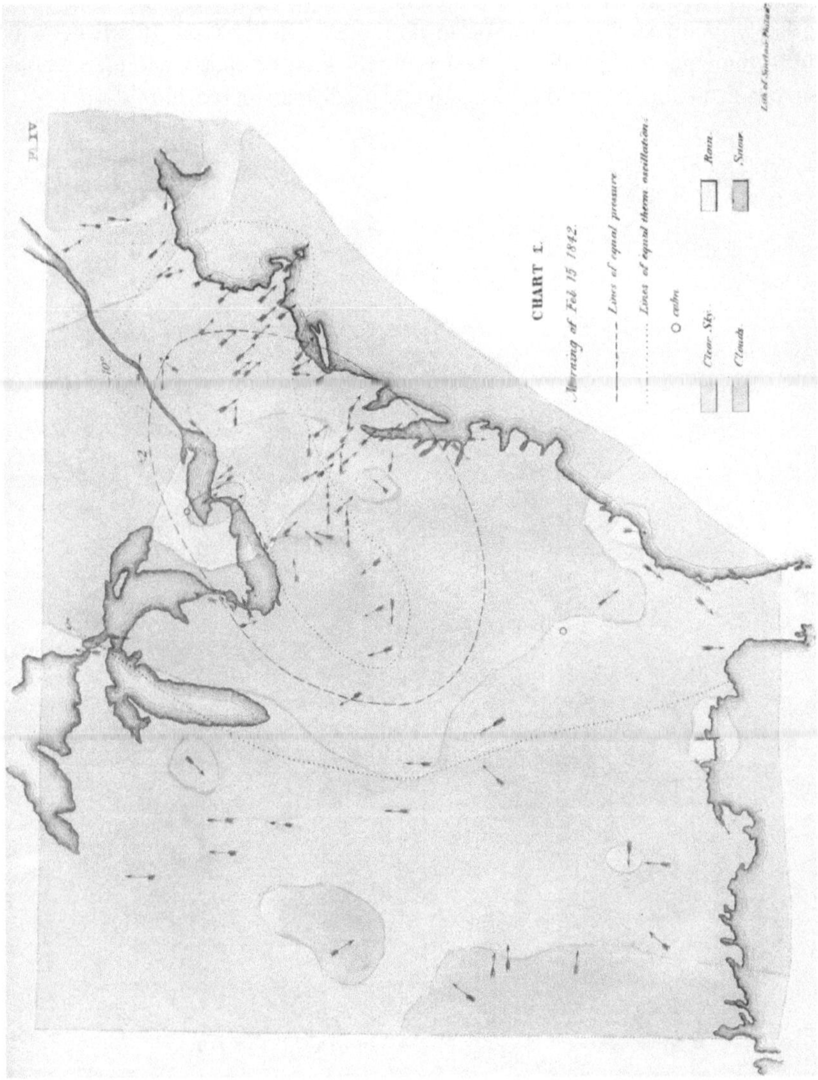
CHART 1.
Morning of Feb 15 1842.
Lines of equal pressure
Lines of equal therm oscillation
calm
Clear Sky
Clouds
Rain
Snow

Chart 1.

observers spread over the entire country at distances from each other not less than fifty miles. This would require five or six hundred observers for the United States. About half this number of registers are now kept, in one shape or another, and the number, by suitable efforts, might probably be doubled. Supervision is needed to introduce uniformity throughout, and to render some of the registers more complete. Is not such an enterprise worthy of the American Philosophical Society? The general government have, for more than twenty years, done something, and have lately manifested a disposition to do more for this object. If private zeal could be more generally enlisted, the war might soon be ended, and men would cease to ridicule the idea of our being able to predict an approaching storm.

1.2

BAROMETRIC WAVES

Editorial Headnote

1.2 Barometric Waves

Echoes of the American storm controversy reached Britain in the early 1840s. Extracts and reviews of the participants in the controversy (William Charles Redfield, James Pollard Espy and Elias Loomis) began to appear in British scientific journals. They were also discussed in elite circles such as the British Association for the Advancement of Science (BAAS) and the Royal Society of London.[1] The 'Report on the reduction of meteorological observations', read by John Frederick William Herschel (1792–1871) at the BAAS meeting in Cork, Ireland, 16–23 August 1843, and published in London by John Murray in 1844, provided British scholars with the opportunity to see how the country's then-leading authority on science introduced the new concept of the *barometric wave*.[2]

Herschel's report sought to settle the storm controversy once and for all by drawing upon a notion that was vaguely floating around in scientists' minds at the time. In his 1832 report on the 'Recent progress and present state of meteorology', read at the Oxford meeting of the BAAS, James Forbes had already suggested that variations in atmospheric pressure were due to 'great atmospheric tidal waves which are perpetually traversing oceans and continents'.[3] But even Herschel himself, in his influential *Treatise on Astronomy*, published as early as 1833, was already speculating about winds and the 'great fluctuations, of the nature of waves' in the atmosphere.[4] In his 1844 report, however, the concept of the barometric wave was defined within the framework of the search for storm laws and his system of inductive universalism, which required large sets of global observations over long periods.[5]

Herschel failed to mention in his report the work of Elias Loomis, whose wave formulation he emulated. Loomis was a decisive figure in the introduction of wavelike notions ('cold air wave') into meteorological science (see Vol. I, 1.1 'Precursors'). In two papers read to the American Philosophical Society on 20 March 1840 and 26 May 1843, Loomis presented his ideas on cold air waves and barometric oscillation waves as a solution to the storm controversy, and in the latter even included an analogy of his waves to ocean waves, just five months before Herschel's report was read at the BAAS meeting in Cork. Given his intimate knowledge of the controversy, it is almost inconceivable that Herschel was unaware of Loomis's ideas.

An empiricist by training and a champion of scientific induction, Herschel proposed that storms were the result of the crossing of two large atmospheric waves moving in different directions. This hypothesis of global wave dynamics was not a mere suggestion, but the result of the study of many and varied barometric records – as can be seen from the tables in the report – from the campaign of simultaneous meteorological observations promoted at British observatories worldwide (including the Cape of Good Hope, to which he travelled and on whose return he presented his report). Sometimes accused of dilettantism for his propensity to work in many fields, and the son of William Herschel, the 'father of stellar astronomy', John Herschel had an able assistant, William Birt (1804–1881), from 1839 to 1843.[6] The two showed an increasing eagerness to engage in the search for diurnal and monthly regularities in ways that involved the reduction, tabulation and representation of barometric data. The conceptual shift from barometric curve to barometric wave is evident in the correspondence between the two colleagues during these years. Thus, we find explicit mentions of the notion of "barometric curves" in 1841, and also of "atmospheric curves" in June 1842.[7] However, it is in Herschel's letter to Birt of 28 July 1843 that the barometric wave is described as a clearly defined concept – a letter reproduced and commented on in detail by Vladimir Jankovic.[8]

An amateur astronomer and variable star researcher, Birt came to Herschel's attention when he submitted his first studies of stellar variations for publication in the *Memoirs of the Royal Astronomical Society*, of which Herschel was president. His work as Herschel's assistant was shaped by the award of a BAAS grant to calculate barometric fluctuations and determine the nature of atmospheric waves.[9] He attracted the attention of Herschel and readers of the BAAS annual reports and journals such as *Philosophical Magazine* for his ability to sort and analyse observational data, an ability he would later apply to the arrangement and reduction of copious series of barometric measurements.[10] It was precisely *Philosophical Magazine*, founded by Alexander Tilloch in 1798 and edited by Richard Taylor in the 1840s, that Birt chose to publish his results, not only because it attracted the most prominent scientists of the day, such as Michael Faraday, Lord Kelvin or Herschel himself, but also because its influence extended beyond Britain and reached an international audience. Over time, Birt's stance shifted from wholeheartedly supporting Herschel's line of research to questioning the reality of barometric waves, as evidenced by the following two articles published in *Philosophical Magazine.*

Birt outlined his support for the wave theory in a study, 'On the Storm-Paths of the Eastern Portion of the North American Continent' (1846), the first of his series of articles published in the *Philosophical Magazine*. The article expresses Birt's concern to acknowledge the work of cyclone researchers such as Redfield, Espy and Loomis in the USA, William Reid in England and Henry Piddington and Alexander Thom in India, while emphasising their failure to explain atmospheric undulations. In particular, he warned of the need to accompany all wind observations with barometric readings, readings which he felt should include the whole system of observations corresponding to the waves involved, both in space and in time – a practice not common among the aforementioned scholars. In his view of storms as the crossing of two large atmospheric waves moving in different directions, Birt followed Herschel *au pied de la lettre* in his explanation of the BAAS report of 1842.

The contrast between Birt's comments in his 1846 study and those in his 1847 study entitled 'On Certain Atmospheric or Barometric Waves Which Traversed Europe during November 1842' illustrates his growing concern at the lack of consistency between the empirical data and the barometric wave hypothesis. In his analyses and reductions of barometric records from 1845 to 1847, Birt found numerous anomalies, such as erratic waves and others that appeared to overlap. Unlike Herschel in his later papers and correspondence, Birt was cautious about the reality of barometric waves. His change of mind about the ontological significance of waves is undeniable: Where in 1846 Birt confidently asserted that evidence had been provided (dating from 1842) of the passage of large atmospheric waves occupying fifteen to seventeen days between successive crests, in 1847 he had to admit that barometric curves do 'not represent the form of any reality in nature'.[11]

Notes

1 For example, William Reid, 'A Statement of the Progress Made towards Developing the Law of Storms; and of What Seems Further Desirable to be Done, to Advance Our knowledge of the Subject', in *Report of the Eighth Meeting of the British Association for the Advancement of Science* (London: John Murray, 1839), vol. 7, pp. 21–5.

2 Aitor Anduaga, *Politics, Statistics, and Weather Forecasting, 1840–1910. Taming the Weather* (London and New York: Routledge, 2020), pp. 48–53.

3 James D. Forbes, 'Report upon the Recent Progress and Present State of Meteorology', in *Report of the First and Second Meeting of the British Association for the Advancement of Science, 1831 and 1832; at York in 1831, and at Oxford in 1832: Including Its Proceedings, Recommendations and Transactions* (London: John Murray, 1833), pp. 196–258, on p. 235.

4 John Frederick William Herschel, *A Treatise on Astronomy* (London: Longman, 1833), p. 26.

5 Christopher Carter, 'Herschel, Humboldt, and Imperial Science', in Raymond Erickson, Mauricio A. Font, and Brian Schwartz (eds.), *Alexander von Humboldt. From the Americas to the Cosmos: An International Interdisciplinary Conference, October 14–16, 2004* (New York: The City University of New York, Bildner Center for Western Hemisphere Studies, 2004), pp. 509–18, on pp. 512–15; Gregory A. Good, 'A Shift of View: Meteorology in John Herschel's Terrestrial Physics', in James Rodger Fleming, Vladimir Jankovic, and Deborah R. Coen (eds.), *Intimate Universality. Local and Global Themes in the History of Weather and Climate* (Sagamore Beach, MA: Watson Publishing International, 2006), pp. 35–67, on pp. 47–8.

6 David S. Evans, 'Herschel, John Frederick William', in Charles Coulston Gillispie (ed.), *Dictionary of Scientific Biography* (New York: Charles Scribner's Sons, 1972), vol. 6, pp. 323–28.

7 Gregory A. Good, 'A Shift of View: Meteorology in John Herschel's Terrestrial Physics', in James Rodger Fleming, Vladimir Jankovic, and Deborah R. Coen (eds.), *Intimate Universality. Local and Global Themes in the History of Weather and Climate* (Sagamore Beach, MA: Watson Publishing International, 2006), pp. 35–67, on p. 52.

8 Vladimir Jankovic 'Ideological Crests versus Empirical Troughs: John Herschel's and William Radcliffe Birt's Research on Atmospheric Waves, 1843–50', *British Journal for the History of Science*, 31 (1998), pp. 21–40, on pp. 28–9.

9 Roger Hutchins, 'Birt, William Radcliff', in H.C.G. Matthew and Brian Harrison (eds.), *Oxford Dictionary of National Biography* (Oxford: Oxford University Press, 2004), vol. 5, pp. 853–54; Marvin Bolt, 'Birt, William Radcliff (1804–81)', in Bernard Lightman (ed.), *The Dictionary of Nineteenth-Century British Scientists* (Bristol: Thoemes Continuum, 2004), vol. 1, pp. 215–217.

10 Eric G. Forbes, 'Birt, William Radcliff'. In Charles Coulston Gillispie ed., *Dictionary of Scientific Biography* (New York: Charles Scribner's Sons, 1970), vol. 2, p. 147.

11 William Radcliffe Birt, 'On Certain Atmospheric or Barometric Waves Which Traversed Europe during November 1842', *Philosophical Magazine*, 30 (1847), pp. 482–502, on p. 491.

3

REPORT ON THE REDUCTION OF METEOROLOGICAL OBSERVATIONS

John Frederick William Herschel

Source: John Frederick William Herschel, 'Report on the Reduction of Meteorological Observations', *Report* of *the British Association for the Advancement of Science, Held at Cork in August 1843* (London: John Murray, 1844), pp. 60–61, 97–100.

Report of the Committee appointed for the Reduction of Meteorological Observations.

By Sir J. F. W. HERSCHEL, *Bart.*

EVERY exertion having been used to complete the series of equinoxial and solstitial observations during the years 1835, 1836, 1837, 1838, whether by writing to parties who have communicated observations, for duplicates of missing series, or by searching the records of observatories, meteorological registers, scientific journals and periodicals, these endeavours have proved so far successful, that at length 334 sets of observations have been collected, made at sixty-nine distinct stations. A synoptic statement of these, with the geographical elements of the stations, the names of the observers or communicators of the observations, and other particulars as far as they could be collected, or are necessary for our present purpose, is annexed to this report. See Appendix (A.).

It will be at once seen on inspection of this synopsis, that although at a few stations (as London, Greenwich, Brussels, Port Louis, Markree, Cadiz) consecutive series, extending over a period of three complete years, have been procured (within the limit assigned to these reductions), yet that this is not the rule but the exception; and that, taken altogether, the observations form anything rather than a connected whole. Under these circumstances, the only point of view which seemed to promise any distinct and definite results, bearing reference to causes prevailing over extensive regions, was that of the barometric fluctuations, considered with a view to the propagation of atmospheric waves, which, it is manifest, can only be traced over any considerable tract of country by this method of inquiry. It is accordingly to these, and to these only, that my attention has been directed; using the observed temperature (where the original observations have not been corrected by the observers) merely as elements of reduction, and

DOI: 10.4324/9781003650294-6

referring to the registered state of wind and weather, whenever such reference has been considered elucidatory of any point suggested by the main branch of the inquiry.

If we consider the trifling depth of the ponderable atmosphere regarded as an envelope of the whole globe, the interruption and obstacles offered to its oscillatory movements as a whole by the configuration of the continents and the distribution of mountain chains, but above all, by the vast and capricious variety of local causes affecting the temperatures of particular districts, and thereby causing partial ascensional and descensional movements and local generations and precipitations of vapour, we shall clearly perceive that, so far as our present purpose is concerned, the particular dimensions and form of our planet have little to do with our inquiry, and that for the immediate purposes of that inquiry we may regard our globe as a plane surface of infinite extent, over particular districts of which systems of oscillation of local origin, and independent of each other, are in progress, and in which we may regard ourselves fortunate if we can now and then succeed in obtaining distinct evidence of the direction, extent, height and velocity of a single wave. The distribution of our stations into groups, grounded on this view of the subject, and the mode of referring the observations of each group to a central station within it, have been described in my report for 1840, and need not therefore here be recapitulated. These groups, it is true, abstractedly considered, are far from those which would be chosen *à priori.* For example, Mauritius and Van Diemen's Land are but ill adapted to form a group with Indian stations. But for this there is no remedy, and the Mauritius observations (of which, owing to the diligence and zeal of Captain Lloyd, Surveyor-General of that island, we possess a nearly complete series) merit and will receive a separate discussion.

Two objects have been chiefly kept in view in the present inquiry. First, the tracing, where it can be accomplished, the course of one particular wave over the whole area embraced within one of our groups; and secondly, where this cannot be done, the observation of connexions between particular localities with a view to the subdivision of the total area into barometric districts, in which the atmospheric fluctuations shall be, generally speaking, similar in their phases. With these objects the projection of the barometric curves, for all the stations of a group, one sheet for each separate term, has been executed with great care and delicacy by Mr. Birt, and on a scale so large as to allow of the minutest corresponding changes, if any, to be distinctly followed out. The number of sheets so projected is fifty-three, on a scale of an inch to the hour in time, and one inch to 0·066 of barometric altitude; and I must not lose this opportunity of acknowledging many valuable remarks received from that gentleman on the subject of particular cases of much interest, which will be given in his own words, under their several heads as they occur.

I proceed now, therefore, to the discussion of the observations of each term, so reduced and projected, seriatim.

As the whole of the curves of the American group, and seventeen others of the European series, together with their tabulated reductions, have been on a former occasion submitted to the inspection of the meeting, to the number of 105, it has not been thought necessary to risk the loss or damage of the remainder by transmitting them herewith, they being in all respects similar.

[. . .]

Concluding Remarks (H.). – Having now discussed in their order the observations of each term, and stated the conclusions which have presented themselves in the course of their examination, it may be expected that I should express some opinion as to the utility of prosecuting a similar series of observations more extensively, and on the objects chiefly to be aimed at in so doing. And here I must first observe, that supposing any such continued series set on foot, it would be highly desirable, by previous communication and concert, to secure a multitude of cooperators on chosen lines of connexion, so as to obtain, at least for those stations, more regularly continued series of terms than those which are comprised in our list, and to narrow in some degree the field of inquiry by limiting it (as I have been compelled to do) to some distinct point of meteorology, such as the tracing of atmospheric waves and the determination of the coefficients of the diurnal periods. To the latter point the *hourly* system of observation *alone* is applicable, and monthly series (on the 21st of each month) of such observations from a vast number of stations in which absolutely nothing else was set down than the hourly march of the barometer for twenty-five consecutive hours (so as to *begin and conclude the twenty-four hours with an observation*), would in itself, if continued for a few years, leave nothing to be desired on this important head. I need not insist on each observer reducing his own observations. The time is arrived in meteorology when *unreduced* observations (at least barometrical ones) can no longer be tolerated, and must henceforward (except at very peculiar stations) be considered as not having been made. The tables for the purpose are in every one's possession who can be considered competent to use the barometer scientifically; and if the correction be applied *at once* at the time of reading off, the trouble is so subdivided as to be reduced almost to nothing. It should, however, be remembered by all who may undertake such observations, that unless made *at each hour of the twenty-five and at the exact hours,* a series loses much of its value; and if two or more hours be deficient, it is hardly worth using, as its comparability with others ceases. We have seen, from the instance of Mauritius (and the remark applies to most inter-tropical stations), that from twelve to sixteen unexceptionable series of the kind in question suffice to afford a perfect solution of the problem of the diurnal movements when mean quantities only are concerned; and in all probability three or four years' continuance of *monthly* terms would lead to a knowledge of the annually periodic variations in their coefficients. For *such* an object, dew-points, clouds, temperatures (other than needed to reduce the barometer to the freezing standard), and even winds, *might* be pretermitted, though I am far from *advising* their neglect.

As regards, however, the exceedingly interesting subject of the atmospheric waves, it is right I should observe that, without setting on foot (except with special views and in highly civilized localities) *any* express series of observations, but simply by comparing and reducing what already stand on record, – or even by projecting them unreduced, where great undulations only are to be traced, – the way lies open to most valuable conclusions. The fluctuations we have been able to trace by projecting the term observations, are those only whose total duration of rise and fall is comprised within or does not much exceed the twenty-four or thirty-six hours over which the term extends. But these are by no means those which either theoretically or indeed practically speaking are to be regarded as most important. I have the pleasure to lay before the Section specimens of barometric projections by Mr. Birt, laid down from his own observation, in which the interval between two successive maxima of pressure occupied in the one case seventeen days without any intervening maximum; in another, a similar period *with* two subordinate maxima interposed; and another where a beautifully symmetrical wave of an extremely remarkable character occupied thirteen days for its complete rise and fall. [See curves designated as "Barometric Waves," respectively annexed. Plates I. and II.]

For such objects (as before observed) we need not travel beyond existing records. In the records of our own magnetic and meteorological observatories, and those which stand in correspondence with them throughout the world, we have ample materials; and when dealing with undulations of such extent, it is by no means a visionary speculation to consider the possibility of tracing them over the *whole* of our globe; nay, perhaps of obtaining evidence of their performing, tide-like, two or more revolutions round its surface. And although the pressure of other avocations will (unfortunately perhaps for myself) entirely preclude my taking any further personal share in this most interesting inquiry, I am authorized by Mr. Birt to state, that should it be the pleasure of the British Association to intrust the subject to his inquiries by appointing him a committee for that purpose, he is prepared to pursue it if provided with a moderate grant to clear unavoidable expense.

Before concluding this report it may not be irrelevant to add a few words on the subject of the winds generally as connected with barometrical movements. In the mode in which I have been accustomed to consider the subject, the winds may be divided into two classes, – winds of translation and winds of oscillation; nearly in the way in which the movements of the ocean may be divided into oceanic currents and tide-streams, and these again (regarded as the result of oscillatory movements) may be referred to the general laws which regulate the molecular movements of water in contact with the bottom of the sea, when under the influence of undulatory agitation The trade-winds and greater aerial currents of a similar character, to which the name of monsoons is given, are winds of translation. They have a distinct barometrical origin, in the diminution of pressure in approaching the equator, caused by the expansion of the equatorial atmosphere and the overflowing of the upper strata outwards towards the poles. But this cause is not oscillatory but permanent. Monsoons also arise in the same way; from local heating and cooling periodically renewed, it is true; but in long periods of six months in duration, so as to give rise to steady currents. With such winds the present research has little connexion, except in so far as their powerful influence mixes with and masks the effect of the other class of winds, – those which arise from barometric *oscillation,* and which are connected with such oscillation in a more direct and intimate manner. The oscillations themselves perhaps take their rise in local and temporary causes prevailing over great areas simultaneously, the principal no doubt depending on the prevalence of cloud or clear sky, rain, or dryness over great tracts for several days or weeks in succession. But once produced, and an extensive atmospheric undulation once propagated, a wind or system of winds dependent on such undulation necessarily arises also. Every wave-like movement in a fluid (see Weber's 'Wellenlehre') consists of two distinct things, *an advancing form* and a *molecular movement,* which latter consists in a two-fold motion of each particle, vertical and horizontal. Laying aside at present the consideration of the vertical movement (which belongs only to the strata not in contact with the ground, and with which probably many interesting particulars respecting the formation and dispersion of clouds, the precipitation of rain, and the generation of lightning are connected), those strata which are in immediate contact with the ground, in obedience to the general laws of fluid movement, have their vertical movement zero and their horizontal a maximum, and for the most part (in periodical waves) alternately progressive and retrogressive.

Now, the *advancing form* is indicated to us by the barometer, the *molecular movement* by the wind, and between these two phænomena there subsists of necessity a close and purely dynamical connexion. And it would be no small meteorological discovery if, by the study of the characters and progress of barometrical fluctuations, we could either make out any law of the greater ones which would enable us even roughly to predict them, or any peculiarity in their physiognomy by which we could recognize them in their earlier stages, as by this we might possibly be led to the prediction of great storms.

Everybody must have remarked the sudden reversion of wind which often accompanies short and brisk storms accompanied with thunder and lightning (I do not now speak of the great so-called "revolving" gales). This phænomenon I consider as quite a case in point. But the horizontal motion of a particle of air in contact with the earth's surface need not be a straight line or even a returning curve. It would be the former only in the case of a straight, cylindriform wave once passing or periodically repeated; it would be the latter in the case of an ospillatory movement revolving round a node in the way in which Mr. Whewell has proved the tide-wave to do round certain definite points in the Channel. Now it *is* a fact, which has of late been a good deal insisted upon, viz. that in certain stations the winds do revolve in one uniform direction. The vane, for example, at Greenwich (as I am informed by the Astronomer Royal) makes five revolutions per annum in one uniform direction. May not this phænomenon, which, I confess, offers otherwise some difficulty of conception, be in effect an indication of some such atmospheric node, where a line perpendicular to the strata of the atmosphere may be regarded as describing a conical surface round the true vertical? If it be true (as the discussion of the term-observations has led to suspect) that Brussels is in effect such a nodal point, the examination in this view of its "Wind-Rose" would be interesting.

On the subject of "revolving" storms I am not fully prepared to speak; but there is certainly one point of view in which some of the principal of their phænomena would seem capable of explanation in this way of conceiving winds of oscillation, and in which they would become traced up, not to "funnel-shaped revolving depressions" in the nature of waterspouts, but simply to the crossing of two large long waves running in different directions thus: – where A A′ A″ represent the progress of one wave, B B′ B″ that of the other, and *t t′ t″* that of the tornado made by their intersection. The way in which a rotary movement in an ellipse or circle, or in some other partly oval and partly rectilinear figure, may result from the combination of two rectilinear movements of advance and recess, will easily be understood by the analogy of the circular and elliptic polarization of light, where rectilinear movements of the æthereal molecules are conceived to be similarly combined. Some features in such storms are strongly in harmony with this view, viz. the fact that in them the direction of the wind at a given locality never makes more than one rotation, and not always that; and that in the central line of the storm's progress there is a simple and sudden reversal of direction. On the other hand, it must not be concealed that some features militate against it; for instance, the fact that such gales are stated always to "revolve" in one direction, whereas on this view of their origin the changes of wind ought to be in opposite directions on opposite sides of the medial line. But for the present it must suffice to have pointed out a mode of considering the subject of at least certain sorts of winds which, being in the nature of a *vera causa,* resulting from dynamical considerations perfectly general and indefeasible, cannot be without some influence, the only question being that of amount.

Fifty-five pounds of the original grant of 100*l*. have been expended: should the Association order the printing of this report, a portion of the remainder may be applied to cover the expense of engraving or lithographing the curves above noticed and accompanying this report, expressing Mr. Birt's observed waves. The publication of all the projected term-curves would incur a more serious additional expense without a corresponding benefit. In furnishing the basis of the above discussions they have performed their office, though as records they should be carefully preserved.

(Signed)
J. F. W. HERSCHEL.

4

JOHN FREDERICK WILLIAM HERSCHEL TO WILLIAM RADCLIFFE BIRT, 28 JULY 1843

John Frederick William Herschel

Source: John Frederick William Herschel to William Radcliffe Birt, 28 July 1843.

Collingwood, July 28, 1843

Dear Sir,

. . .

I have already come to the conclusion that our conception of a barometric wave can be easily enlarged (i.e. if we set out with my notion of their analogy to sea-waves) so as to take in vast (?) flattened (blanket-like if I may be allowed such a homely comparison) heavings more like the tide waves of the ocean [. . .] And if we consider (which is already well-known of the phenomena of the barometer) that we have often a rising glass for whole month and vice versa – and sometimes a general high or low level . . . for one or two months, it will be clear that our atmosphere is a vehicle for wavelike movements which may embrace in their single swell & fall a whole quadrant of a globe.

My view of the matter to which I have alluded is that every barometric movement is essentially and intimately connected with an appropriate and corresponding wind – as . . . every wavelike movement in a fluid consists of two distinct things – an advancing form and a molecular movement – . . . the advancing form is indicated by the barometer – the molecular movement by the wind, and between these two phenomena there exists of necessity a close and purely dynamical connection. [Knowing barometrical fluctuations one would be able to predict wind] for this could lead to prediction of great storms.[1]

Editorial Endnote

1 Herschel to Birt, 28 July 1843, RS:HS, 19.106, as reproduced by Jankovic (1998, pp. 28–29).

DOI: 10.4324/9781003650294-7

5

ON THE STORM-PATHS OF THE EASTERN PORTION OF THE NORTH AMERICAN CONTINENT

William Radcliffe Birt

Source: William Radcliffe Birt, 'On the Storm-Paths of the Eastern Portion of the North American Continent'. *Philosophical Magazine*, 28, 1846, pp. 379–382.

To the Editors of the Philosophical Magazine and Journal,
GENTLEMEN,

IN the opening articles of the first and second numbers of the second series of the American Journal of Science, Mr. Redfield has contributed some further important information to our present stock of knowledge respecting the storm-paths of the eastern portion of the North American continent and its adjacent seas. The chart No. 1, illustrating the articles, exhibits the tracks of sixteen hurricanes; eight of these form an interesting group distinguished by this striking characteristic: the paths are semi-orbital, presenting in each case the half of an ellipse of greater or less eccentricity; the apices are confined to the thirtieth parallel of latitude, and are more acuminated as they recede from the fiftieth degree of longitude west of Greenwich. The remaining eight depart considerably from this type: four appear in some measure to approximate to it, but most probably with very acuminated apices. A most remarkable one, of a parabolic form, with its apex on the twenty-sixth parallel, was observed in October 1837 (XV.). In October 1842, the north-western portion of an elliptic path with the apex still lower was traced (XIII.). The storm pursuing this path is particularly discussed in the continuation of Mr. Redfield's paper, and its identity with the northers of the Mexican coast insisted on. Mr. Redfield also speaks of other storms that had exhibited the character of northers in the Gulf of Mexico, and afterwards presented all the features of Atlantic storms. Speaking of a storm that occurred in October 1837, he says, "This norther of the Mexican coast had become in due course of progression an Atlantic storm" (see American Journal of Science, second series, No. 2, March 1846, p. 166). In October 1844, a storm passed nearly in a direct line from the Gulf of Honduras to Newfoundland (XIV.); and to the gale which is first discussed in the articles

 DOI: 10.4324/9781003650294-8

before us (XII.), a nearly direct westerly course has been assigned from all the observations that have come to hand.

It is not my intention in the present communication to enter into any examination of the particular gales above enumerated, or to attempt to substantiate or refute either the one or the other of the rival theories which have been offered as an explanation of the phænomena. Those of your readers who are acquainted with Col. Reid's work, are aware that he has most ably discussed the rotatory theory, and for the centripetal theory, I beg to refer to various papers in the American Journal of Science. I apprehend that the labours of Redfield, Loomis, and Espy in the United States, Col. Reid in England, and Piddington and Thom in India, have brought the inquiry to that point at which it becomes essential to connect it with some kindred branch of science, in order to see our way clear in resolving the interesting problems that suggest themselves, to strike out a path for working energetically in surmounting the obstacles that still retard our progress in becoming acquainted with the dynamical system of our atmosphere, and in successfully removing the desiderata as they arise.

Among the desiderata of these phænomena, and by far the most important, will be found their origin and final disappearance. Sir John Herschel has suggested, in his Report on Meteorological Reductions,[1] that they may be produced by the crossing of two large atmospheric waves moving in different directions. Some interesting evidence of the existence of such atmospheric waves has been brought forward at late Meetings of the British Association. This evidence rests entirely on the barometric affections of the atmosphere over a large tract of country. In order, therefore, to extend our knowledge of the rotatory gale (Redfield) or the centripetal hurricane (Espy), especially with regard to the desiderata abovementioned, I apprehend it will not only be important, but absolutely necessary to accompany all the observations of the direction, force, and variation of the wind with barometric readings. These readings, however, must not be confined to the mere period of the passing of the gale; evidence has been adduced of the passage of large atmospheric waves occupying from fifteen to seventeen days between the anterior and posterior troughs, or between successive crests; so that in order to detect the origin of a gale arising from the intersection of two waves, to trace it throughout its destructive course and to observe its final disappearance, it will be essential to discuss the entire system of observations appertaining to both waves, not only in time, but also in space. Our meteorological observations are approaching a degree of uniformity and system that bids fair for uniting these kindred inquiries. I have now before me Prof. Loomis's interesting and ably conceived charts for exhibiting the principal phænomena during the passage of two storms over the United States in February 1842, in which he clearly shows the barometric, thermometric, and anemonal phænomena, and exhibits in a very striking manner the extent of rain, snow, cloud, and blue sky over the whole of the United States twice in the course of each day that the storms prevailed. The barometric phænomena are shown by *lines of equal pressure*; and I apprehend that these lines of equal pressure indicate, especially in one of the two cases, that *two* waves passed over the United States; that as the posterior slope of one wave passed off, the anterior slope of a wave of a different system approached; and that in the point of intersection the storm raged. Should the entire barometric observations taken over the United States on that occasion, February 1 to 4, 1842, support the theory of atmospheric waves, I apprehend Sir John Herschel's suggestions will be partly realized.

I cannot close this notice without adverting to a most important desideratum in this interesting inquiry. Mr. Redfield's storm-paths are confined to the westward of the fiftieth meridian, and he is only able to give one-half of the ellipse in any case. Nearly the whole breadth of the Atlantic is a perfect blank in any storm-map. Surely several captains may be found who will be happy to make observations on their passages out and homeward, and transmit them to head quarters, either in London or some principal city in America; the only additional information required to that furnished by their logs, will be the altitudes of the barometer at given hours of the day and night, with its attached thermometer. The American steamers might furnish important and valuable information.

I have the honour to be,
Gentlemen,
Your very obedient Servant,
W. R. Birt.

Cambridge Heath, April 22, 1846.

Source Endnote

1 Report of the British Association for the Advancement of Science, 1843, p. 100.

6

ON CERTAIN ATMOSPHERIC OR BAROMETRIC WAVES WHICH TRAVERSED EUROPE DURING NOVEMBER 1842[1]

William Radcliffe Birt

Source: William Radcliffe Birt, 'On Certain Atmospheric or Barometric Waves Which Traversed Europe during November 1842'. *Philosophical Magazine*, 30, 1847, pp. 482–493.

AT the sittings of the Geological Society on the 3rd and 17th of May in the year 1816, a paper was read by Dr. Macculloch On the Employment of the Barometer in measuring heights. In this paper the author endeavoured to show that the *differences* indicated by different instruments, placed at various stations on the earth's surface, are *real* differences of atmospheric pressure, and that the changes which take place at several stations are far from being simultaneous: he also expressed an opinion, that the differences of pressure, combined with their progressive character, are in some way or another connected with the direction of the wind, especially the two cardinal currents N.E. and S. W. of our island; but the imperfect state of the barometer at that time was such as to prevent an efficient investigation of the subject.

In the year 1832, Professor Forbes, in his Report on the Recent Progress and Present State of Meteorology, presented to the British Association for the Advancement of Science during its sitting at Oxford, alluded to the *accidental* variations of barometric pressure as greatly influenced by latitude, the amount of variability increasing towards the poles. He appeared to regard these variations as resulting from *great atmospheric tidal waves*, which perpetually traverse our oceans and continents, and he viewed a future and more advanced state of meteorological science, as furnishing data from which the most interesting and important conclusions might be drawn in immediate connexion with such atmospheric waves.[2]

On the 21st of March in the year 1835, Sir John Herschel made a series of hourly meteorological observations at Feldhausen Wynberg, near the Cape of Good Hope. These were accompanied by a simultaneous set at the Royal Observatory near Cape Town. The observations thus set on foot were continued at numerous

DOI: 10.4324/9781003650294-9

stations during the following three years; and the discussion of the whole, collected by Sir John, forms the subject of a report presented by that gentleman to the British Association during its sitting at Cork in the year 1843. One of the objects to which Sir John directed his attention in this discussion, was the barometric fluctuations considered with a view to the propagation of atmospheric waves. For this purpose the stations were distributed in groups, in one of which (the European) Sir John succeeded in detecting and tracing two well-defined waves; one on the 21st of September 1836, the other on the 21st of December 1837. These waves were so vast in their dimensions, that to take in (says Sir John) an effect of this nature, we must enlarge our conception of an atmospheric wave till it approaches in some degree, in the extent of its sweep and the majestic regularity of its progress, to those of the tide-waves of the ocean.[3]

The principle on which the discussion of these hourly observations was conducted, was that of projecting the barometric curves for all the stations of one group on a single sheet for each separate term, the scale being one inch to an hour in *time,* and one inch to $0^{in}\cdot066$ of barometric altitude. From the nature, however, of the observations, no one curve was continued longer than twenty-four hours, except on the occasion of a remarkable storm. During the progress of this work, it was deemed desirable to obtain the complete rise and fall of the barometer; so that a complete wave might be observed at least at one station. Observations undertaken with this view, and projected in curves, illustrate Sir John Herschel's report. As a test of any conclusions that may be drawn from a discussion of observations on this principle, the observations may be so combined as to exhibit, not only the barometric relations with regard to *time*, but also with regard to space: in other words, the distribution of pressure over a tract of country, at any given epoch, will indicate the real character of the waves then transiting.

The last report which I had the honour to present to the British Association during its sitting at Southampton, contains an examination of certain barometric observations exclusively on the last-named principle. These observations were made during November 1842, the same month in which the observations were made from which some of the curves illustrating Sir John Herschel's report were projected. The result of the examination has been the apprehension of several large atmospheric waves of so distinct a character, that the barometric movements over an area included by the angular points, the Orkneys, St. Petersburgh, Geneva and Cork, have been fully explained, especially during the earlier portion of the month. In order clearly to place these barometric waves before the readers of the Philosophical Magazine, I shall at once proceed to make various extracts from the report bearing immediately upon the subject, commencing with the

DEFINITION AND PHÆNOMENA OF AN ATMOSPHERIC WAVE.

When a number of barometric observations are projected on paper according to a suitable scale, and continued for months and years, the eye on contemplating them will recognise a variety of curved *forms*, some of large and some of small amplitude; some rising to a considerable altitude, others sinking far below the level, representing the mean barometric pressure at the station of observation. At first there appears but little regularity in these curvilinear records of the ever-shifting state of our atmosphere, but here and there the attentive observer will notice some similarity existing between two or more *individual* curves, and he may notice some which possess a certain symmetrical arrangement of the ascents and descents. In consequence of this similarity and symmetrical arrangement, he examines more carefully the records of barometric pressure, and not only discusses the observations at *one* station, but compares those observations with others made at *various* stations; and here again he finds apparent irregularity and confusion. The curves to a certain extent agree, but in many minor points they differ often very considerably, in some cases rising at one station while falling at another; this induces a still more minute and careful investigation: the distribution of pressure over the largest area he can command is carefully examined; and whether his stations are few or many at any given time, he finds on this area a point of maximum pressure and a point of minimum pressure; between these points he finds various pressures, generally increasing from the point of least pressure to the point of greatest pressure. On some occasions he finds a *line of high* pressure, stretching quite across the area, and on others a *line of low* pressure. By continuing his inquiries for successive epochs, he finds these lines of high and low pressure move across the area, or in other words, the high pressure or low pressure is gradually transferred from one point to another. He also finds at still more remote epochs other lines of high and low pressure, some having the same direction with the lines originally noticed, and others crossing the direction of the original lines at various angles.

The questions which now suggest themselves are the following: – What are these movements? How can they be represented? In what manner can they be explained? A simple consideration of the curves suggests the idea of waves as explanatory of the phænomena, and the term atmospheric wave has been used to designate that *ideal individuality* which the mind attributes to the process which it observes of the successive change of place which the barometric maxima and minima undergo, and by which they regularly succeed each other over the area under examination; this ideal individuality has been employed as a mean of examining the movements just alluded to.

Mr. Scott Russell, in his admirable report on waves, presented to the British Association during its sitting at York in 1844, has given the elements of an aqueous wave. These elements appear so clearly to express the relations of the barometric phænomena (simply considered as such), that I shall avail myself of Mr. Russell's terms to indicate the distribution of pressure over the surface of the earth, as in figure 1.

In reference to this figure, the line of high pressure stretching across the area (the figure being supposed to cut this line transversely) has been termed the *crest,* W; the line of low pressure in advance of the crest, the *anterior trough, a* (the

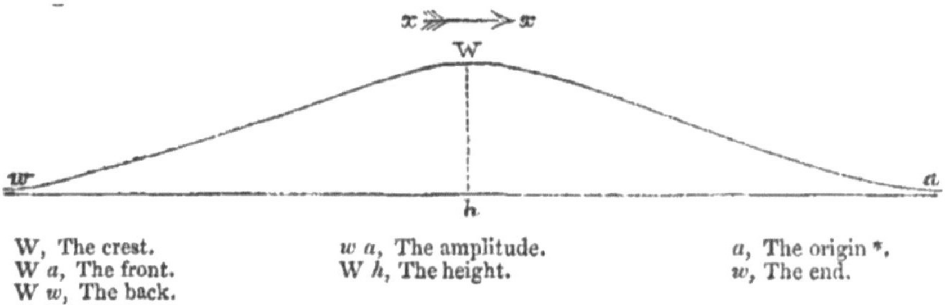

Fig. 1.[4]

origin of Mr. Scott Russell's water wave); the line of low pressure succeeding the crest, the posterior trough, *w* (the end of Mr. Scott Russell's water wave); the line *w . . . h,* as measured by the mercurial column, the altitude of the wave; the slope W *a,* the anterior slope or front of the wave; the slope W *w*, the posterior slope or back of the wave; *w a* constitutes the amplitude of the wave, and $x - x$ in the same direction, the axis of translation.

The observations to which allusion has been made, were published by Mr. William Brown in a paper that appeared in the Philosophical Magazine for April 1846. This paper is entitled "On the Oscillations of the Barometer, with particular reference to the Meteorological Phænomena of *November* 1842." The object of the author is to show that the barometric oscillations are produced by the *meeting* of opposite or nearly opposite aërial currents; that one current thus meeting or impinging on another, deflects it, and under some circumstances produces a rise of the mercurial column, but under others occasions a fall in many cases of considerable magnitude. In order to elucidate his views, Mr. Brown has collected barometric observations from eleven stations, which are scattered over an area included by the following angular points: – The Orkneys, Christiania in Norway, Paris, Plymouth and Cork, These observations are in most cases given as read off from the scale. In addition to these the paper is accompanied by six plates, in which the direction of the wind at numerous stations is indicated for every day during twenty-six days in the month by arrows. The aneraonal observations published in the body of the paper not being in all cases for consecutive days, a comparison of them with the plates is rendered difficult; nevertheless the plates form a very valuable portion of the communication, and if they have been laid down from accurate observations, they furnish us with an important addition to our knowledge of the *arrangement* of the aërial currents, especially with respect to the distribution of pressure. It is a matter of regret that Mr. Brown did not so arrange his observations and plates, that the accuracy of the latter could have been seen by inspection.

This paper is peculiarly interesting at the present time, when the attention of meteorologists is directed to the important and interesting problem of the barometric oscillations. I have already noticed the views which have been taken of these oscillations as the results of waves. In this paper they are regarded as the effects of currents; and it appears that meteorologists are now divided as to which of these hypotheses they shall be referred. Professor Dove remarks, in his letter to Colonel Sabine on the Magnetical and Meteorological observations,[5] "the so-called irregular variations of the barometer are regarded by some persons as only the effects of currents of air of unequal temperature and moisture; other persons distinguish the effects of currents from the effect of undulations *progressing in the manner of waves of sound, and propagating themselves with great velocity over large portions of the earth's surface.*"

It is worthy of notice, that these movements have been referred to two very distinct kinds of waves. Prof. Forbes speaks of them as great atmospheric *tidal* waves, and Prof. Dove speaks of a class of individuals who regard them as progressing in the manner of *waves of sound.*

It is not my intention to enter into an examination of the conclusions and results which Mr. Brown has arrived at; as the question is open, I apprehend I have not committed an injustice towards that gentleman by having employed a rather different process to that which he has used, and further discussed the observations he has given. I beg to acknowledge the obligations I am under to him for these observations, and especially for the plates, of which I have before spoken: they are extremely interesting in the present inquiry.

In accordance with these remarks, I have selected the following stations from Mr. Brown's list: – the Orkneys, Belfast, Shields, Cork, Bristol, Plymouth, London, Paris, and Christiania, The reason I have omitted Glasgow and Armagh will be apparent from Mr. Brown's notes. As I have discussed these observations with especial reference to the wave hypothesis, I have most cautiously avoided in my remarks any thing that may at all bear on Mr. Brown's views. The plan I have proceeded on is as follows. I have selected the middle observation of each day; at those stations where only two are given morning and evening; I have taken a mean of them. These observations I have so arranged as to exhibit the distribution of pressure over the area for each day – the line or lines of the greatest diminution of pressure – and the relation of such distribution and of such lines to the aërial currents or winds. As a convenient method of readily expressing these various relations and giving to the discussion that completeness which otherwise it would want, I have adopted the wave hypothesis, and to every line of barometric maxima applied the term *crest* and to every line of minima the term *trough.* In a word, I have regarded the progress of the barometric and anemonal phænomena as the progress of *waves*. The observations remain the same both in Mr. Brown's and my own discussions, the results only are different; and it remains for other philosophers, by more closely investigating the subject, and submitting the observations to a more rigorous and searching discussion, to advance this interesting inquiry and to become more intimately acquainted with the causes of these interesting phænomena.

In addition to the discussion of these observations making us acquainted with several large waves similar in character to the definition before given, it also reveals to us the arrangement of the aërial currents with respect to such waves, or rather by means of these observations and the plates accompanying them, we are able to ascertain, not only the general direction of the aërial currents, but also the barometric phænomena resulting from their transit.

Prof. Dove, in his second letter to Col. Sabine relative to the magnetical and meteorological observations, has announced his opinion that the equipoise of the atmosphere is maintained in the temperate zone by currents *on the same level flowing in opposite directions*[6]; thus we have a bed or stratum of air moving from the S.W., and on each side of this are strata of N.E. winds. We may here inquire, how are these alternate aërial currents related to the waves before alluded to? It is one of the objects of the discussion to exhibit this relation, which may be thus briefly expressed, at least in so far as the examination of the observations has yet extended.[7] Let the strata *a a a' a'*, *b' b' b b*, fig. 2, represent two parallel aërial currents, *a a a' a'* being from S.W. and *b' b' b b* front N.E., and conceive them both to advance from the N.W. in the direction of the large arrow, that is the strata themselves will advance with a *lateral* motion. Now conceive the barometer to commence rising just as the edge *b b* passes any line of country, and to continue rising until the edge *b' b'* arrives at that line, when the maximum is attained. The wind now changes and the barometer immediately begins to fall, and continues to fall until the edge *a a* coincides with the line of country on which *b b* first impinged. During this process we have all the phænomena exhibited by an atmospheric wave; when the edge *b b,* fig. 2, passes the line of country, the point *a*, fig. 1, of the wave (the anterior trough) transits that line of country and the barometer begins to rise with a N.E. wind. During the period the stratum *b' b' b b*, fig. 2, transits the line the anterior slope W *a,* fig. 1, passes; when the conterminous edges of the strata *a' a' b' b'*, fig. 2, pass, the crest W, fig. 1, extends in the direction of

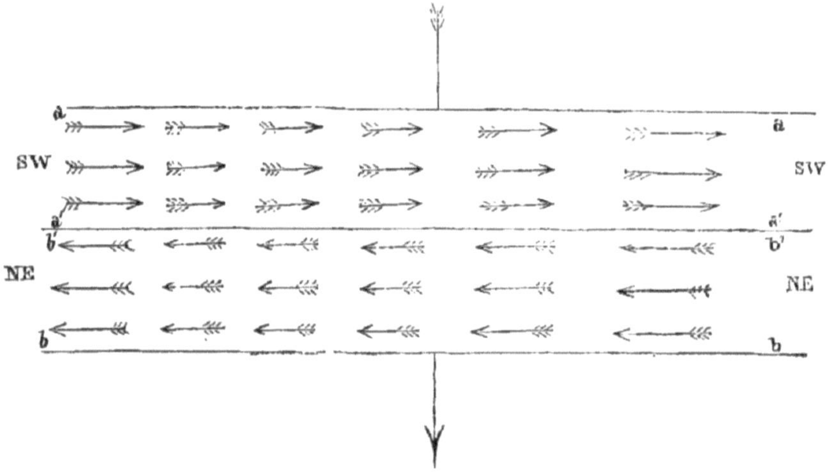

Fig. 2.

the preceding trough: the barometer now begins to fall, and when the edge *a a*, fig. 2, occupies the place of *b b*, also fig. 2, the descent of the mercurial column is completed; the posterior slope W *w*, fig. 1, has passed, and the posterior trough *w*, fig. 1, now occupies the line in which the anterior trough extended.

This arrangement of the aërial currents in connexion with a barometric wave may be termed the *air-motion* of the wave. In contemplating the transference of the barometric maxima and minima, we regard only the *wave-motion*. In contemplating the direction and force of the wind, we regard also the *air-motion*. The *wave-motion* consists in the *lateral* transference of the parallel beds of aërial currents; the *air-motion* being these oppositely directed parallel currents themselves.

From these considerations, we readily see that the *wave* is a convenient method of representing the barometric fluctuations. We have seen that the rise is due to the anterior slope and the fall to the posterior; and we now further learn that the direction of the aërial current on the anterior slope is at right angles to the axis of translation directed towards the *left-hand*, while on the posterior slope it is the reverse; still at right angles to the axis of translation, but directed towards the *right-hand*.

Having thus noticed the *wave-motion* with its accompanying *air-motion*, these interesting questions suggest themselves: – How are the forces of this *air-motion* arranged? Do all the particles move with the same velocity? Are there different velocities in different parts of the wave? Our anemometers will answer these questions. In the troughs, the edges *b b a a*, the forces are strongest; as the barometer rises, the force gradually subsides; when the crest passes, it is zero; and as the barometer falls, it increases until the trough passes, when it is again strongest.

M. Dove has suggested that the parallel currents, by which the equipoise is maintained in the temperate zones, may be shifting ones, and we have supposed that the parallel currents of N.E. and S.W. winds may advance from the N.W. with a lateral motion. The same cause that produces the well-known opposite and superposed equatorial and polar currents, will also give rise to the same opposite *but* parallel currents in the temperate zones, namely, the ascending column of heated and consequently rarefied air. Now it is well known that in stormy weather, when the wind is blowing with great force, the barometer being nearly at its minimum, upon the *wind changing* the barometer *commences rising*; the wind however continues to blow with about the same force as it did with the previous falling barometer. Upon M. Dove's view of parallel and opposite currents, somewhere in or near the line forming the boundary between the currents, towards or in the torrid zone, we ought to find the point of rarefaction, and to this point the N.E. current would rush with the greatest force to supply the ascending column of heated air.[8] This N.E. current would be compensated by a S.W. current of nearly or quite the same force, situated just to the S.E. of it, as in fig. 3, in which let a point of rarefaction, *a* for instance, exist in any locality, so that a N.E. current may be established to supply the ascending column; suppose the greatest force to exist along the line of crossed arrows *b b*, the air would be drawn from the end of this line to fill up the vacuum at *a*, and a compensating S.W. current, *c c*, established. This S.W. current would be established partly by the descent of the overflowing

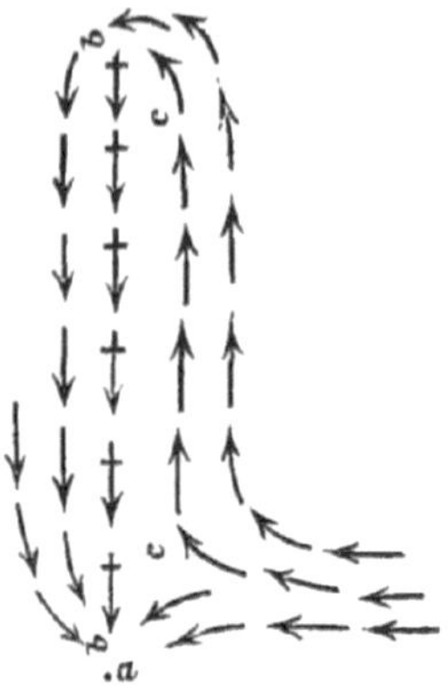

Fig. 3.

current at *a*, and partly by the rush to supply the air constantly drawn off to feed the ascending column. When however it is once established, the velocity of the line of S.W. current nearest to the N.E. would probably be equal, or nearly so, to that of the N.E. current itself.

In this way it is easy to conceive that a complete barometric wave may be produced; the lines of greatest velocity of the parallel currents will indicate the trough; the rapidity until which the currents pass in opposite directions greatly diminishes the pressure, and according to this view somewhere near the direction of the trough and to the S.W. of it, we ought to find the point of greatest rarefaction; the velocity decreases on each side this trough, and with this decrease of velocity the pressure increases, so that we have a distribution of pressure of a *wave form* gradually rising on each side the trough, the *pressure* being dependent on the velocity of the parallel currents.

The constant ascent of air at the point of rarefaction would continually draw off a quantity of air from the S.E. side of the line of greatest velocity *b* – *b*, fig. 3, and this would be attended with two results; first, there would be a *real* hollow or trough formed in the line of junction of the parallel currents; and secondly, this line would gradually advance towards the S.E.; for as more air would be drawn off from that side, the whole body of air would advance in that direction to supply the deficiency; and should the rarefying process cease, we can readily conceive that not only will the *waveform* be continued, but also *wave-motion.* The establishment of the parallel currents will give the *air-motion*; the diminution of pressure towards the lines of greatest velocity will give the *waveform*; and the drawing-off of air from the S.E. will induce the *wave-motion.* The wave thus generated is *negative*; it consists of a hollow produced by the ascending current of heated air carrying off a considerable portion of air set in motion by this ascending column, and its direction of motion is determined by more air being drawn off from the S.E. slope than the N.W.

CROSS WAVES.

The examination of the transit of a single wave by means of barometric and anemonal observations, would be comparatively easy, but it seldom happens, from the operation of natural causes, that an isolated or solitary wave is produced. In almost every instance (except in those in which the generating power is very much greater than any which occasions the production of smaller waves) the wave is contemporaneous with others of equal, if not of greater magnitude, so that different systems are in motion at the same time, each individual pursuing its own course, and although perfectly independent of every other, yet greatly modifying the *resulting phænomena* as exhibited by the barometer and anemometer. When therefore we proceed with the examination of certain barometric and anemonal phænomena in the manner above alluded to, we are speedily perplexed with the barometric and anemonal effects of *cross waves*; the flowing of one set of waves in a certain direction is apparently interrupted and interfered with by another in a different direction, and before the first set can be exhibited with its proper proportions, and the true altitudes, amplitudes, velocities, and directions of its individual waves assigned, all the phænomena of the other set must be carefully disentangled and separated from the aggregate phænomena presented by the contemporaneous systems. The barometric curve, including a complete rise and fall at any one station, is *not* the curve resulting from the transit of any *one* wave; it does *not* represent the *form* of any *reality* in nature; but it *does* represent, and is an exponent of the effects resulting from the contemporaneous transits of waves, or systems of waves, such as have been described.

The discussion of Mr. Brown's observations has clearly brought to light a set of parallel and opposite currents at right angles to those we have been contemplating, namely, from N.W. and S.E. with a *wave-motion* towards the N.E., producing the cross waves which occasion the complexity before alluded to. The late Professor Daniell has remarked that the curves increase in range towards the N.W., and in general the neighbourhood of *water* presents curves remarkable for the boldness of their contour and the large extent of their range. In venturing a speculation on these cross waves from the S.W. with parallel and opposite currents from N.W. and S.E., I should be inclined to attribute them to the effect of the solar influence on the terrestrial surface, extending from Cape Verd in Africa to the extreme north of Lapland in Europe. This surface extends from S.W. to N.E., or somewhat in that direction. It may be remarked, that to the *northeast* of Cape Verd is situated the Sahara or Great Desert of Africa, and here we have a great rarefying surface. To the north-west or west-north-west of this extensive rarefying surface, the broadest part of the Atlantic ocean is situated. The relative positions of the Great Desert and the broadest extent of the Atlantic will produce a great indraught of cool air from the ocean; the direction of this wind will be W.N.W. or N.W. To the north-east of this current, probably in the neighbourhood of Morocco, Fez, Algiers, Spain and Portugal, and the north-west portions of the Mediterranean sea, we ought to find the counter current from the S.E. or E.S.E., the two portions in juxtaposition moving with the greatest velocity. Somewhere in the Atlantic the turning-point of these oppositely directed currents should exist. The line of junction of these parallel currents will determine the trough of the wave, and as before shown, in consequence of the air being drawn off from the north-east to supply the ascending current, the wave will progress towards that quarter; the barometer first descending with the S.E. wind as the trough approaches stations to the N.E., and rising with the N.W. as the current produced by the rarefaction approaches, until the crest passes, when the new counter current or slope of the next wave would set in.[9]

Pursuing this idea further, there can be no question that Ireland and Scotland become points, or unitedly constitute a great point of rarefaction, forming as they do the nearest land to the northern part of the Atlantic, the land becoming hotter than the neighbouring water, and in consequence a N.W. current with its compensating current from the S.E. is induced. Not only will the rapidity of the currents reduce the pressure, but the ascending column from the land will transfer some of the air into the general current of the atmosphere, and there will be a real difference in the distribution of air as well as pressure; a section transverse to the line of greatest velocity will exhibit a hollow or trough, and the same phænomena will result from this arrangement of the aërial currents as we noticed arising from the N.E. and S.W. currents, the only difference being in direction.

In thus considering these rectangularly posited systems of parallel and opposite currents, many complex anemonal and barometric phænomena receive an easy explanation, particularly the revolution of the vane in one uniform direction, and the barometric wind-rose, When the conterminous edges of any two currents pass a station, the barometer is either at a maximum or minimum with respect to that particular system of currents; the wind also changes at this time. If the barometer has previously been rising with a north-easterly wind, it now begins to fall with a south-westerly: the cross currents are however passing at this time with a lateral motion towards the N.E.; in this set of cross currents the barometer will rise with a north-westerly wind and fall with a south-easterly. Suppose while the posterior slope of a N.W. wave transits, wind S.W., and before its trough passes, the trough of the cross wave from the S.W. also transits, and is immediately succeeded by the following anterior slope with its N.W. current, the wind will pass from S.W. to W. Now while this slope continues, upon the trough of the N.W. system passing, the wind changes to N.E., and the resultant of the two currents is N. It is easy to pursue this reasoning, and thus trace the changes of the wind arising from these two cross systems completely round the compass.

The two systems of cross currents naturally divide themselves into four beds of opposite currents, namely, N.E. S.W., N.W. S.E.; with the first of each system, N.E. N.W., the barometer rises, and with the last of each, S.W. S.E., it falls, so that in the barometric wind-rose the maximum is found about the N.E., the prevailing system, and the minimum near the S.W., the opposite current of this system.

The extent of arc which the wind-vane frequently describes, especially in stormy weather, also receives an explanation from these systems of cross currents.

A contemporaneous S.W. with a N.W. wind will occasion large arcs to be described between these points; the south-westerly gusts prevailing, directing the vane to that quarter; and the north-westerly immediately following, instantly occasions a change carrying the vane towards the N.W. These sudden and extensive changes are rendered more distinctly perceptible by means of a small kite flown with about 250 or 300 feet of string, or even more; the distinctness and independence of the direction of the two currents are readily seen, as well as the difference in their strength.

Source Endnotes

1 From Report of the British Association for the Advancement of Science for 1846.

2 Report of the First and Second Meetings of the British Association for the Advancement of Science, p. 235.

3 Report of the Thirteenth Meeting of the British Association for the Advancement of Science, pp. 60–103.

4 Mr. Scott Russell designates the point *a* the *origin*; a better term I apprehend would be *commencement.*

5 Report of the Fifteenth Meeting of the British Association for the Advancement of Science, p. 30.

6 Report of the Fifteenth Meeting of the British Association for the Advancement of Science, p. 61.

7 For this knowledge I am indebted to Mr. Brown's plates.

8 In attributing the greatest force to the N.E. current, I do not by any means wish to put forward or support any hypothesis that would at all interfere with the well-known fact, that the greatest force is usually manifested by S.W. winds. The point to which I wish more particularly to solicit the attention of the Association is this, *the cause which induces the south-westerly current itself.* This must reside in or near the torrid zone. Here we have a sufficient cause; we are presented with phænomena fully adequate to explain an influx of *cool* air from the N.E. This is the current that must first be established, and in the first instance its force will be *greatest.* We have however only to turn to Prof. Dove's letter to Col. Sabine (Report of the British Association, 1845, p. 61), and we shall at once find the reason why S.W. winds manifest by means of our instruments the greatest force. The N.E. currents are *narrower*, and the force *soon abates* as they pass over towards the S.E.; while on the other hand the same station is not only *oftener*, but *longer* in the S.W. currents, and as the line of greatest force approaches, the force increases, on some occasions very rapidly, until the wind changes. The line of greatest force soon passes the station, so that upon a mean of numerous observations the southwesterly wind exhibits the *greatest* force.

9 In the above suggestion I have considered the northern portion of the African continent as inducing the N.W. current, but of course, the entire surface, as far as the extreme north of Europe, including Great Britain and Ireland, will act as a rarefying surface.

1.3

MAPPING ATMOSPHERIC WAVES

Editorial Headnote

1.3 Mapping Atmospheric Waves

Following on from the focus on Elias Loomis, John Herschel and William Birt and the conceptualisation of atmospheric waves in the 1840s, the extracts in this section survey a key work by the man who took up their baton, Adolphe Quetelet (1796–1874), and examples of two of his key achievements, the application of statistical methods to meteorology and isobarometric mapping. It is a story of the introduction, development and application of statistical methods in meteorology, which went hand in hand with the founding of Belgium as a nation-state after the Revolution of 1830 and the objectives that Quetelet set himself as director of the national observatory then created. By promoting international observing campaigns in the 1840s, which to some extent took over from those promoted by Herschel in the 1930s, Quetelet sought to put Belgium on Europe's map of science.[1] This section explores a methodological representational achievement of significant influence: the identification of atmospheric pressure isolines with barometric wave crests. Atmospheric wave mapping became, in fact, the main tool of weather science until the 1860s.

Mathematician, statistician, founder of social physics, astronomer and director of the Royal Observatory of Brussels until his death, Adolphe Quetelet published *Sur le climat de la Belgique* (On the climate of Belgium) in two volumes between 1845 and 1857, although the different parts were printed separately by M. Hayez, the official printer of the Royal Belgian Academy.[2] The work brings together several works on meteorology already published in the *Annales de l'Observatoire* and the *Mémoires de l'Académie*, as well as new works. Thus, Volume I comprises the following parts: I. *Rayonnement solaire et températures de l'air et du sol* (1845); II. *Direction, intensité, durée et caractères distinctifs de vents* (1848); and III. *De l'électricité de l'air* (1849); and Volume II includes: IV. *Pressions et des ondes atmosphériques* (1851); V. *Des pluies, des grêles et des neiges* (1852); VI. *De l'hygrométrie* (1856); and VII. *De l'état du ciel en général* (1857).[3] Consisting of two sections, 'De la pression atmosphérique' and 'Des ondes atmosphériques', with nine and three chapters respectively, Part IV, which is reproduced in part here, involved the systematic application of graphical representation techniques to the study of variations in atmospheric pressure. Quetelet identified these variations with the atmospheric waves defined by Herschel and Birt (see 1.2 'Barometric Waves'). However, unlike Loomis, who identified the isolines with cold air waves, Quetelet introduced isochronic lines representing both the maximum pressure values and the crests of atmospheric waves. This comes across clearly in the extract reproduced here (including the map), which reveals the plotting, on a map of Eurasia, of curves showing the barometric variations around twenty stations from Brussels to Beijing and the lines linking the points where the barometric value is at its maximum. The reviewing tone of the extract is typical of the attitude of acknowledgement and correspondence which Quetelet adopted towards his source material. His extensive explanations are in line with the physical arguments adduced mainly by Herschel and Birt, although he also mentions Espy, Redfield and Dove. Surprisingly, he omits Loomis, perhaps because their focuses were different: whereas Loomis emphasised cold air waves as the origin of storms, Quetelet was interested in the advance of pressure waves propagating through the atmosphere.[4]

Notes

1 Christina Helena da Motta Barboza, *As viagens do tempo. Uma história da meteorologia em meados do século XIX* (Rio de Janeiro: E-papers Serviços Editoriais Ltda., 2012), pp. 176–78; Fabien Locher, *Le savant et la tempête. Étudier l'atmosphère et prévoir le temps au XIX siècle* (Rennes: Presses Universitaires de Rennes, 2008), pp. 24–7.

2 Jean-Guy Prévost and Jean-Pierre Beaud, *Statistics, Public Debate and the State, 1800–1945: A Social, Political and Intellectual History of Numbers* (London: Pickering & Chatto, 2012), pp. 49–61; David Aubin, 'Observatory Mathematics in the Nineteenth Century', in Eleanor Robson and Jacqueline Stedall (eds.), *The Oxford Handbook of the History of Mathematics* (Oxford: Oxford University Press, 2009), pp. 273–98, on pp. 287–91.

3 Nicolas Édouard Mailly, *Essai sur la vie et les ouvrages de L.-A.-J. Quetelet* (Bruxelles: F. Hayez, 1875), pp. 148–50.

4 Aitor Anduaga, *Politics, Statistics, and Weather Forecasting, 1840–1910. Taming the Weather* (London and New York: Routledge, 2020), pp. 82–98.

7

SUR LE CLIMAT DE LA BELGIQUE. QUATRIÈME PARTIE. PRESSIONS ET ONDES ATMOSPHÉRIQUES

Adolphe Quetelet

Source: Adolphe Quetelet, *Sur le climat de la Belgique. Quatrième partie. Pressions et ondes atmosphériques* (Bruxelles: M. Hayez, 1851), pp. 73, 75, 91–92.

PART FOUR

SECTION 2. – ATMOSPHERIC WAVES

CHAPTER

ATMOSPHERIC WAVES IN GENERAL

The variations in the barometer show us that atmospheric pressure undergoes continuous changes. At more or less distant intervals, this pressure reaches a maximum state through a series of oscillations and then passes into an opposite state.

Experience also teaches us that this maximum pressure does not occur in a single place, but that it is generally observed at the same time at a number of points which are connected by the law of continuity and thus form a more or less extensive line on the surface of the Earth.

This line of maximum pressure is mobile, moving in directions and at speeds that have not been studied until recently. By analogy with what happens at sea, the interval between two lines of minimum pressure can be called an atmospheric wave. In this sense, the crest of the wave is the line of maximum pressure.

It is important, as I shall soon show, not to confuse these waves indicated by the barometer with what I shall call atmospheric currents, which are generally indicated by wind directions. This is an important distinction, and if it is not made clear, serious mistakes can be made.[1]

I intend here to examine various issues relating to this interesting subject, but I think I should first go into some detail about the nature of the observations I have used and the way in which they were collected.[2]

DOI: 10.4324/9781003650294-11

During his stay at the Cape of Good Hope, Sir John Herschel had already turned his attention to the phenomena associated with the general movements of the atmosphere. In order to study this type of phenomena more closely, the illustrious English scientist had encouraged the Philosophical Society of Southern Africa to make hourly observations of the barometer, thermometer, etc., four times a year (at the solstices and equinoxes); these observations were also to be made at other points on the globe. Early in 1855 he kindly invited me, through our mutual friend Mr Ch. Babbage, to assist him in his work.[3]

The requested observations began in Brussels in June this year and continued, as requested, hour by hour from one o'clock in the morning on 22 June until midnight the following day. From the following equinox, Mr Crahay agreed to take part in the observations, and his example was successively followed by several observers in Belgium and abroad.[4]

However, towards the end of 1858, I received a letter from Sir John Herschel which contained the following passage: "I should be sorry to cause you and the observers who assist you fruitless trouble (and I must ask you to express my thanks to them) by drawing your attention any longer to observations of this kind. If, however, you see in them sufficient interest and hope to derive from them material likely to increase our knowledge, I would not wish to discourage you from continuing them, especially as you seem to devote so much of your attention to meteorological observations in general: But, after a careful examination, the result of all the communications I have received has confirmed me in the opinion that, beyond the 40th degree of latitude, the accidental variations of the atmosphere are too great, and too much obscure the regular variations, to permit clear conclusions of a general nature to be drawn from observations made for only four days in the year. My own observations ceased on my departure from the Cape, and although I have received several communications from different countries during the course of this year, I do not expect any more after the end of 1858, and I shall not really be able to reduce and combine them after that time.[5] I therefore ask you, at least out of consideration for me, not to continue with such observations until next December".[6]

I had already learned from experience how useful it was for the study of meteorology and the physics of the globe in general to continue and extend such a system of observations. On the other hand, I understood the motives of delicacy which had dictated the letter of the learned English astronomer, and I sought with even greater eagerness to find collaborators in the various parts of Europe and, above all, to arrange the observation points in such a way as to be able to follow the atmospheric waves step by step throughout their course. At the same time, I combined this great meteorological enterprise with the study of the *periodic phenomena* of animals and plants; and I left no stone unturned in order to extend as far as possible this great scientific crusade, which I placed under the auspices of the Belgian Academy. I was fully aware of everything I needed to accomplish my mission and to inspire the confidence necessary for the success of such an association.

ATMOSPHERIC WAVES

3. MAIN CONCLUSIONS

From what we have just seen, we can draw the following conclusions, at least for the periods covered by our research:

1 In general, several different wave systems pass through the atmosphere. These waves interfere with each other and create a specific pressure condition for each part of the Earth.

2 In the midst of all these particular movements, there is a predominant wave system that seems to remain more or less constant for the same climate.

3 In both Europe and Asia, atmospheric waves propagate from north to south, but not at the same speed; they travel faster in the Asian and Central European systems than in Russia or the Urals.

4 Atmospheric waves appear to propagate with fewer obstacles at the surface of the oceans than inland. In general, the unevenness of the earth, especially mountain ranges, reduces their speed and also modifies their intensity.

5 The inequality of speed on the continent, on the one hand, and near the sea, on the other, explains the inflections that the line representing the general movement of the wave in our hemisphere undergoes throughout its extent.

This line bends so that it is pushed forward in the direction of maximum speed: the wave thus penetrates the European continent almost simultaneously along the various coasts of the North Sea, the ocean, and the Mediterranean; on the other hand, it comes to rest almost simultaneously along the Urals and the Tyrolean Alps.

6 The speed at which barometric waves propagate is very variable; it can be estimated at an average of 6 to 10 leagues per hour from France: it is slightly higher in central Europe and lower in Russia.

Moreover, this speed varies from wave to wave, and even between different parts of the same wave. As we have already said, it is greater towards the coasts and in all places where the propagation of the movement seems freer. On the other hand, in the vicinity of mountains and plateaus, this speed decreases considerably; in the Urals it is sometimes reduced to less than 2 leagues per hour.

7 The directions of the winds have no apparent relationship with the directions of the barometric waves. This important fact seems to favour the hypothesis of compensating currents at the bottom of the atmosphere, in directions opposite to those of the currents flowing from the pole to the equator (see page 80). It should also be noted that the air can also be condensed by lateral pressures without there being inflows of new air and, consequently, significant winds in the directions of these pressures. On the contrary, the prevailing winds may well remain unchanged while the air masses they displace change their density appreciably.

Certain barometric waves must be like sound waves, which are transmitted in all directions, despite the obstacle of the winds, which can indeed modify their intensity and speed.

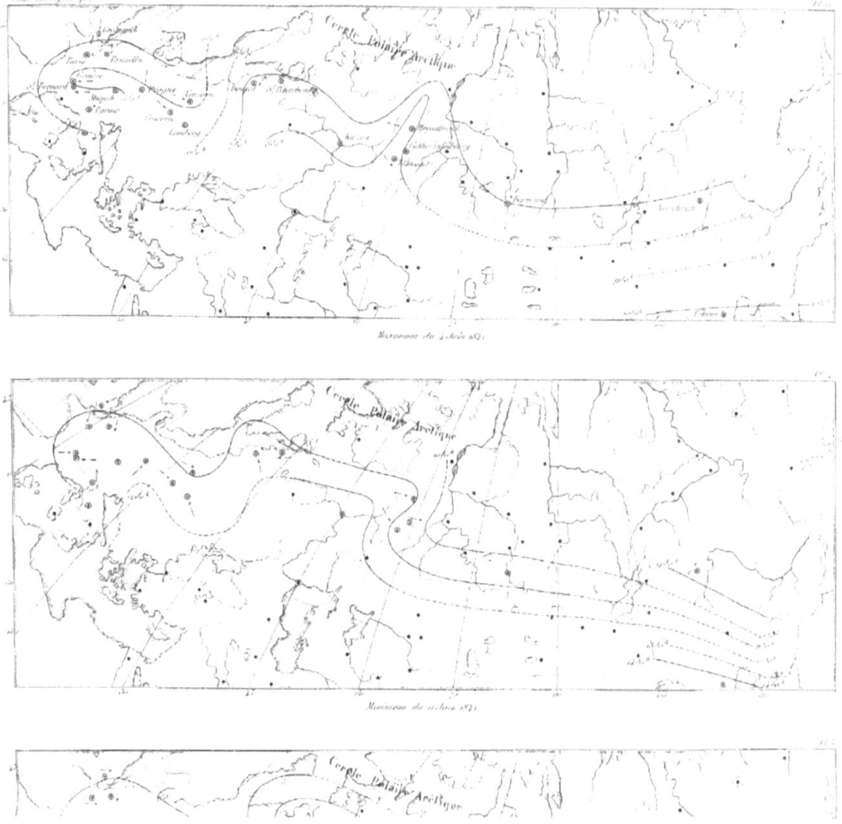

Editorial Endnotes

1 This distinction was also made by Sir John Herschel, in his report on the reduction of meteorological observations, presented to the British Association meeting in Cork in 1843. *In the mode in which I have been accustomed to consider the subject, the winds may be divided in two classes, winds of translation and winds of oscillation; nearly in the way in which the movements of the ocean may be divided into oceanic currents and tide-streams, and these again (regarded as the result of oscillatory movements) may be referred to the general laws which regulate the molecular movements of water in contact with the bottom of the sea, when under the influence of undulatory agitation . . . Every wave-like movement in a fluid (see* Weber's Wellenlehre) *consists of two distinct things,* an advancing form *and a* molecular movement, *which latter consists in a two-fold motion of each particle, vertical and horizontal . . . Now, the advancing form is indicated to us by the barometer, the molecular movement by the wind, and between*

these two phenomena there subsists of necessary a close and purely dynamical connexion. (Report of the British Association for 1845, p. 99).

2 In 1843, I stated the aim of my researches into air waves as follows: The study of these atmospheric oscillations, if undertaken by a sufficiently large number of observers, would reveal the most curious facts; it would tell us the size of the atmospheric waves, their average speed, the general direction of their movement; the places where they are formed; those where they disappear; the influence that mountains or certain localities may have in modifying them, and an infinity of other circumstances that we cannot foresee (*Rapport sur les travaux de l'Académie royale de Belgique,* during the year 1842–1843, vol. X, 2nd part of *Bulletins,* p. 551 et seq.).

3 *Bulletins de l'Académie royale de Bruxelles,* May 1855, vol. 2, p. 125.

4 In December 1830, the observations from Alost and London were added to those from Brussels and Leuven, volume IV, pages 5 and following of the *Bulletins de l'Académie*. Observations from Ghent began at the autumn equinox in 1838, volume V, page 575 of the *Bulletins*. However, it was not until 1840 that this system of observations developed considerably.

5 The report of the thirteenth meeting of the British Association, held at Cork in 1845, contains an examination of all the observations made by Sir John Herschel himself. This initial discussion led to a number of interesting observations, which I shall mention briefly.

6 *Bulletins de l'Académie royale de Bruxelles*, session of 3 November 1838, volume V, page 631.

1.4

EMBRACING WAVES FOR STATE INTERESTS

Editorial Headnote

1.4 Embracing Waves for State Interests

As we showed in the introduction to this volume, much of the importance of atmospheric waves to nineteenth-century science of weather forecasting lay in the way in which the analysis of their nature and propagation, through barometric curves and oscillations, led to the prioritisation of a scheme and view in which the barometric pressure field and fixed terrestrial coordinates took precedence. This atmospheric wave scheme paved the way for the so-called Eulerian approach: in this framework, the observer had to follow the patterns of air movement recorded at fixed (usually ground-based) observation stations, and then project them onto a synoptic map. The Eulerian approach created clearer understanding of the storm as a living, dynamic system, the result, according to John Herschel, of the crossing of two great atmospheric waves, the progress and evolution of which could be foreseen to some extent. Atmospheric waves thus contributed in various ways to the gradual reconceptualisation of the technique and scope of weather forecasting. By focusing on the atmospheric pressure field at level 0, the atmospheric wave framework led weather scientists to investigate how these waves arise and interact with storms, how they propagate through the atmosphere, and, most provocatively, how their progress can be foretold. This section and the following one deal with the adoption of the Eulerian approach in France and Italy. In particular, it deals with a dynamic process with significant consequences: Although the Eulerian approach had few adherents in Britain in the 1850s, it represented a powerful *protégé* in the thought of Western meteorology, which would later influence British science. In fact, the Eulerian approach became the mainstream of weather forecasting practice until the beginning of the twentieth century.[1]

This section deals with Urbain Le Verrier (1811–1877), the controversial and authoritarian director of the Paris Observatory, whose adoption of the Eulerian approach in the interests of his leadership and national security and order set the pattern for the subsequent direction of meteorology and, in particular, weather forecasting.[2] Practical needs and aims, rather than theoretical motivations, were the reason for adopting such an approach. This occurred in a context of political urgency under Napoleon III, caused by a maritime tragedy – the fatal Black Sea storm of 14 November 1854 – in the Crimean War.[3] Although John Herschel, William Birt and especially Adolphe Quetelet anticipated some elements of his work, and although Emmanuel Liais, his young assistant at the Observatory and the main proponent of the mathematisation of meteorology in France, contributed substantially to his studies on the famous storm of 1854, Le Verrier is undoubtedly regarded as its most significant founder.[4] He was amongst the first to plan a vast international meteorological network, centralised in Paris, to warn sailors of approaching storms, and was a pioneer in organising a daily bulletin distribution service from the Paris Observatory, dispatching notes on atmospheric conditions to a good number of French and foreign weather stations.

Two extracts from his work are included. The first, 'Note sur le développement des études météorologiques en France' (1855), is itself the précis concerning meteorology of his then-unpublished work entitled *Mémoire sur l'état actuel de l'Observatoire de Paris, et projet d'organisation scientifique* (December 1854), which was to be published as "Rapport sur l'Observatoire impérial de Paris et projet d'organisation" (Report on the Paris Imperial Observatory and organisational project) in the *Annales de l'Observatoire Impérial de Paris* in 1855, as part of the first issue of that journal. Although Le Verrier's name appears in the heading of the latter, his assistant Liais was to claim authorship of the section on meteorology many years later. This project for the scientific reorganisation of the Observatory was born in a climate of political urgency, after Minister of War Jean Baptiste Vaillant urged Le Verrier to analyse the conditions of the Black Sea storm in order to anticipate the arrival of similar phenomena in the future. The précis summarising part of the 1855 Report (pp. 55–59) was published in the *Comptes Rendus Hebdomadaires des Séances de l'Académie des Sciences de Paris*. In the précis, Le Verrier introduces some new considerations that do not appear in the report of the Commission set up in January 1854 to reorganise the Observatory, of which he was a member. In particular, he highlights the practical function of meteorology, which was geared to the progress of navigation, agriculture, public works and hygiene. But most importantly, he also emphasises the importance of observing barometric variations in order to announce the approach of storms and to warn of their arrival by electric telegraph. The Black Sea storm had a knock-on effect: storm warnings became an essential service in the reorganisation plans.

The second extract, a report on Liais's work at the Imperial Observatory on the storm in the Black Sea, also published in the *Comptes Rendues* of the Paris Academy in 1855, illustrates well Le Verrier's ability to adapt to the context of political urgency of the time. As Liais was not a member of the Academy, Le Verrier presented it on his behalf at a session on 31 December 1855. Before an audience full of sceptical academics, including the anti-empiricist mathematicians Jean-Baptiste Biot and Augustin Louis Cauchy, Le Verrier and Liais aimed to win academic approval for the project to reorganise the Observatory. In his presentation, Le Verrier sought to convince the rationalist and anti-positivist currents in the Academy of the practical usefulness of the science of meteorology. Although he had the emperor's endorsement, the explicit approval of the members of the Academy was essential for his project.[5]

In the report read by Le Verrier, Liais gives the results of his study carried out in 1855 on the course of the Black Sea storm, based on data on pressure and wind direction sent to the Observatory from various countries. After describing the data in detail, he argues that there is no relationship between the progress of the storm and the direction and speed of its winds. He then distinguishes two types of pressure lines: *atmospheric waves* (here he is inspired by Quetelet waves), which

represent displacements of air masses with barometric maxima at their crests; and *wave transport lines*, which represent lines of absolute values of barometric minima and maxima in a wave (this is an original idea of his).[6] Confident that the audience would appreciate the advantages of monitoring the storm's path in real time, Liais draws a map illustrating the passage of the disturbance across Europe, a resource that Le Verrier had hitherto used sporadically from isolated data and without following any methodology.[7] By showing Liais' map to the audience, Le Verrier seeks to demonstrate that the graphical representation of barometric iso-lines is the most appropriate method for studying the approach of storms. Thus, he postulates the existence of a correlation between changes in atmospheric pressure and the path of a storm. The strategy used in the presentation has an immediate effect: as Le Verrier notes, 'I certainly hope it will be possible to analyze the main circumstances of propagation, and I note with pleasure that M. Cauchy makes a sign of assent' (p. 1203). However, despite this tacit approval, the evidence suggests that the thorny issue of wave propagation was not mathematically analysed.

Notes

1 Aitor Anduaga, *Politics, Statistics, and Weather Forecasting, 1840–1910. Taming the Weather* (London and New York: Routledge, 2020), pp. 106–27.

2 John D. Cox, *Storm Watchers: The Turbulent History of Weather Prediction from Franklin's Kite to El Niño* (Hoboken, NJ: John Wiley & Sons, Inc., 2002), pp. 85–90; Fabien Locher, 'L'empire de l'astronome: Urbain Le Verrier, l'Ordre et le Pouvoir', *Cahiers d'histoire. Revue d'histoire critique*, 102 (2007), pp. 33–48; James Lequeux, *Le Verrier – Magnificent and Detestable Astronomer* (New York: Springer, 2013).

3 Helmut Landsberg, 'Storm of Balaklava and the Daily Weather Forecast', *The Scientific Monthly*, 79 (6) (1954), pp. 347–352; Aleksandr Khristoforovich Khrgian, *Meteorology: A Historical Survey*, 2nd ed. (Jerusalem: Israel Program for Scientific Translations, 1970), p. 154.

4 Christina Helena da Motta Barboza, 'Liais, Emmanuel-Bernardin', in Noretta Koertge (ed.), *New Dictionary of Scientific Biography* (Detroit, MI: Scribner, 2008), vol. 4, pp. 293–94; Jacques Ancellin, 'Un homme de science du XIXème siècle: l'astronome Emmanuel Liais', in M. Maurice Durchon (ed.), *Mémoires de la Société Nationale des Sciences Naturelles et Mathématiques de Cherbourg*, vol. 57 (Coutances: Imprimerie OCEP, 1975–1978).

5 Aitor Anduaga, *Politics, Statistics, and Weather Forecasting, 1840–1910. Taming the Weather* (London and New York: Routledge, 2020), pp. 120–27.

6 Christina Helena da Motta Barboza, *As viagens do tempo. Uma história da meteorologia em meados do século XIX* (Rio de Janeiro: E-papers Serviços Editoriais Ltda, 2012), pp. 183–84.

7 Fabien Locher, *Le savant et la tempête. Étudier l'atmosphère et prévoir le temps au XIX siècle* (Rennes: Presses Universitaires de Rennes, 2008), pp. 39–41.

8

NOTE SUR LE DÉVELOPPEMENT DES ÉTUDES MÉTÉOROLOGIQUES EN FRANCE

Urbain Jean Joseph Le Verrier

Source: Urbain Jean Joseph Le Verrier, 'Note sur le développement des études météorologiques en France', *CR,* 40, 1855, 620–626

"The Academy has read with interest a Note from its illustrious Associate, Mr Alexandre de Humboldt, concerning meteorology. M. de Humboldt expresses the wish that the Government, which is always inclined to favour institutions closely connected with the progress of agriculture, should permanently endow a certain number of well-chosen stations in terms of latitude and height throughout France and its colonies. "We have these stations from Memel to the Rhine," adds Mr. de Humboldt.

I must inform the Academy, and Mr. de Humboldt will be pleased to know, that the Government has responded to its wishes and has taken steps on its high *initiative* to ensure that meteorological studies are established in Paris, in France and in our overseas possessions *on the widest possible scale.* These are the very expressions of the orders given to the administration in the interests of science.

These orders do not only concern meteorology, but the entire work plan which, according to article 10 of the decree of 31 January 1854, we had to submit to the Government for approval. On a recent occasion, I told the Academy that I would have the honour of presenting the various parts of this organisation in turn. Mr. de Humboldt's letter, which will be a great encouragement to us, encourages me to begin with meteorology.

I therefore extract what concerns meteorology from a work entitled *Mémoire sur l'état actuel de l'Observatoire de Paris, et projet d'organisation scientifique,* written in December 1854 and never published. (As it is impossible to reproduce this document in its entirety here, we shall confine ourselves to a brief analysis).

After having explained the relationship between astronomy and meteorological phenomena, M. Le Verrier reviews the progress made since the 17th century, when the two most valuable meteorological instruments, the barometer and the thermometer, were invented. Then, after examining the state of meteorological

DOI: 10.4324/9781003650294-13

research in France and abroad, he regrets that our country had not always kept pace with the state of the art: We must not be afraid to agree provided we are determined to put an end to this situation.

Of all the scientific establishments owned by the State, the observatories in Paris and Marseilles are the only ones where meteorological observations are made on a regular basis. In Marseille, observations are made six times a day: at 6am, 9am, midday, 3am, 6am and 9pm. In Paris, the weather was observed only four times a day: at 9am, midday, 3am and 9pm. Since last November, observations have been added at 6 pm and midnight, but this is the limit of what can be achieved with a small staff, specially trained in astronomy and entrusted with a tiring task. At Greenwich a special staff of four makes twelve observations a day, from two o'clock in the morning to two o'clock in the afternoon, covering the whole twenty-four-hour period; and the observatory is also equipped with instruments which record their own indications by means of photography. In the many magnetic and meteorological observatories in England, which has a model of this kind in Kiev, observations are everywhere made from two hours to two hours, the two-hour series having been recommended by the Royal Society of London. In Russia and the United States of America, observations go even further and are made every hour. The same applies to Brussels, most observatories in Germany and some in Italy. In addition to their astronomical observatories, these countries have even more magnetic observatories. The whole surface of England, Russia, Germany, and especially Austria and the United States are covered with these observatories, intelligently distributed in the most interesting physical stations; and France does not have a single one!

It must be said, however, that the French nation has an eminently scientific spirit, and commendable efforts are being made in our departments by a few private individuals, whose zeal makes up for the inadequacy of the public institutions. Meteorological series, in some places more complete than those of the Observatory, have been made for several years at Versailles, Vendôme, Toulon, Bordeaux, Dijon, Lyon, Metz, Rodez, Rouen, Orange, Cherbourg, Goersdorff, Nantes, Bourg, Saint-Hippolyte-de-Caton, Le Puy, Privas, and so on. However, most of the instruments used for these series are not known; many observers were unable to obtain suitable stations for their instruments, which casts doubt on several series that were nevertheless carried out with such dedication to science.

It should be added that private meteorological observatories have not been able to collect all the necessary observations. Humidity is measured at very few points. Atmospheric electricity is not observed anywhere in France. As for incidental observations, only a few private observers keep meteorological diaries. Magnetic observations, which require special constructions at a distance from inhabited places containing iron, are beyond the reach of private observers. For this reason, France remained outside the great association created to solve the curious problems posed by terrestrial magnetism. While instruments with a continuous display by means of photography are becoming widespread in all observatories, and Greenwich had them since 1847, France has not yet seen them.

And yet meteorology is an eminently practical science. Navigation, agriculture, public works, and hygiene have a special interest in its progress, and it is important that such serious interests should no longer be neglected.

The usefulness of meteorological research for navigation is unquestionable. It is thanks to the study of winds that the length of crossings has been considerably reduced in recent years. The various seafaring nations owe a great debt of gratitude to Lieutenant Maury, whose skilful plans have contributed greatly to this result. The average crossing time from the United States to Cape Saint-Roch has been reduced from 41 to 22 days, and from California from 180 to 100 days . . .

The importance of the compass required that the equator and magnetic meridians be determined for the entire globe. France, through several expeditions, had the glory of contributing to this fine work. But it was not enough: the elements of terrestrial magnetism are constantly changing; after a few years, the magnetic meridians have shifted considerably, and it is important to combine the results in such a way as to determine the true state of the magnetic forces at any given time and in any place on Earth.

While lightning conductors protect ships from the effects of lightning, which are often so terrible on board, the barometer, through its variations, tells the sailor when a storm is approaching and warns him, if he is near a dangerous coast, to move away from it, or to seek refuge in the ports it can offer. Krusentern attributes his ability to anticipate gales to the constancy with which he observed the barometer. the barometer. Scoresby claims to have predicted storms *seventeen* out of *eighteen* times by consulting this instrument. And yet the indications deduced from isolated observations of the barometer are of little value compared to that which can be obtained from the simultaneous study of all the meteorological instruments, many of which are still so infrequently observed. A good series of research studies will provide a better understanding of the forecasts that can be made from these observations combined with those of the sky.

By linking the various stations where meteorological observations are made, by means of electrical telegraphy, we will be able to know at any moment the direction and speed of the propagation of storms. We will be able to warn our coasts several hours in advance of certain gales, especially the most dangerous ones; for the history of shipwrecks tells us that almost all these events occur with winds blowing on land, and we know that almost all hurricanes propagate by suction, i.e. in the direction from which they blow.

Rainfall will also be useful. While it provides useful information for the inhabitants of high, dry areas on the dimensions of cisterns and reservoirs, it can also enable the inhabitants of low-lying areas near our great rivers to save their crops, their herds, and even their lives by warning them of impending floods. A hydrometric commission set up in Lyon after the terrible floods on the Saône in 1840 has already rendered outstanding services. On several occasions, by monitoring the rainfall in the Saône basin, it was able not only to warn of impending floods, but also to predict to inform a few decimetres the height to which the water would rise. At present, thanks to the zeal of some engineers of the Ponts et Chaussées, pluviometric observations are being made in conjunction with the observations of the levels of several of our rivers; and the curious relations discovered by Mr Belgrand, chief engineer of the Ponts et Chaussées, between the floods of the various rivers of France, give us the certainty of obtaining in this way results of great importance for the inhabitants of our watercourses.

Consequently, M. Le Verrier considers that it is necessary, as soon as possible:

1. To equip the Paris Observatory with standard instruments;
2. To organise in this establishment a complete system of regular meteorological and magnetic observations;
3. To endeavour to establish a meteorological journal, as complete as possible, for random phenomena;
4. To establish relations with the various private observatories in France and with some of the principal foreign establishments; to encourage the establishment of new meteorological stations; to verify their instruments; to publish instructions on the choice of exposure for thermometers, hygrometers, etc . . .;
5. To publish daily in the newspapers the observations of Paris and of the principal stations of France, which would be transmitted to the Observatory by electric telegraph (1);
6. To present, at the end of the year, in a general publication, a summary of all the observations and of the discussions to which they had given rise;
7. To discuss the observations previously made in Paris and to collect documents on all that has been done in France.

Magnetic observations, in particular, require special constructions which private observers are generally unable to carry out. In France, no one had undertaken such work. It is therefore up to the government to set up magnetic observatories. However, this question, apart from that of the astronomical observatories, which came under the responsibility of the Ministry of Public Instruction, is of particular concern to the Ministry of the Navy, which has a special interest in this matter and has trained personnel on the staff of each military port and in the colonies, who can be put in charge of these observations. In addition, the Ministry of the Navy could distribute the magnetic observatories in our various colonies and select seven stations which, together with the Paris Observatory, would be sufficient to define the magnetic state of the globe at any given moment. Magnetic observatories in Algiers, Saint-Louis (Senegal), Saint-Pierre or Miquelon, in one of our West Indies or in Cayenne, in Tahiti or in the Marquesas Islands, in Reunion Island or in Pondicherry would be admirably located.

(1) At the time of the General Exhibition of Industry in England, the state of the atmosphere in the various cities of the United Kingdom was posted every day at the door of the Exhibition. In May 1855, shall we be less advanced than the English in 1851?

After this presentation, Mr Le Verrier added that the plans, which had been submitted to the higher Administration last December, had been approved, and that the orders for their execution had been given in February. The organisation will continue with all the speed compatible with the proper execution of the work, much of which is of a very delicate nature. The special participation at the Observatory of three of the most eminent meteorologists, Messrs. du Moncel, Liais, and Renou, will be a precious guarantee for the friends of science.

The special installation of the Paris Observatory should leave nothing to be desired in terms of both the nature and precision of the equipment. We have been fortunate to find at the Conservatoire des Arts et Métiers models of the recording apparatus, which General Morin, with the utmost generosity, will make available to the Observatory at a time when this method of observation will be used for the first time in France.

At the same time, the regular organisation of meteorological studies on the French surface, an organisation which the Central Observatory in Paris does not separate from its own, is progressing. The production of atmospheric maps of France using the information collected by the Administration of Telegraphic Lines, maps which the Academy had just seen, was only a first attempt to show how the organisation to be set up could be definitively and usefully established. It was decided that a fairly large number of meteorological stations, at least twenty-four, already selected, established by the Administration and equipped with precise instruments by the Paris Observatory, should be set up immediately and should transmit their observations every day.

These stations, added to those founded by honourable scientists to which we have already given the justice they deserve, will complete a respectable meteorological network, which is likely to render great service. They will avoid duplicating the work of private observatories, which work admirably and whose observers have kindly agreed to contribute to the common effort. The faculties of science, even the lycées, would intervene if necessary. The organisation of colonies and studies at sea will follow immediately afterwards.

It would be of little use, however, to collect a considerable number of observations and to assemble the originals at the Paris Observatory, if they were not published immediately, in a suitable order, and in such a way that they could be discussed as a whole, not only by the Observatory itself, but by all meteorologists. Measures have been taken to ensure that all observations are published, some on the day they reach the Observatory, others within a month at most. The new publication will by no means duplicate existing publications, in particular the excellent *Annuaire de la Société Météorologique*. These publications, which serve the same purpose, will complement each other.

As can be seen, the Imperial Observatory of Paris has no other ambition than to contribute, as far as it can, to the dissemination of meteorological studies by developing observations and making them immediately available to scientists. It will leave no stone unturned to achieve this objective, seeing the interest shown by all sides in such undertakings.

9

M. LE VERRIER PRÉSENTE UN TRAVAIL FAIT À L'OBSERVATOIRE IMPÉRIAL, PAR M. LIAIS, SUR LA TEMPÊTE DE LA MER NOIRE, EN NOVEMBRE 1854

Urbain Jean Joseph Le Verrier

Source: Urbain Jean Joseph Le Verrier, 1855. 'M. Le Verrier présente un travail fait à l'Observatoire impérial, par M. Liais, sur la tempête de la mer Noire, en novembre 1854', *CR*, 41, 1855, pp. 1197–1204.

We have not forgotten the hurricane which, on 14 November 1854, caused so many disasters in the Black Sea and led to the loss of *the Henri IV*. On the same day, or a day apart, depending on the places, storms broke out in Western Europe, over Austria and Algeria, so that the phenomenon seemed to have spread over a vast area. This remarkable circumstance attracted the attention of our illustrious colleague, Marshal Vaillant, Minister of War, who kindly wrote to me inviting me to undertake a study of the conditions under which the phenomenon had occurred and assuring us of his support.

In order to be able to respond to Mr. Maréchal's intentions, I sent a circular to astronomers and meteorologists in all countries, asking them to send me any information they could gather on the state of the atmosphere during the days of 12, 13, 14, 15 and 16 November 1854. I limited my request to these five days so as not to tire our correspondents, and because my sole aim was to obtain initial indications which would make it possible to draw up a plan for later discussion.

In response to this circular, the Observatory received more than two hundred and fifty documents. I regret that it is not possible for me to give a complete list of the persons who have been so kind as to supply us with information in this way, and I must confine myself to mentioning the names of those who have sent observations covering the entire surface of a country: in England, Messrs. James and Glaisher; in Belgium, Mr. Quetelet; in Holland, Mr. Buys Ballot; in Prussia, Mr. Dove; in Austria, Mr. Kreil; in Sweden, Mr. Aghardt, etc. We received the comments from India through Mr. Secretary of the Asiatic Society and those from our colonies through Mr. Mestro, Director of Colonies at the Ministry of the Navy.

 DOI: 10.4324/9781003650294-14

We entrusted the discussion of all these documents to the head of the Observatory's meteorological division, Mr Liais, and it is his excellent work that we are now going to report on.

The simultaneity of the storms in Eastern and Western Europe initially led people to believe that the phenomenon was continuous. But as more information became available, it soon became clear that the storms in France and the Crimea were different. In Central Europe, including Vienna, the atmosphere was calm on 14 November, although the wind in the Austrian capital was still quite strong on the night of 13–14 November. The storm that began in Paris on the 14th and reached its peak in France and England during the days of the 15th and 16th November was therefore in no way related to the storm that raged in the Crimea on the morning of the 14th.

Looking at all the evidence, we can see that the two storms did not remain stationary, but moved from west to east. Thus, the storm that crossed the Black Sea on the 14th was felt, albeit weakly, in France around the 11th, more strongly in Austria from the 12th to the 13th, in Transylvania on the 13th, and finally, with increasing intensity, in the Crimea. From there we saw the phenomenon move on to the Caucasus.

The movement of the storm should not be confused with the movement of the wind. The direction of the wind has been determined at weather stations, sometimes by means of weather vanes, sometimes by the movement of clouds. The directions obtained in this way were almost always different, but neither of them seemed to have any direct relationship with the direction of the storm, or at least none that could be discerned from the documents available.

Let's move on to the indications provided by barometric observations.

We know that in every place the barometer oscillates continuously; since these movements are due to the displacement of masses of air, it follows that there must exist in the atmosphere types of waves characterised by a barometric maximum and necessarily having a translational movement. Observations have confirmed the existence of these waves and shown that they are generally of immense length. M. Liais was able to detect and follow several of these waves, which crossed Europe from west to east in November 1854. The graphs on the maps that I have the honour of presenting to the Academy clearly show the progress of the phenomenon. These few maps summarise the mass of documents. This example will prove to those who are afraid of the multiplicity of meteorological material that, with a good graphic representation, can be reduced to a form that allows conclusions to be drawn at a glance.

In general, atmospheric waves do not travel in isolation. The interval between two successive waves necessarily forms a trough. It was one of these troughs that crossed the Black Sea at the time of the storm on 14 November 1854. Initially quite weak, it crossed Spain and France between 10 and 11 November. By the 12th, it had reached the Danube provinces, and on the 13th, at the time of the first gale mentioned in the *Pluton* commander's report, it began to reach the Crimea.

Throughout its voyage it continued to increase, accompanied and followed by storms.

The wave which preceded this depression was small and crossed the Caucasus on the 12th. The subsequent wave, however, was very strong and reached the western coast of France on the same day. It was this latter wave that we were able to follow precisely, thanks to the information gathered by the Observatory.

At noon on 12 November, the track of the wave passed over the east coast of England, which it crossed towards the 55th parallel, and then moved southwest. Towards the 6th degree of longitude, it turned south to cross the Bristol Channel and the tip of Cornwall, crossed the English Channel and Brittany, and passed the mouth of the Loire, from where it proceeded in a straight line towards Narbonne. It then crossed the Mediterranean and reached the coast of Algeria, which it intersected by *one* degree of west longitude. All along this line, the barometric pressure, reduced to sea level, was close to 770 millimetres and even exceeded this limit in some places.

Twelve hours later, the wave was developing on the coasts of Holland and Belgium, passing east of the Zuiderzee, heading towards Lille, passing a little east of Paris and Lyon, and entering the Mediterranean near the mouth of the Rhône.

At noon on 13 November, according to the information available, the wave was tracked from the entrance to the Gulf of Finland, from where it headed towards the south-east coast of Sweden near the island of Oland, then passed near the island of Rugen, then a little west of Berlin and Dresden. There it turned east and crossed the 50th parallel at 13 degrees east longitude. It then turned south again until it reached the Alps, over which it fell, partly following their contours. Finally, at the borders of France, it entered the Mediterranean Sea, where we can no longer follow it.

Thus, while in the centre the wave had travelled from the coast of Brittany to Berlin in twenty-four hours, in the south it had only crossed the Rhône valley and come to rest on the Alps. This explains the anomaly in the observations from the Rhône valley, which we first noticed in the information sent to us from Orange by our learned colleague M. de Gasparin. The barometer there remained more or less stationary on the 13th and 14th, while in Paris it fell rapidly.

At midnight on the 13th, the wave was still over the Alps. However, it had crossed the border between France and Italy and was entering the Mediterranean at the bottom of the Gulf of Genoa. To the north, it passed the mouth of the River Oder.

At noon on 14 November, the wave passed just west of St Petersburg and headed straight for Dantzick. There it turned south and headed straight for the Dalmatian coast, forming only a sinuosity that carried it a little to the west and brought it closer to Vienna. It crossed the Adriatic Sea, intersected the Italian coast at about the 13th degree of longitude, and re-entered the Mediterranean in the middle of the Gulf of Taranto.

From midday on 14 November, the part of the wave in the centre of Europe moved more slowly than its extremities. Its shape then resembled that of a hyperbola. At noon on the 15th, the southern branch of this curve undulated around the Crapacks Mountains, reached Kronstadt, and from there moved towards the Bosphorus. By midday on the 16th it had already crossed the Black Sea. We saw it in the Urals, near Katherinenburg, from where it headed west-south-west towards Kaluga. There we lost track of it due to a lack of observations, but it must have retreated to the south-east because we found it near Tbilisi.

If we trace the waves along the 45th parallel, the latitude of the Crimea, we see that they decrease as they cross the centre of Europe and increase again as they approach the Black Sea. Thus, on the 12th, the maximum was 771 millimetres on the western coasts of France[1]; on the 13th, over the Alps, it was only 767 millimetres, but the wave abruptly decreased like a wave about to break; on the 14th, the maximum dropped to 762 millimetres; on the 15th, it rose again to 764 millimetres, and on the 16th, it seemed to increase again as we approached the Caucasus. Another section towards the 49th parallel also shows that the wave weakened in the centre of Europe.

The central part of Europe was colder than the western and eastern parts. Below the same parallel, the difference between west and centre was as much as 15 degrees.[2]

So, Mr Liais concludes, the light wind accompanied the atmospheric wave; the strong wind and the squalls, the depression. Probably the effect on the vapours caused by the cooling due to the drop in pressure, is not entirely unrelated to this phenomenon. The vapour may also have had an effect on the increase in intensity that the depression experienced as it crossed the Black Sea. Finally, the temperature difference between the continent and the ocean must also have had an influence. New records, covering a longer period, will be needed to disentangle the specific influence of each of these causes. We would also need observations from North America, not all of which have been received, to know whether the wave originated in the ocean or on the coasts of Europe, or whether it already existed on the new continent.

The wave has always shown a marked tendency to retreat and stop over mountains and high ground. In this way, it took almost twenty-four hours to cross the Alps. It would be very interesting to trace a large number of atmospheric waves. Their relationship with the unevenness of the ground would suggest that they have predominant forms in each locality. The same is undoubtedly true of the transport lines of maxima and minima. Perhaps there is also a certain periodicity in the direction of the waves, depending on the season. This seems to be the result of some studies carried out by M. Quetelet on the June waves. Very useful practical applications would then result from these studies.

In a nutshell:

The storms of November 1854 were caused by atmospheric waves that crossed Europe from west to east.

The wave of 14 November covered the whole of Europe.

It crossed Europe in about four days.

The direction of the meteor and its speed have no definite general relationship to the direction and speed of the wind, any more than the waves propagating on the surface of flowing water depend on the direction of that water.
In the region occupied by the top of the wave, the atmosphere was quite calm. The storms corresponded to the lows.
The later low caused the storm in Paris on 14, 15 and 16 November.
The previous low-pressure system, weak at Gibraltar on the 10th, quite weak at Malta on the 11th, stronger at Vienna on the 12th, stronger still at Vienna, Gorfu and Kronstadt on the 12th and 13th, finally reached the Black Sea on the 14th.

This raises a series of questions:

Where did these waves come from? Where did the phenomenon that created them occur?
Is their translation always independent of the direction of the wind? Would their speed of translation offer any law in all conditions?
Would the action of these waves produce only an oscillation of the atmosphere, without translation, and would the translation of the phenomenon be due to the mean wind?

As can be seen from the above, despite the mass of documents already collected, new information is needed. Even the observations from India and the colonies could not be used for the time being because of the lack of intermediate points. Observations from North America, Southern Russia, Siberia, Tartary and Greece have not all been received. Although we hope to be able to fill these gaps, we did not wish to delay any longer in informing the Academy of the results already obtained, in order to show our correspondents that the information they have sent us has not been useless, and to make them realise, before they ask for more, the need for new details relating to a greater interval of time.

In the presence of such widespread phenomena, and in the face of the generality with which small local actions disappear, we certainly hope it will be possible to analyze the main circumstances of propagation, and I note with pleasure that M. Cauchy makes a sign of assent. But before we can undertake this study mathematically, it is first necessary to understand the conditions of the phenomenon, and to do this we must increase the number of observations, which are already so numerous.

Finally, when we see this regular transmission of the November storm, we wonder whether the presence of an electric telegraph between Vienna and the Crimea could not have been used to warn our armies and fleets.

When we learned in Vienna that the storm had raged on the coast of France at such and such an hour, in Paris at such and such an hour, in Munich at such and such an hour, and always increasing in intensity, could we not have foreseen that it would reach the Black Sea? We do not hide the fact that we will encounter great practical difficulties in obtaining results of this importance, but we will undoubtedly be able to overcome them. The Observatory is working on it and I will soon be able to present to the Academy the first steps taken to make progress on this issue.

Editorial Endnotes

1 All these pressures are reduced to sea level.

2 After tracing the general course of the wave, Mr Liais examines how the maximum and minimum are transported on this wave.

"The line, he says, which separates the points at which the barometer rises and those at which it falls, does not depend on the absolute value of the pressure; the isobarometric line, on the other hand, is not defined by the movement of the barometer, but only by its absolute value. In addition to these two systems of lines, we will consider a third, which can be called the *wave transport lines*, and which depends both on the absolute value of the pressure and on the movements of the barometer.

To define this system of lines, we need only look at the pressure variations at the top of a wave. If we follow this peak, we can see a series of increases and decreases in pressure. These differences are carried with the wave. If we connect by a line the series of points which successively occupy the same maximum or minimum on the top of a wave, we shall have the system of lines we are considering, which cut the waves at variable angles.

In the wave of 12 November, there was a maximum in the English Channel. On 13 November, at noon, this maximum had moved near Dresden, and from there it seemed to move towards the Bosphorus. On the same wave of 12 November, there was a minimum in southwestern France. It moved near Vienna, passed north of the Alps, and from there seemed to be heading towards the archipelago. Finally, a maximum seems to have accompanied the wave over the Mediterranean.

Another minimum undoubtedly existed in the north, but the wave of the 12th can be extended beyond the east coast of England for lack of observations. This minimum had arrived on the 13th near the island of Oland, reached the Prussian coasts near Dantzick, advanced eastwards to the 25th parallel, and from there moved south-eastwards. The wave therefore seemed to have a tendency to move towards the Black Sea and the archipelago. Unfortunately, the total lack of records for Turkey and the small number of records for southern Russia that have reached the Observatory do not allow us to extend these lines with certainty.

1.5

A NEW METHOD

Editorial Headnote

1.5 A New Method

In the 1850s, the government of the Papal States made an important contribution to the development of telegraphic meteorology and storm warnings. Following the presentation of Urbain Le Verrier's project (see Vol. I, 1.4 'Embracing waves for state interests') and the Maritime Conference held in Brussels in 1853, which recommended the adoption of uniform and standardised observation systems to meet the needs of maritime safety, the Pontifical government became convinced that its own security and state cohesion was linked to advances in meteorology. Through the pages of the journal *Corrispondenza scientifica in Roma*, edited by her husband Erasmo, the astronomer Caterina Fabbri Scarpellini from the Observatory of Campidoglio proposed to the government the creation in Rome of a network similar to that established by Le Verrier in France. Pope Pius IX entrusted the mission to Father Angelo Secchi (1818–1878), director of the Observatory of the Jesuit-run Collegio Romano and one of the most important promoters of observatory sciences in the nineteenth century.[1] In 1855, Secchi led a plan for the telegraphic transmission of simultaneous meteorological observations, centralised at the Pontifical Observatory and with three stations (Ancora, Bologna and Ferrara) as data providers.[2] In doing so, the Papal State claimed the importance of the storm warning service and used it to strengthen its internal cohesion, both territorially and administratively.[3]

A Jesuit astronomer who pioneered stellar spectroscopy and was exiled to Stonyhurst in England and Georgetown University in Washington, Secchi was imbued with the works of Captain Matthew F. Maury (architect of the Brussels Conference of 1853) and his vision of dynamic meteorology and telegraphic communication.[4] He became interested in atmospheric waves as reflectors of storm motion through his friendship with Adolphe Quetelet, with whom he corresponded extensively from 1853 to 1863. Secchi published numerous studies on dynamic meteorology and terrestrial magnetism in the journals of the Osservatorio Romano (which he directed), such as the *Bullettino meteorologico* and the *Memories,* and also in publications of Italian and foreign academies, such as the *Comptes rendus* of Paris and the *Giornale Arcadico* of the *Accademia degli Arcadi*.[5] For many years he presided over the *Accademia Pontificia dei Nuovi Lincei* (founded in 1603 and revived by Pope Pius IX in 1847 to integrate religion and science), where he published a total of ninety-six articles between 1847 and 1877. The extracts from Secchi's two studies presented here were published in the *Atti* (Proceedings) of this academy.[6]

In the first extract from a study on the results obtained from telegraphic meteorological correspondence and the balance barometer (1858), Secchi proposes a new graphical method for atmospheric waves. It is based on the availability of near-real-time international weather information, and, in particular, Secchi's collaboration with Le Verrier's *Bulletin quotidien*. The intention is to show that plotting barometric curves at major stations in different parts of Europe can reflect the true movement of storms. Each curve plotted on a piece of paper can be compared with those of its companions, and by distributing these pieces over the geographical positions of latitude and longitude of the stations, it is possible to see at a glance how the disturbance has propagated. The observations discussed are exclusively barometric. Secchi argues that the reason for preferring this instrument is that it is 'least affected by the local conditions of its location': 'The thermometer and the wind direction are too much subject to particular influences' (p. 389).

The second extract is taken from Secchi's study of a storm that occurred in 1859–60 and was published in the *Atti dell'Accademia dei Nuovi Lincei* (1860). Secchi argues for the importance of a comprehensive and global view of all atmospheric phenomena and the installation of complete graphical instruments to understand how storms form and propagate. The ability of the instruments devised by Secchi to concatenate all the phenomena involved contrasts with the reductionist mentality prevailing in many observatories, which tended to use instruments that did not order or compare phenomena. Secchi was here attacking a practice of graphical recording in many observatories, which used separate instruments unsuitable for the object being studied. He praised Maury's inclusive vision, whose works *Explanations and Sailing Directions to Accompany the Wind and Current Charts* (1851–1859) and *Investigations of the Winds and Currents of the Sea* (1851) are acknowledged in the extract (see Vol. II, 2.3 'Sailing directions'). He shared Maury's view that the knowledge of nautical meteorology was still imperfect, not only as regards the origin of storms, but above all as regards the nature of air movements, despite mentioning the work of the main proponents of the two theories (mechanical and thermal) on this subject – William Redfield, William Reid, Henry Piddington and Heinrich W. Dove on the one hand, and James Espy on the other.

Chief among the instruments that Secchi devised at the Roman Osservatorio was the meteorograph, which was essential for the search for physical correlations (see Vol. III, 1.1 'Secchi meteorograph').[7] Integrating graphical and electromechanical techniques, he created a device that automatically recorded all phenomena on a single sheet. He argued that this allowed the observer to 'see at a glance the course of any storm and recognise all the circumstances that accompany it' (p. 234). Under these premises, he was able to establish a general relationship between barometric variations and wind direction and speed. This relationship was empirical and easily perceptible insofar as his meteorograph showed *d'un coup d'oeil* the succession and concatenation of all meteorological phenomena.

Secchi's proposals in these two studies are revolutionary in several respects. On the one hand, he breaks with a stream of statistical thought, carved out from Alexander von Humboldt to Quetelet, which is based on the representation of mean values rather than on the objectification of individual phenomena. On the other hand, by considering the barometric curve as the primary instrument of analysis, Secchi prioritises the pressure field over other fields. However, unlike Le Verrier, who does so for political and military reasons, Secchi adopts the Eulerian approach for fundamentally technical reasons.

Notes

1 Luigi Iafrate, 'Padre Angelo Secchi: l'astronomo italiano che inventò le previsioni del tempo', *Nova Historica: rivista internazionale di storia*, 36 (2011), pp. 110–30, on p. 120.

2 Francesco Zantedeschi, 'Dei documenti comprovanti la proposta e l'applicazione da lui fatte negli anni 1849, 1850, 1853 e 1854, del telegrafo elettro-magnetico alla meteorologia e agli avvisi delle burrasche', *Atti dell'I.R. Istituto veneto di scienze, lettere ed arti*, 23 (1864–1865), pp. 1414–23, on pp. 1420–21.

3 Beniamino Besso, *L'elettricità e le sue applicazioni* (Milano: E. Treves, 1871), pp. 447–49; Aitor Anduaga, *Politics, Statistics, and Weather Forecasting, 1840–1910. Taming the Weather* (London and New York: Routledge, 2020), pp. 127–37.

4 Edoardo Proverbio and Letizia Buffoni, 'La meteorologia dei fenomeni Estremi alla metà del XIX secolo e il contributo di Angelo Secchi allo studio dei rapporti tra fenomeni meteorologici, magnetici e solari', *Physis; Rivista Internazionale di Storia della Scienza*, 41:1 (2004), pp. 125–65, on pp. 131–37.

5 Aldo Altamore and Sabino Maffeo (eds.), *Angelo Secchi. L'avventura scientifica del Collegio Romano* (Vaticano: Specola Vaticana, 2012); Agustín Udías, *Jesuit Contribution to Science: A History* (New York: Springer, 2015); Ileana Chinnici and Wiktor Gramatowski, 'Le carte di Angelo Secchi S.J. (1818–1878) conservate presso la Pontificia Università Gregoriana. Un inventario inedito rivisitato', *Nuncius*, 16 (2) (2001), pp. 571–627; Ileana Chinnici, *Decoding the Stars: A Biography of Angelo Secchi, Jesuit and Scientist* (Leiden, Boston: Brill, 2019).

6 Luigi Iafrate and Maria Carmen Beltrano, 'Angelo Secchi and Meteorology in Italy', in Ileana Chinnici and Guy Consolmagno (eds.), *Angelo Secchi and Nineteenth Century Science. The Multidisciplinary Contributions of A Pioneer and Innovator* (Cham, Switzerland: Springer, 2021), pp. 153–74.

7 Aitor Anduaga, 'Epistemic Network: The Jesuits and Tropical Cyclone Prediction, 1860–1900', *Isis*, 113:3 (2022), pp. 513–36, on pp. 529–31.

10

DI ALCUNI RISULTATI OTTENUTI DALLA CORRISPONDENZA METEOROLOGICA TELEGRAFICA, E DAL BAROMETROGRAFO A BILANCIA

Angelo Secchi

Source: Angelo Secchi, 'Di alcuni risultati ottenuti dalla corrispondenza meteorologica telegrafica, e dal barometrografo a bilancia', *Atti Accademia Pontificia dei Nuovi Lincei*, 11, 1858, pp. 389–390.

It is now a year since I definitively installed the balance barometer at the Observatory, and for almost the same period meteorological observations have been telegraphed from Rome and many other major European cities to Paris, which are then lithographed at the Imperial Observatory and communicated in a special bulletin to the various collaborating observatories. This correspondence extends in France to the towns of Dunkerque, Mézières, Strasbourg, Tonnerre, Paris, le Havre, Brest, Napoleon-Vendee, Limoges, Montauban, Bayonne, Avignon, Lyon and Besancon; and outside France to Madrid, Rome, Turin, Geneva, Brussels, Vienna, Lisbon, Petersburg, Algiers and Constantinople. Observations from Athens are also sent to our observatory via a special bulletin every 10 days. The results which may be obtained from such a correspondence, continued for a number of years, are not in doubt, and will succeed in establishing over Europe the rule of the propagation of meteorological phenomena, from which will also arise the possibility of practical benefit to agriculture and commerce; its fruits have already been abundantly reaped in the discussion of observations of the same kind made on board ships, the result of which has been to save many lives by the discovery of the law of the circular motion of storms, and to give a new impetus to navigation itself, shortening voyages by more than a third of their former duration, showing new courses to be taken, etc., etc.

In spite of the short time that has elapsed since the Parisian correspondence has become regular, I wished to discuss the results by combining them with those of the barometer to see if I could determine the speed of the atmospheric waves. The observations which I discussed were exclusively barometric, and the reason

DOI: 10.4324/9781003650294-16

for preferring this instrument to all others was that it is not only the fundamental instrument having the closest general relation to the state of the atmosphere, but is also the one least affected by the local conditions of its location. The thermometer and the wind direction are too much subject to particular influences, to see which it is necessary to know well their exposure and a thousand other peculiarities which can easily disturb their course. I have also preferred the last quarter of March, April and May to the rest of the year, because it is a season which is neither too stormy, as is the winter, nor too calm, as is the state, so that it has been possible to trace clearly the course of many singular events which are quite distinct from each other and very instructive. The observations recorded in the bulletins, which are made only once a day, at 7 a.m., leave a large gap in the 24-hour period. We shall see how this can be remedied and how desirable it is to have more frequent observations.

The method used in this discussion was the graphical one, which is the only useful one in this matter. I have therefore plotted the barometric curves of the principal stations scattered over various points of Europe, collected in the above-mentioned bulletins or obtained by other means, using, therefore, small square lined paper, which facilitates the operation immensely. Each curve plotted on a piece of paper equal to the others, can be easily compared with all its companions, and by distributing these pieces on the respective geographical latitudes and longitudes of the various countries, the main consequences can be seen at a glance.

The fundamental result deduced from this operation and which can be seen from the simple trend of the curves, is as follows: "The great atmospheric events extend over the whole of Europe, propagating successively in such a way as to cross it from north-west to south-east in little more than a day, decreasing in strength and increasing in number as they advance southwards, and decreasing in number and excursion as the summer season approaches, so that an advance in the season is equivalent to a geographical advance towards the equator".

11

ALCUNE RICERCHE METEOROLOGICHE SULLE TEMPESTE OCCORSE NEL 1859–60

Angelo Secchi

Source: Angelo Secchi, 'Alcune ricerche meteorologiche sulle tempeste occorse nel 1859–60', *Atti Accademia Pontificia dei Nuovi Lincei*, 13, 1860, 231–237, 249.

Of all the meteorological studies, the study of the formation and progress of storms is perhaps the most important, not only from a scientific point of view, but above all because of its practical usefulness; it rightly places meteorology alongside astronomy, for if the former gives the sailor a way of navigating the immensity of the oceans, the latter shows him how to choose the shortest or least troublesome course, taking advantage of the winds and currents, and thus avoid and escape the most terrible dangers that threaten him.[1] This study has recently been given a new impetus by the great American work *Sui venti e le correnti dell' Oceano* (Investigations of the Winds and Currents of the Sea), published by Maury, director of the Washington Observatory, in which, by perusing a large number of nautical journals, the crossing of the widest and most difficult seas is reduced to certain and reliable rules, and voyages are immensely shortened. So great was the activity and zeal of this illustrious scientist, who truly deserves the name of benefactor of humanity, that the opusculum entitled *Sailing Directions*, which began in 1817 with four printed sheets, grew in 12 years to two large volumes, and is still growing and being improved by the zeal of his compatriots and all the nautical and navigational peoples who contribute without rival to its enrichment.[2]

In this great work, however, the author himself admits that our knowledge of several major points of nautical meteorology is very imperfect, not only as regards the true origin of storms, but even (what is more important) as regards the nature of the air motions that govern them and their mode of rotation. Much progress has certainly been made and much is known, but even at the most studied point, that of whirlwinds or cyclones, there remains a fundamental doubt that leaves the sailor in cruel perplexity. In fact, two are the most reasonable theories that have been proposed in this regard:

DOI: 10.4324/9781003650294-17

> one, whose authors and defenders include Redfield, Reid, Piddington and other illustrious navigators, as well as the famous meteorologist Dove, reduces all storms to great cyclones or whirlwinds, attributing the barometric drops observed in them to centrifugal force. The other is that of Espy, according to which the vortex occurs only in a few cases and the common storms are nothing more than great atmospheric imbalances caused by ascending currents generated in the atmosphere, which, like huge chimneys, suck in the air around them: the shape of the partial vacuum would be more longitudinal and in the form of a long furrow rather than a circular one with translation.

The theoretical solution to this problem is not an insignificant point for practice; indeed, it is the essential one, because according to one principle or another, the instruction given to the sailor to escape from the centre of the storm must be different. In the case of a cyclone, by running perpendicular to the radius of the circle connecting the centre of the vortex, the wind must take a path to escape from it; in the other case, as the wind is along the radius, it has to take a different course, so that, so that according as the captain shall rule on either principle the ship may be saved or lost.

But doubt lies not only in the theory: the facts themselves are not yet well established, and much remains to be studied, and none of the theories, according to Maury himself, satisfies all the facts observed so far.[3] The motions, according to him, are spiral rather than circular, and this must be due to the law of the composition of the translatory motion of the wind with the earth's rotation, to which recent mechanical research on problems of this kind has shed so much light.[4] This movement must also be accompanied in the centre by an ascending current, such as is seen in the small whirlpools of dust or snow, and thus we also find true what Espy supposes about the progressive rise of the air, not only by the heat developed in the condensation of vapours, but also by mechanical action. The wind must always blow towards this centre, but it cannot go there in a straight line, and theory teaches that it must go there in spiral curves. However, in view of the small thickness of the atmospheric layer in motion, cyclones as large as they should be seem improbable when they bend their course from west to east after the 25th or 30th degree of latitude; and the attempts made to reduce various storms in the higher latitudes to the same vortex have been in vain, because there may be more than one vortex in the same storm. One of the fundamental points to be clarified *de facto* is whether the course of the air is directed *towards* the centre of the maximum barometric depression, as Espy wants, or normally towards it, as the other theorist wants, or obliquely, as Maury wants. There is no doubt that in the equatorial seas the air turns in cyclones, but whether it does so constantly at high latitudes is still uncertain, and new observations should be addressed in this research. Such a study is of great importance because of the complete similarity between the barometer movements here and those described by navigators as so capricious around Cape Horn in the southern hemisphere,[5] and the law found in Europe would also apply in the southern hemisphere with the usual modifications.

One of the reasons for all these uncertainties about the true visceral movement of storms, especially outside the tropics, is that the observations made at sea are too incomplete and inaccurate. In fact, the seafarer's position in the face of such enormous dangers is too critical for him to be able to deal with all the minute circumstances that can shed light on the matter with the necessary coolness. The position of the ships is also often uncertain, and it is always the case that several ships are in relative positions suitable to shed light on the true movements of the air and their succession. Consequently, I believe that this will be the time when it will be necessary to resort to observations made on land. It is true that the storms are very modified there, but I see that the marine experts believe that they propagate across the continents with sufficient regularity to make trends easy to identify.[6] Except that, even in the land, their study is not easy. Attention is usually drawn to the phenomenon only when it is already raging and devastating, by which time it is already late and has certainly been in the making for several days. Moreover, in order to clearly distinguish the phases of a storm, it would be necessary to make frequent and minute observations, especially near the height of its fury; this is often done at inconvenient hours, so that even in the places where the people who have this as their duty are present, they are not observed; consequently, owing to these difficulties, little progress can be made in science. Ordinary observations made three or four times a day are valuable, but they leave too many empty intervals which interrupt the course of the meteors. All the more so for those observations that are made only once a day, which can hardly give comprehensive information about the longest and most violent phenomena. It would be desirable at least for those who are interested in meteorological observations to multiply their observations when storms are approaching, and it would be a case of organising observing societies like

> But even where there is no lack of observations, as in certain observatories that are made at every hour, their discussion is an immense work, and the end result is that the meteorological archives are filled with figures and nothing more. Therefore, it seemed to me that it would be a blessing for mankind and science if an instrument could be made available that fulfilled the following two conditions:
>
> 1 Complete description of all phenomena associated with storms and, more generally, capable of generating records of all atmospheric events:
> 2 That this register is set up in such a way as to make it easy to reduce and compare at a glance the trends of all atmospheric phenomena, and storms in particular.

This is what I have been trying to do for the last two years by constructing an instrument which I have called the *meteorograph,* which automatically records all the atmospheric phenomena on the same sheet of paper, one after the other, so that one can see at a glance the course of any storm and recognise all the circumstances that accompany it. Its description with figures can be found in the *Memorie dell' Osservatorio del Collegio Romano* for 1859. No. I. and the improvements introduced are to be found in No. XIX. Here I will only say that it has two panels, each with a vertical movement and regulated by a clock, and that in the first, which takes ten days to make its journey of 36 centimetres, 1.° the curve representing the atmospheric pressure is marked by means of a balance barometer;[7] 2.° The hourly wind speed with as many lines for each hour as the ordinates of a curve, each 5 millimetres long representing a geographical mile with a wind speed of 1852 metres. 3.° The direction of the wind, directly for the 4 main rhumb lines and indirectly also for the 4 intermediate rhumb lines. 4.° The variation in temperature using a Kreil metal thermograph. 5.° The rain in the hour in which it falls and in duration, and shortly also in quantity.[8]

These indications, juxtaposed with abscissae corresponding strictly to the same hours and collected on the same sheet for 10 days without interruption, make it possible to see at a glance the succession and concatenation of all the phenomena which would be difficult to recognise if they were marked on different sheets, as is the case with most other instruments. In my travels and visits to various observatories where the graphic register method has been introduced, I have always been struck by the great defect of such instruments, which, although they *record all* the phenomena, do not *order* them or compare them, so that if one wants to discuss them, one must first reduce everything to figures and then return from them to comparative curves.

However, since many phenomena require a more minute and detailed study, and since it is not easy to detect them in the short curve space occupied by a day of 36 millimetres in the frame above, this problem is solved by a second frame, which can collect the observations of one or two days, depending on whether the curves are to be lengthened or shortened. It contains the following records.

1. A repetition of the barometer, because in great storms the oscillations are often rapid within a few minutes and it is interesting to know the exact time. 2.° A repetition of the thermograph, but on a larger scale of ordinates. 3.° Repetition of the rain. 4.° Complete recording of the psychrometer, i.e. the two thermometers, dry and wet, that make up this instrument. 4.° Rainfall on a larger scale.

The wind and the psychrometer are marked by electricity, because this medium has the great advantage of transmitting movements over great distances, so that the measuring instruments can be placed in the most favourable positions: so that their indications are reliable and not dominated by local causes, which is all too easy in the case of the wind: together with this, the recording machine can be placed in a place that is most convenient for the observer, and that is easily accessible to monitor the instrument and make that study that serves to form the true concept of the course of material phenomena, and that no machine but intelligence can provide. So here the windmill (which is Robinson's) and the wind vane are on top of the tower of the old observatory, a place that is not at all free from all sides, 48 metres above street level, while the machine is in the centre of the new observatory, 90 metres away, and the thermometers are 6 metres away, facing north in a very good position. The care required to maintain the pile and its cost is reduced to a minimum thanks to a successful modification introduced in Daniell's construction, which makes it absolutely constant for a period of 50 days or more, always maintaining a higher strength than that of the local pile used in telegraph offices.

It is already a year since this system of observations has been continued without interruption, and although there were some difficulties in the early days, the paperwork has now been reduced to a simple matter, and we have a 37-sheet atlas containing the whole meteorological history from 10 to 10 days, with the details for the other days clarified in the other sets of sheets, two by two. It is clear, however, that this is only an imperfect study of these valuable data, since in order to make proper use of them it would be necessary to be able to compare our curves with those recorded by similar machines in other places, since it is a general condition of this class of facts that they manifest themselves over a vast area, while at single points they never show themselves in all the phases of their development. The only way to ascertain the law of weather patterns would be to place several such instruments at great distances around a central station, so that storms could be studied simultaneously in time and space. If the distances are too great, there may be doubt as to the identity of the eddies and gales, and if they are too close together, they are almost useless. But even 130 or 150 miles is a distance at which the phases differ in time, and I have proof of this even from the simple observations made four times a day in Loreto by Father Ferrari, Professor of Physics in that College, which now advance and now retard noticeably in comparison with Rome.

Then the reduction of all observations concerning a great storm will consist in comparing the curves traced on the various sheets of paper belonging to the different places, and thus, if the turn they usually take is understood, the first indications of them at one station can be telegraphed to the neighbouring stations, and the much studied problem of the turn of storms can be solved with certainty.

I. GENERAL RELATIONS BETWEEN WIND SPEED AND BAROMETER MOTION

Coming now to our subject, the relation between the motions of the barometer and those of the wind, it seems really incredible that so little has been studied with precision, although it is such a fundamental motion in the action of storms, for the catastrophes they produce are no more grievous than the mechanical force that the air in them acquires. It is precisely on this subject that we lack precise data: we know only that major hurricanes and strong winds are accompanied by rain and a fall in the barometric pressure, but the reciprocal relationship is only vaguely indicated even in the best meteorological works. In everyday observations, wind force is recorded in such vague terms that it is impossible to draw any conclusions. Kaemtz rightly states that for a long time there were no reliable means of assessing it, and it is only in recent years that some have actually been introduced into observation, but with such imperfect methods that anemometric and barometric observations are difficult to compare without laborious reduction work. A simple examination of our meteorographic curves gives the following laws.

1. Atmospheric storms of any kind are accompanied by a corresponding drop in barometric pressure, forming a *hollow wave* and a wind proportional to the speed of the barometric upward or downward motion. In order not to make this law too general, it is necessary to define the meaning of gale or storm. Under this name fall, in the first place, hurricanes, whirlwinds and the great succession of several consecutive days of bad weather, which prevails in winter, in which heavy rain, gusty winds and rough weather reign together. The small rainfalls and variations, which are mostly local, are also included, but to a lesser extent and in proportion to their minor importance. We will deal mainly with the major ones, especially those that occur in winter, which are the most terrible for navigation.

CONCLUSION

From what has been said so far, if I am not mistaken, it is clear what great benefit the navy and navigation can derive from land-based observations of the course of storms. Admittedly, these means of registration are a little expensive, but the expense will be paid a hundred times over by the fruits. If I may make a comparison, it seems to me that meteorology is still in its infancy, as astronomy was three centuries ago, and that all certain science is reduced to a few mean values and a few variabilities, as was then the case with the planetary motions. To perfect science, it was first necessary to improve the instruments of observation, and Tycho took this step at great expense, resulting in the true laws of planetary motion discovered by Kepler. Later, Bradley, with even more sophisticated instruments, founded exact astronomy, which is now used only for calculations. It is true that these instruments were difficult to use and expensive to acquire, but science was derived from them. As the ordinary instruments have been observed for so long without yielding more fruit than either Theodore Palatine or Toaldo drew from them in their time, it must be said that with so many observations and so much effort, as is fortunately found today in the cultivation of meteorology, the defect is on the side of the instruments, which are inadequate to the need. If my love for an invention in which I have a part does not deceive me, it seems to me that the time has come to change the system, and to erect complete graphic instruments for everything that can be done. A number of these, placed from 2 to 300 miles apart, would be sufficient, and then the other intermediate observatories, making their observations by the present methods, would be most useful.

The present complete meteorograph machine is rather expensive, but having calculated all its parts, I have already begun to consider its economical construction, and I am sure that I can reduce its cost to such an extent that it will be within the reach of private amateurs, and yet equally useful, though not so complete.

In another paper I will give a more detailed account of the average results similarly derived from this instrument for as long as it has been in operation. Below I give a list of the storms observed during the year, which I hope will inspire many to study them better in the future.

Editorial Endnotes

1 It is an old custom that astronomical observatories also keep meteorological records: this custom is now being abandoned on account of the principle of division of labour, but I think it is a bad idea, and it is certain that many sailors would be much more interested in knowing the rotation of storms than in certain subtleties of the course of the moon.

2 In the *Annali di Scienze Fisiche e Matematiche* by Prof. Tortolini Vol. I, we give a mention of this work then about to begin.

3 See Maury's *Sailing Directions*. 8 ed. Vol. I p. 262, 264 and Vol. II. 450, 451 and especially T. I. p. 266. §. 252. and p. 275. §. 261.

4 See *Comptes Rendus de l'Ac. des Sciences* 1859 a long and interesting discussion on this subject.

5 See what Maury reports from Jansen op. cit. tom. II. p. 451 lin. 7.

6 See Reid. Of the Law of Storms etc. Maltese ed. p. 306.

7 See my description. *Atti dell'accad. De'N. Lincei.*

8 The latter is best measured in an ordinary rain gauge.

Part 2

CONFLICT

The Eulerian vs. Lagrangian Approach

2.1

SYNOPTIC METHOD

Editorial Headnote

2.1 Synoptic Method

As weather scientists moved away from climatology and towards "weatherology" to explain non-periodic atmospheric phenomena, they turned their attention to the motion of air past fixed observation points and its temporal patterns. These patterns were then projected onto a synoptic map. Atmospheric pressure oscillations were represented by both barometric curves and atmospheric wave propagation. The study of the atmospheric pressure field and its correlation with the movement of storms expressed the adoption of the Eulerian approach. There was, nonetheless, a shift in the form and purpose of representation as the 1850s wore on, from the graphical method of showing the motion of atmospheric waves to the cartographic method of foreseeing the approach of storms.[1] An article published in 1862 by the Jesuit Father Angelo Secchi (1818–1878), who was helped by his young assistant Giuseppe Serra-Carpi, articulated this act of evolution. There is a sense of progression in these extracts through which cartographic techniques and perspectives are discussed.

Secchi was a Jesuit physicist and astronomer who had been trained in Stonyhurst and Washington during his exile. Upon taking over the direction of the Osservatorio del Collegio Romano, he turned his attention to publishing studies on what he called an integral view of atmospheric phenomena and their mutual connections (relating to heat, electricity and magnetism).[2] Serra-Carpi was a twenty-one-year-old engineer and inventor, familiar with electrical telegraphy, who was to publish several books on electrostatic induction and heat conductivity. Upon joining the young assistant at the Osservatorio, Secchi became a mentor figure, guiding him in his research and engaging in discussions on the best way to visualise the synoptic maps. By the early 1860s, they had developed a relationship as cartographic pioneers, propounding the synoptic method, a method of representation that combined isothermal and isobaric line drawing, inspired in part by Adolphe Quetelet – though not Emmanuel Liais – as they admit in the article, excerpted here. Isoline tracing was central to this method that showed the succession and concatenation of atmospheric phenomena. Serra-Carpi was working on this method in Secchi's 1862 paper, and continued to explore the isothermal line technique in subsequent essays.[3] Years later, Secchi would say that the 1862 results probably led Urbain Le Verrier and his assistants at the Paris Observatory to emulate their method and subsequently publish maps (for Secchi and Serra-Carpi did not publish any maps), using the 'resources available in a rich nation like France'.[4] Secchi's article was published on 31 March 1862 in the *Bullettino Meteorologico* of the Osservatorio – one of his favourite forums, in which he published 74 studies between 1862 and 1877. The rest of the results appeared in the *Bullettino* of 15 August 1862.[5]

Secchi's article sheds light on the new synoptic method and indicates the path followed by Serra-Carpi to show the distribution of pressure and temperature in Europe. On a large map of Europe laid out on a table, he drove a twenty cm stylet over the points where the weather stations were located. In this way, he represented the barometric and thermometric heights. He drew upon different coloured pencils to trace the isobaric lines and the peaks and troughs (which he called *pozzi* or holes) of the atmospheric waves. Instead, he represented the direction and strength of the wind with a series of banners and lines of varying lengths and arrowheads, while he plotted the isotherms with thicker lines. In his own way, Serra-Carpi transformed the graphic technique of Secchi's barometric curves into a cartographic method of synoptic representation.

Notes

1 Aitor Anduaga, *Politics, Statistics, and Weather Forecasting, 1840–1910. Taming the Weather* (London and New York: Routledge, 2020), pp. 106–58.

2 Ileana Chinnici, *Decoding the Stars: A Biography of Angelo Secchi, Jesuit and Scientist* (Leiden, Boston, Brill, 2019); Ileana Chinnici and Guy Consolmagno (eds.), *Angelo Secchi and Nineteenth Century Science. The Multidisciplinary Contributions of a Pioneer and Innovator* (Cham, Switzerland: Springer, 2021).

3 Giuseppe Serra-Carpi, *Sulle linee isotermiche dell'Italia de'suoi mari ed isole adiacenti* (Roma: Tipografia delle belle arti, 1865).

4 Angelo Secchi, *L'astronomia in Roma nel Pontificato di Pio IX. Memoria* (Roma: Tipografia della Pace, 1877), p. 44.

5 Angelo Secchi, 'Sulle onde atmosferiche', *Bulletino Meteorologico dell'Osservatorio del Collegio Romano*, 1:12 (1862), pp. 93–5.

12

DISTRIBUZIONE DELLA PRESSIONE ATMOSFERICA SULL' EUROPA DURANTE IL GENNAJO E IL FEBRAJO DEL CORRENTE ANNO 1862

Angelo Secchi

Source: Angelo Secchi, 'Distribuzione della pressione atmosferica sull' Europa durante il gennajo e il febrajo del corrente anno 1862', *Bulletino Meteorologico dell'Osservatorio del Collegio Romano*, 1(3), 1862, pp. 19–20, 24.

Among the fine institutions currently flourishing for the advancement of meteorology, there is the telegraphic-meteorological Correspondence, founded at the Paris Observatory by M. Leverrier, who was able, with all the means at the disposal of a great empire, to do what we had tried to do on a microscopic scale, connecting the extremes of the Papal States with corresponding simultaneous observations. However, to my knowledge, no one has yet taken it as a point of departure for studying the course of storms over Europe, so some discussion of the same observations will not be unwelcome. I say some discussion because I do not pretend to exhaust the subject, nor to draw from it all that this great mine is rich in, but only to give an essay so that its richness may be better seen, and also to point out the weaknesses of the institution so that it may be improved.

1. QUALITY AND DISTRIBUTION OF OBSERVATIONS

In this Correspondence, the meteorological observations made at many stations are telegraphed every morning to Paris, where they are published the same day in a lithographed bulletin, which is sent free of charge to correspondents and collaborators, and from where it is distributed to many other newspapers. The observations are made at 7 or 8 a.m., according to the season, and include the

DOI: 10.4324/9781003650294-20

barometer with its thermometer to reduce all the readings to normal pressure, i.e. to 0 degrees, temperature and sea level; the external thermometer, the direction of the wind and to estimate its strength; the state of the sky and also of the sea for locations close to it.

The stations included in France are Paris, Dunkerque, Mezieres, Strasbourg, La Havre, Cherbourg, Brest, Lorient, Napoleon – Vandée, Rochefort, Limoges, Montauban, Bayonne, Montpellier, Cette, Lyon, Marseille, Toulon, Antibes, Nice, Avignon, Besanzone and Algiers in Africa.

In Italy; Rome, which is the southernmost point of the peninsula, Livorno, Florence, Turin.

In England; Greenwich, Scarborough, Penzance (at the southernmost tip of the island), Yarmouth, Portland.

In Ireland; Galway, Valentia (on the western tip of the island), Queenstown near Dublin.

In Belgium and Holland; Brussels, Den Helder, Groningen.

In Denmark Copenhagen, in Norway Christiania, in Sweden Stockholm and Haparanda near Tornio at the bottom of the Baltic Sea in the Gulf of Bothnia. In Spain and Portugal; Madrid, Bilbao, Porto, Lisbon, San Fernando near Cadiz, Alicante, Barcelona, Palma in Mallorca. In Germany and Austria; Leipzig and Vienna. In Russia; Petersburg, Riga, Libau, Helsinki, Moscow, Kiev, Warsaw, Nikolaev (on the Black Sea near Odessa).

These stations are so numerous in certain places that it is no inconvenience to miss one or two of them: all of them have been generally selected with full intelligence, and are very convenient for giving an idea of atmospheric events over Europe. They include the whole of the western coastline, and extend from the Arctic Circle in the north to Africa, for 32° latitude and 47° longitude, from Moscow to Valentia and Lisbon, and cover an area which, with Vienna almost at its centre, has a radius of 18° and forms 1/40th of the whole globe. This miracle would not have been possible without the electric telegraph, and a few years ago it would have been considered a dream.

But although this network of stations is so vast, it would be desirable to extend it to a few more points, which the discussion has shown to be extremely important, and a few would be enough to complete it. One would be Palermo or Malta; another in Greece, such as Athens; a third in Constantinople (which has been contributing for some time); a fourth in Transylvania; and a fifth in Dalmatia, such as Trieste or Ragusa. If the stations nearest to Asia, where such observations are already made in connection with the Central Meteorological Observatory of Russia, i.e. Ekaterinburg, Kazan, Yakutsk, Nerchinsk, etc., could be brought together there, nothing more would be desired. If these stations are too remote to be sent by telegraph, the value of the observations would not be diminished by sending them by ordinary mail at the end of the month, and we would be glad to receive them to join their discussion with that of the others.

2. MODE OF DISCUSSION

In order to draw some conclusions from such a useful collection of elements, we first had to form a material idea of the course of events that would guide us in a scientific discussion. For this purpose we spread out a large map of Europe on a large table of suitable thickness, one metre high and one metre and a quarter wide. For each of the weather stations, a hole was drilled in the table and a 20 cm long wire or sliding metal rod was inserted to represent the barometric or thermometric height, with the remaining part protruding from the table. In practice, for the barometer, we found it sufficient to give the rod a height of 5 mm per 1 mm of pressure.

Therefore, if you want to get an idea of the distribution of atmospheric pressure over Europe at a glance, simply place the bars high or low on a scale according to the number given in the lists, starting with an arbitrary zero, e.g. 730.mm as a base. It would be useful to use such a system in a public educational institution to give an idea of the phenomena. By adding a paper ribbon to the rods, the rotation of the winds can be studied to great advantage. In order to determine these configurations, they are shown each day on a small, specially lithographed map of Europe, on which the main meteorological stations are indicated and on which the corresponding figures are written, and then the isobaric lines and the great crests or depressions (which we call wave *troughs* for brevity) are drawn with different coloured pencils, red for maximums, blue for minimums and green or yellow for averages. The wind direction is indicated by a line about 20 mm long with an arrowhead indicating the estimated strength (-> = moderate) (▶ = strong) (-< = weak).

Once the eye has become accustomed to the configuration of the rods, it is no longer difficult to recognise their atmospheric states even from the drawn maps alone, and so the material configuration was retained only in a few singular cases.

After this brief outline of how these observations were discussed, we will now present some of the most important results obtained, adding only that the execution of these maps was carried out by the distinguished young Mr Dr. Giuseppe Serra-Carpi with singular diligence and care, who also took care to note the relationship of some of the phases that we are going to expose.

CONCLUSIONS

It seems clear to us from this discussion essay that the two rival theories of storms, one claiming that they are large waves, the other that they are whirlwinds or cyclones, can both be true within their proper limits. In large events, the mere wave structure seems to dominate, without any significant vortex motion. In small events, on the other hand, the whirling motion is certain from other well-established facts; this motion, on a free surface such as the sea, can progress regularly over large parts of the earth's surface, but on the continents, where it encounters immense obstacles, it cannot propagate over such a distance, so that it must happen that many storms of the latter kind turn into those of the former, especially in Europe. A few essays on one or two of the storms Loomis had already written led him to admit that European storms were independent of American ones, so that generally speaking they did not seem to cross the Atlantic.

2.2

EULERIAN APPROACH

Editorial Headnote

2.2 Eulerian Approach

The previous section showed how, as weather scientists moved away from climatology and towards "weatherology" to explain non-periodic atmospheric phenomena, they turned their attention to the patterns of air motion past fixed observation points, which they then projected onto a synoptic map. The atmospheric pressure oscillations expressed both the nature and the propagation of the so-called "atmospheric waves". In fact, Angelo Secchi and Giuseppe Serra-Carpi had just shown the direction of cyclonic storms in Europe by identifying atmospheric waves with pressure fronts. There was, nonetheless, a push to replace atmospheric waves with the new concept of "low-pressure areas" as the 1860s wore on. Two works written by Edme Hippolyte Marié-Davy (1820–1893) – a paper on equinox storms published in the *Comptes Rendus Hebdomadaires des Séances de l'Académie des Sciences de Paris* (1863) and his treatise *Météorologie. Les mouvements de l'atmosphère et des mers considérés au point de vue de la prévision du temps* (1866) – articulate this act of replacement. In these texts, there is a strong sense of adopting the Eulerian approach insofar as the barometric pressure distribution becomes the main cause of all weather changes.

Marié-Davy was a French physicist who had obtained his doctorate in 1851 with a study of the climate of Montpellier. When he joined the Paris Observatory in 1862 under Urbain Le Verrier, he put aside his early interests in electricity, cells and the physiology of vision to head the geomagnetism section.[1] Emmanuel Liais had resigned as assistant at the Observatory in 1857 because of serious disagreements with Le Verrier. With his resignation and the subsequent appointment of Marié-Davy, it was not at all clear whether the storm warning service would continue to rest on the conceptual framework of atmospheric waves.[2] Eight months before Marié-Davy read his study before the Paris Academy (12 October 1863), the president of the *Société Météorologique de France*, Eugène Grellois, had published a laudatory description of Secchi and Serra-Carpi's synoptic method in the *Annuaire* of this society (February 1863, volume 11).[3] This method combined the plotting of isothermal and isobaric lines and was partly inspired by Adolphe Quetelet – though not by Liais – as the authors themselves acknowledged. Given that Marié-Davy himself confessed that he had read 'what European meteorologists had achieved' and that it was only then that he 'began to be captivated by a service that had such marvellous potential', there is every indication that the French physicist had read Grellois's article by the time he defended his study before the members of the Academy.[4] The synoptic method proposed by Marié-Davy for the study of storms owes much to the cartographic proposal of Secchi and Serra-Carpi (see 2.1, 'Synoptic Method').

Marié-Davy's 1863 article sheds light on his synoptic method, its potential and limitations, and points to the influence of Robert FitzRoy and the British storm warning service to ultimately present his isobaric line-based method (see 2.3, 'Lagrangian Approach'). Like Le Verrier, Marié-Davy was concerned that his observatory and the French Navy itself – which was keen to receive FitzRoy's warnings – were dependent on the British service.[5] Earlier in the same year, 1863, Marié-Davy had worked on his method in two articles, in the context of a great international rivalry in storm warnings (in addition to FitzRoy, the Dutch Weather Service had been issuing telegraphic warnings based on the simultaneous barometric pressure difference at several meteorological stations since 1860).[6] In April 1863, he discarded the atmospheric wave hypothesis because, according to him, it was 'difficult to trace any isolated wave with distinct patterns of movement'.[7] In the article excerpted here, dated October 1862, he continued to explore the isobarometric method and reported the results of storm warnings to the Academy. He argues that the first signs of approaching storms can be seen in the inflection of the barometric equal pressure curves. Marié-Davy draws on the storm maps published in the Paris Observatory's daily bulletin – up to six successive and distinct storms in a month – to convince the academic audience that it is possible to predict their arrival 24 or 48 hours beforehand by analysing his weather maps. However, he also points to issues that appeal to the limitations of the synoptic method – for example, it says nothing about where the storm comes from and how it forms – to demonstrate its nascent and rudimentary nature, which does not detract from the importance of constructing daily maps covering the entire northern hemisphere.

Marié-Davy's treatise on meteorology, published in 1866 by Victor Masson et Fils in Paris and reprinted in 1877, is a more illustrative compilation of the same weather forecasting perspective. In the excerpts from Chapter XIV "Weather Forecasting" reproduced here, he envisages a "predictable atmosphere" in which it will be possible to determine a few days in advance where and when storms will occur. In tune with his earlier writings, he argues that the characteristic deformation of barometric equal pressure curves heralds the general motion of the disturbance. Finding the "low-pressure areas" and plotting them in the form of quasi-circular isobaric curves on the synoptic map was fundamental to the Eulerian perspective of weather forecasting. On a noteworthy note, it should be added that in the preface to this first edition, dated April 1866, Marié-Davy expressed his respect for Le Verrier, although at the end of that month he wrote a letter expressing his complaints, a letter that was later sent to the Emperor. Yet, in the preface to the second edition of the work, dated 16 April 1876, he lamented the slow progress of meteorology in France since 1865.[8]

The synoptic weather map took many distinctive graphic forms between 1850 and 1865. Quetelet's atmospheric waves (traced by parallel and equidistant lines), Liais's various types of lines, and Secchi's correlations between waveforms and barometric depressions were some of its precursor forms. These interrelated examples nevertheless illustrate, in a compressed manner, how mid-nineteenth-century weather scientists chose to adopt a Eulerian approach to storm warning and forecasting. Through Marié-Davy's writings, we can trace a growing acceptance of the notion of "low pressure area" as an analytical tool for weather forecasting. Marié-Davy gave the Eulerian approach its most complete formulation, transforming the undetectable atmospheric waves into a traceable and foreseeable object: a depression encircled by isobaric curves.

Notes

1 Edme Hippolyte Marié-Davy, *Notice sur les travaux scientifiques de M. Marié-Davy* (Paris: Gauthier-Villars, 1868).

2 John L. Davis, 'Weather Forecasting and the Development of Meteorological Theory at the Paris Observatory, 1853–1878', *Annals of Science*, 41:4 (1984), pp. 359–82, on pp. 368–69; Fabien Locher, *Le savant et la tempête. Étudier l'atmosphère et prévoir le temps au XIX siècle* (Rennes: Presses Universitaires de Rennes, 2008), pp. 109–10.

3 Eugène Grellois, 'Analyse des bulletins météorologiques de l'Observatoire du Collége Romain', *Annuaire de la Société Météorologique de France*, 11 (1863), pp. 20–6.

4 Aitor Anduaga, *Politics, Statistics, and Weather Forecasting, 1840–1910. Taming the Weather* (London and New York: Routledge, 2020), pp. 137–44.

5 Pauline Halford, *Storm Warning. The Origins of the Weather Forecast* (Stroud: Sutton Publishing, 2004), p. 205; J. Malcolm Walker, *History of the Meteorological Office* (Cambridge: Cambridge University Press, 2012), pp. 42–5.

6 C.H.D. Buys Ballot, 'On the System of Forecasting the Weather as Pursued in Holland', *Report of the British Association for the Advancement of Science*, 33 (1864), pp. 20–1, on p. 20.

7 Edme Hippolyte Marié-Davy, 'Sur la météorologie'. *Revue des sociétés savantes*, 3 (1863), pp. 232–235, p. 233 – as translated by Fabien Locher, 'Atmosphere of Globalisation. Depressions, the Astronomer and the Telegraph (1850–1914)'. *Revue d'histoire moderne et contemporaine*, 56:4 (2009), pp. 77–103, on p. 86.

8 Robert Fox, *The Savant and the State: Science and Cultural Politics in Nineteenth-Century France* (Baltimore: Maryland: The John Hopkins University Press, 2012), pp. 123 and 325.

13

SUR LES TEMPÊTES DE L'ÉQUINOXE

Edme Hippolyte Marié-Davy

Source: Edme Hippolyte Marié-Davy, 'Sur les tempêtes de l'équinoxe', *CR*, 57, 1863, pp. 640–644.

The autumnal equinox was marked by frequent atmospheric disturbances.

"Since our last communication to the Institute, on the 17th of August, we can count up to six successive and distinct storms, separated by an interval of a few days of more or less complete calm.

All these storms have had common features in the way they have appeared and the way they have moved.

The first signs of these storms can be seen several days in advance on the western coasts of Europe, in the inflection of the curves of equal barometric pressure; then the wind rises more or less rapidly on the north-western coasts of France and England, with a very marked tendency to turn around a centre of depression which forms the centre of the storm. The centre itself moves, sometimes regularly and progressively, from west to east, first rising towards the north and then descending towards the south after crossing England, and sometimes the opposite, with a few hesitations that seem to take it backwards for a moment.

The study of these disturbances is of great interest, either from a purely scientific point of view, or from the point of view of the probabilities that can be deduced from them in relation to the points threatened by a storm that is preparing itself or has already begun to rage. At the Imperial Observatory of Paris, this study is regularly followed by means of our maps; but until now, these maps have remained handwritten: we thought we would do something pleasant and useful for meteorologists by inserting them in the daily Bulletin of the Observatory.

In the Bulletin of the 11th of September, we had already reproduced the maps of the 7th and 10th of that month to show the nature of the storm that was then raging over Europe and to indicate the direction and speed of its progress.

Our regular publication began on the 16th of September; it seemed to us that a new storm was brewing on the ocean, and we wanted to be able to follow it as soon as the first signs appeared.

On the 17th, there was a very marked change in the pressure distribution over Western Europe. The 0.765 barometric curve, which had risen to the north-west

DOI: 10.4324/9781003650294-22

over Ireland the previous morning, was folded southwards along our coast and that of Portugal as far as San Fernando, near Cadiz. The likelihood of an impending storm seemed so great that at 3 a.m. we were able to send the following telegraph dispatch to our correspondents in Germany: "*Threat to the west over the ocean*".

On the 18th, the same 0.765 curve had closed, retreating to the north-east and covering only France and part of Germany. At the same time, curve 760 had followed a parallel course and formed a depression in the Bay of Biscay, which for us is one of the most characteristic signs of the arrival of the whirlwinds in central Europe, and which we believe we can attribute to the fact that the smooth surface of the ocean offers less resistance to the transmission of the impulse of the air covering it than the uneven surface of the ground. The atmosphere still seemed calm; a very small number of remote stations, Brussels, Leipzig, Barcelona, and Livorno, indicated a fairly strong wind, while everywhere else the wind was light or moderate. At the same time, the winds over France and Spain were still blowing in a northerly or easterly direction, which in these conditions is only due to the eddies created by the opposing winds tending to settle in the atmosphere.

In this situation, whatever the probabilities of the imminent arrival of a whirlwind, there is still some uncertainty as to the exact point at which it will approach our coasts, and consequently as to the direction and extent of the rotation of the winds. Nevertheless, at 3 a.m. on the 18th, we telegraphed our correspondents the following probabilities for the following day:

Baltic Sea: northwesterly to southwesterly winds, cloudy or overcast.
From Dunkirk to Hamburg: west to south wind, cloudy or overcast.
From Brest to Dunkirk: wind northwest to south-southwest, overcast or cloudy.
From Brest to Rochefort: wind southeast to south veering west or northwest, overcast or cloudy.
From Rochefort to Bilbao: wind northeast to southeast or southwest, fair or cloudy skies.
From Barcelona to Alicante: wind northeast to southeast or southwest, fair skies.
From Port-Vendres to Antibes: wind northeast or southeast, fair skies.
From Antibes to Livorno: wind from north to east or southeast, fair skies.

The wind should freshen everywhere except in the Rochefort and Bilbao areas.

Indeed, on the 19th, we saw the wind gradually turn to the indicated directions and freshen over the Channel. However, only in Brest was it strong at 3 o'clock in the evening. It was over England that the centre of the whirlwind had moved rapidly; but, exceptionally, some of the English dispatches were missing on the 19th, and we could only judge by induction the strength of the storm already raging over England, and the exact position of the centre of the movement. At 3 a.m. on the 19th we gave as probabilities for Sunday the 20th winds from south to west, increasing in intensity from *moderate* or *fairly strong* to *strong*, over the Channel and the Ocean, with overcast skies, and for Monday the 21st more accentuated winds extending to the Mediterranean, with rainy weather. A storm was forecast for today in the Bay of Biscay.

The storm continued its course with some irregularities in its progress on 20, 21, 22, 23, 24 and 25 September. At 8 a.m. on the 25th, winds were light to moderate over most of Europe, with strong winds only in Le Havre and Helgoland. On the 26th, the calm was even more complete, and by 8 a.m. there was no sign of the next g on gales on our western coasts. Only the Mediterranean remained very turbulent and under a persistent threat.

The next morning, Sunday the 27th, the situation changed completely; a second storm announced its imminent arrival, and that evening it began to sweep across Europe, from the north of England to the Mediterranean, disappearing the next day as quickly as it had arrived.

The calm that followed this storm was not to last long.

On the 29th, we saw the characteristic inflection already reported for the 18th of September repeated on the morning map, and we were able to inform our correspondents of the threat that we were facing. The progress of this second major storm was greatly accelerated by the passage of the intermediate storm from the 27th to the 28th. Thus, instead of taking two or three days to invade Western Europe, we found it established over England and the west coast of France at 3 a.m. on the 29th. We should take this opportunity to point out that the meteorological dispatches which reach us every day between 11 a.m. and 2 a.m. are all relative to 7 or 8 a.m.; that the information which can be sent to us about the state of the sky at the time when we draw up our charts and forecasts, i.e. between 2 and 3 p.m., does not reach us until the next morning; and that, in particular, Admiral Fitz-Roy's Bulletin and forecasts do not reach us until the next day by post.

This second storm was extremely energetic. It was still raging over Europe when, on the 5th of October, we saw for the third time the inflection in the barometric lines already reported over the Bay of Biscay. On the same day, 5 October, for the third time since the 17th, we issued a warning of the *threat of gusts of wind.*

On the 6th, England began to feel strong winds.

On the 7th, the atmosphere shook over Europe, and we warned of strong to very strong winds for the 8th on the French coast.

The course of this last storm, in the midst of which we are still today, Monday 12th, presents some peculiarities about which we would like to say a few words.

On the 6th and 7th, the phenomenon was only evident in the inflection of the equal pressure curves along the coasts of the Bay of Biscay; however, the very marked drop in pressure at Valentia, on the western tip of Ireland, showed that the phenomenon was approaching that point. On the 7th, the centre of the storm was fairly clear between Valentia, Penzance and Brest; on the 8th, it moved eastwards and was between Penzance, Brest and Cherbourg; on the 9th, it moved back westwards. The situation continued to worsen until Monday 12 October, when the centre of the storm returned to the west coast of Ireland. This time the depression was strong, the equal pressure curves very steep, and the winds very strong. All indications are that this will be one of the strongest storms of the season.

Examination of our weather charts shows that it is generally possible to foresee the arrival of a fairly long-lasting storm on our coasts twenty-four or forty-eight hours in advance. However, we regard them as no more than a rough guide to the probabilities for the next day or two. All too often, our information is late or incomplete. However, since today, Monday the 12th, the documents that have reached us from England have been helpfully supplemented by Admiral Fitz-Roy with the addition of Nairn and Greencastle.

The undeniable usefulness of this type of work for meteorology makes us very keen to extend it on a wider basis. Although our maps can give us a premonition of a storm and enable us to follow it as it moves across Europe, they tell us little or nothing about where it comes from and how it is formed, and yet this is one of the essential elements not only of the science but also of its applications. We would attach the greatest importance to the construction of daily maps covering the whole of the northern hemisphere, even if it took a year to assemble the elements for each of them. In the midst of the incessant mobility of atmospheric phenomena, there are undoubtedly some great general laws which must be identified and which can be looked up in previous years. The most important of these could be published in our Bulletin, in the form adopted for the maps of the day.

14

MÉTÉOROLOGIE. LES MOUVEMENTS DE L'ATMOSPHÈRE ET DES MERS CONSIDÉRÉS AU POINT DE VUE DE LA PRÉVISION DU TEMPS

Edme Hippolyte Marié-Davy

Source: Edme Hippolyte Marié-Davy, *Météorologie. Les mouvements de l'atmosphère et des mers considérés au point de vue de la prévision du temps* (Paris: Victor Masson et Fils, 1866), pp. 416–417, 434–436, 463–465

CHAPTER XIV

Weather forecast

Short-range weather forecasting, as we understand it, i.e. the indication of likely changes in the state of the sky and winds due to the current state of the atmosphere, is a relatively simple problem in the intertropical regions. There, major changes occur at specific times and are preceded by very noticeable changes in the appearance of the sky and the operation of instruments. Meteorological accidents themselves are either very regular or very rare, and in the latter case the more constant the daily oscillation of the barometer, the more likely it is that the severity of the disturbance will be a warning. The problem is more complex in our regions, where the sky is constantly changing. All the data, all the signs, whatever their nature, must be taken into consideration. Each of them plays its part in the daily discussion. If we establish a certain hierarchy between them, if we attribute a dominant importance to the movements of the barometer,[1] it is because these movements are already like a summary of all the other signs. The latter, however, are the subject of a special and detailed study. There are even certain phenomena that we have not yet discussed because the instruments that accompany them are not within everyone's reach, but which we nevertheless consult with care: such are the movements of the hand needle (see Chapter XV).

DOI: 10.4324/9781003650294-23

APPLICATIONS

The consequences of all the foregoing are: that the changes in the state of the atmosphere in a given region of Europe are the direct result of the displacements of the bed of the great air current coming from the Atlantic, and of the passage of the rotational movements which take place there; that the problem of weather forecasting, therefore, consists in spying on these displacements, in catching the first signs of the arrival of each rotational movement, in determining the extent and intensity of the meteor, the distance at which it must pass from the region under consideration, the direction it must follow, the speed at which it is transported. This is, for the moment, where the power of weather science ends. A certain periodicity in the magnitude of the equatorial current seems undeniable to us as long as we remain within the bounds of generality. As soon as we try to specify the return of each period a little in advance, we lack the bases for assessment, and the frequent errors to which we are exposed show the impotence of statistics based on collections of facts observed at a distance from the primary causes to which they relate. We do not intend in any way to prejudge the future of science; we note its present state, but we have full confidence in its progress. Statistical studies are of great interest; it is important to make use of all that science possesses in the way of well-established facts; the intelligent comparison of these facts should lead to very valuable results, especially when extended to the whole surface of the globe.[2] In addition to ascertaining the facts and comparing them, the most important aim, which must never be lost sight of, is the search for their causes. The synoptic maps, extended first to the surface of Europe and then to the Atlantic, are a great step forward in this direction: they are already not enough. It is essential that, in the future, they cover the Americas, then the Pacific, and finally Asia for the northern hemisphere: one day they will cover both hemispheres. With the present state of science, we are convinced that telegraphic dispatches from America or Siberia will enable us to foresee major changes in the weather eight or ten days in advance: detailed forecasts will necessarily be increasingly limited.

The meteorologist who wants to forecast the weather must carefully monitor the operation of all his instruments and remain attentive to the slightest signs presented by the appearance of the sky.

The information sent by telegraph to the stations corresponding with the Observatory is reproduced in full in the *Bulletin international* and, in the part concerning France, in the newspapers. The Observatory is able to provide daily information on the general state of the atmosphere, the situation and intensity of the equatorial flow, the direction in which this flow tends to move, the rotating movements with which it is strewn, the points threatened by each tourbillon, the random variations that the meteor must cause in the state of the sky and the winds. This information will undoubtedly become more precise and cover a longer period of time as the science of weather becomes more advanced. For the time being, it could serve as a basis for interpreting the indications given by meteorological instruments and by the *signs of the weather*, but in no case should it replace direct or local observation.

As the nature of the information differs depending on whether it is intended for sailors or farmers, we will examine these two points separately.

A new science is being developed, a science which is extremely useful in its purpose and eminently general in the number of interests it can serve. We do not wish to prejudge the means that will be used to ensure that its fruits penetrate to the remotest corners of the territory; we will not even say anything about the organisation that we would consider the most favourable for achieving this: this is a question of administration that is not our place to deal with in this work. But we can, with the confidence of being heard, appeal to the support of all those who are friends of science among our companions. Their interests, as well as their pleasures, depend on the variations of the atmosphere; the state of the sky is often the object of their attention, its changes a cause of their concern; their individual practice, together with the confused experience of the past, gives them a meteorological tact which is often very remarkable. Let them all bring to their observations the spirit of method which is found in many of them, and let them take as the basis of their comparisons and studies the general summaries of the state of the atmosphere published by various newspapers. In this way they will achieve a twofold aim: "They will always be sure," wrote one of their colleagues, M. de Gasparin,

> to find a sweet and useful occupation in meteorological observations, in their comparison with those of other places and other times; and by charming their leisure, they will prepare the time when meteorology will acquire the certainty it lacks, and when conjectures concerning future phenomena will become probabilities.

We would add that by working for the progress of science in general, they will hasten the progress of their own instruction. The *signs of the weather* very often lead to vague and uncertain judgments, because the field of observation is too limited for any isolated observer. The observatory is like a watchtower, whose eye extends over the whole surface of Europe, and even beyond, over the Atlantic, from where bad weather comes to us; knowledge of the distant facts which it signals can be a singular help to the trained eye of the country dweller. Identifying the causes, which are always simple, together with the complex effects, in which we sometimes get lost, helps to classify the signs of the times, to determine their individual or total value, and to give greater certainty to their interpretation. There are, moreover, phenomena which, because of their suddenness, will always elude the forecasts of the meteorologist left to his resources alone, but which the Observatory can foresee from a distance and which it can alert to public vigilance through its advice. The warned meteorologist will be able to focus his attention on signs that would have gone unnoticed.

For its part, the Observatory can derive the greatest benefit from observations multiplied over the surface of France, even without the aid of instruments.

Our general synoptic maps do not deal with the details, which form a separate branch of our studies. The study of these details has been carried out with regard to the storms of 1865; and although the study is not yet complete, important results have already been obtained with regard to the forecast of these dreaded meteors. It is important to go further and construct daily synoptic maps specifically for France, designed to follow the movement of clouds and rain over the surface of France in each particular movement of the atmosphere. With the help of these maps, as soon as a general movement is announced by the characteristic deformation of the curves of equal barometric pressure on the surface of Europe, it will be possible to determine a few days in advance the place and time when a given phenomenon will make its appearance. This statement may appear to be out of the ordinary reserve of a man who has only the interests of science to serve; it is the result of a profound faith inspired by the progress made in a few years and by the immense resources that public favour brings to the new science.

We have spoken of observation without instruments. The study of the sky does not in fact require any instruments, and we prefer this pure and simple inspection to the exclusive use of the barometer, with forecasts recorded beforehand on a scale. But the meteorologist deprives himself of a very powerful aid if he neglects to combine the observation of the instruments with the state of the sky and the movement of the clouds or the lower winds.

The barometer is the first of all meteorological instrument. At present, the market offers very portable and inexpensive ones. The day will come when every village will have its own communal barometer, just as it has its own clock, and the farmers will consult it every day when they come in from the fields or before starting work. These new instruments are perfectly adequate for ordinary purposes. The needs of science require more precise instruments, controlled with great care and observed by trained men. A single barometer meeting these requirements is sufficient for each department, and all the primary schools have been equipped with them. The same remarks apply to thermometers and hygrometers, which are available at even lower prices, as long as they are used to determine the general course of the weather. From this point of view, they are second in importance to the barometer, but since the temperature varies more from place to place than the barometric pressure, they regain the advantage from the point of view of the study of the French climate, and their number should be increased.[3]

Editorial Endnotes

1 *Instructions sur l'usage da baromètre pour ta prévision du temps,* par Marié-Davy *(Bulletin international*, 8 septembre 1864 et suivantes)

2 We are referring in particular to a work of this kind carried out by Mr. Charles Sainte-Claire-Deville, the remarkable results of which have not yet been published.

3 *Instructions sur l'installation et l'usage des instruments de météorologie,* par H. Marié Davy. – Chez V. Masson.

2.3

LAGRANGIAN APPROACH

Editorial Headnote

2.3 Lagrangian Approach

The Eulerian approach became the preferred method for storm warning services in France and Italy in the 1850s and 1860s, largely as a result of the quest for security, order and centralism in state building. However, there was an alternative tradition centred on the Lagrangian approach to local weather forecasting systems, which acted as a counterweight to the Eulerian approach in the same period. Weather forecasting in Lagrangian form, as approached by Robert FitzRoy, developed in Britain and was characterised by the pursuit of autonomy and individual judgement in the face of continental rationalism.[1]

Vice-Admiral Robert FitzRoy (1805–1865), a skilled seaman and hydrographer forged in Mediterranean and South American waters, first came to public notice with his participation in the Beagle expeditions of 1828 and 1836 with the young Charles Darwin.[2] Appointed Meteorological Statist to the Board of Trade (the forerunner of the Met Office) in 1854, he went down in history as the man who coined the term "weather forecast" on the maps he drew after the 1859 storm that caused the loss of the Royal Charter.[3] His *Notes on Meteorology* (1859) sought to introduce to the public a method of forecasting that did not rely on the detection of atmospheric waves. By this time, he had launched his programme of fishing barometers for sailors and fishermen, and published the second of four editions of his book *Barometer and Weather Guide* (1st ed. 1858), all printed by Eyre and Spottiswoode for Her Majesty's Stationery Office in London.[4] In the eyes of official and elite science, FitzRoy was regarded as a radical empiricist and anarchist, a lover of speculation and rule-of-thumb forecasting.[5] In this extract from the sections on atmospheric waves, he distinguishes between the 'supposed undulations of the atmosphere' (as postulated by John Herschel, William Birt and Urbain Le Verrier, among others) and the reality represented by barometric curves, two things often conflated by weather scientists. According to FitzRoy, the fact that the curve representing the oscillations of a barometer corresponds to a mechanical, wave-like undulation of the body of the atmosphere has not been sufficiently proved. The lack of evidence for atmospheric waves does not mean that there are no other alternatives for weather forecasting. Instead, by analysing the variation of temperature and pressure in relation to wind changes on an empirical basis, he concludes that the most appropriate approach to weather forecasting is to combine barometric and thermometric readings with wind observations.

FitzRoy pioneered a Lagrangian approach to weather forecasting; his empirical method, based on simple rules and instrumental readings, prioritised the search for an overview of wind flow patterns over atmospheric waves and the isobar technique. His forecasting method is described in detail in his 1863 book *The*

Weather Book: A Manual of Practical Meteorology – translated into French by M. MacLeod as *Livre du Temps. Manuel pratique de Météorologie* (Paris, 1870) – and in his 1862 *Report of the Met Department of the Board of Trade*. The part of Chapter VIII 'Synoptic charts' reproduced here gives an account of the series of synoptic charts made by FitzRoy and his assistant Thomas Henry Babington to show accurately 'the consecutive simultaneous states of the atmosphere' (p. 103). By discarding the plotting of isobaric shapes and introducing techniques such as shading for clouds and rain, symbols for precipitation and wind vectors, the reader could find FitzRoy's charts, such as that of the 1859 Royal Charter storm, quite confusing, with such a level of visual complexity that it was difficult to forecast the weather from his charts. His synoptic charts also tended to show the confrontation between two types of air currents (warm and cold); this view was the result of the influence of the ideas of Heinrich Wilhelm Dove, whose *The Law of Storms* had just been translated from the German original (*Ueber das Gesetz der Stürme*) by Robert Henry Scott and published by the Board of Trade in 1858 (see Vol. II, 2.2 'Opposing current theory'). The distinctly empiricist image of FitzRoy and his work was particularly widespread among his detractors, such as Francis Galton and other members of the Royal Society of London, who were to write a devastating report against him in 1866. This somewhat caricaturing of his forecasting method, reducing it to a set of simple rules of thumb, is clearly seen in the 1863 opuscule by the Dutch Captain F.H. Klein entitled *The Foretelling of the Weather*, in which FitzRoy's method is compared with that of Christoph H.D. Buys Ballot, Director of the Royal Netherlands Meteorological Institute in Utrecht. For Klein, while FitzRoy's rules 'are useful to isolated observers', Buys Ballot's so-called system of deviations (see 3.1, 'Dutch statistical method') is 'more serviceable to promote the interests of meteorology'.[6]

George Jinman's *Winds and Their Courses; or a Practical Exposition of the Laws Which Govern the Movements of Hurricanes and Gales* (1861), included in the *Publishers' circular and booksellers' record* of 1861 (a reference periodical for professional publishers and booksellers), established him as a pre-eminent pioneer of the Lagrangian cyclone model. In some ways, his work symbolises the concerns of those naval officers and mariners who proposed theories of cyclones and storms based on flow models. Probably based on the experience of the Atlantic expedition that led from Newfoundland to Brazil in 1858, this English master mariner was able to propose a flow model of mid-latitude cyclones. Published by George Philip & Son in London in its first and third editions (1861 and 1865) and reprinted by P.M. Cranenburgh in Calcutta (1863), Jinman's model rests on three pillars: his empirical knowledge of winds and clouds, his extensive experience of tropical and extra-tropical cyclones at sea, and his familiarity with the storm theories of William C. Redfield, William Reid and Henry Piddington. By studying storm behaviour in the temperate North Atlantic, Jinman concludes that the circular theory and Piddington's rules are 'not only wrong, but dangerously misleading'.[7] His storm model is founded upon the meeting of 'two distinct currents of air flowing in opposite directions, and crossing each other on either side of the centre'

(p. 1). Although clearly inspired by Dove's theory of opposing currents, nowhere in the book is the German physicist mentioned. However, in stark contrast to the circular theories he so strongly criticises, Jinman's model offers only vague explanations for the physical processes that cause the formation of air currents.

These three texts condense the fundamentals of the adoption of the Lagrangian approach to weather forecasting. FitzRoy adopts the most groundbreaking forecasting perspective, proposing a synoptic method based on rules of thumb and instrumental readings. His approach seeks to prove that weather changes are not solely determined by the distribution of barometric pressure. While Jinman develops a mid-latitude cyclone model in which the trajectory of the airflows is the element to be considered. Their perspectives anticipate aspects of storm formation, frontogenesis and dynamic meteorology in the early twentieth century.

Notes

1 Aitor Anduaga, *Politics, Statistics, and Weather Forecasting, 1840–1910. Taming the Weather* (London and New York: Routledge, 2020), pp. 159–200.

2 Katharine Anderson, *Predicting the Weather: Victorians and the Science of Meteorology* (Chicago: The University of Chicago Press, 2005); Derek Barlow, 'From Wind Stars to Weather Forecasts: The Last Voyage of Admiral Robert FitzRoy', *Weather*, 49:4 (1994), pp. 123–32; James Burton, 'Robert FitzRoy and the Early History of the Meteorological Office', *British Journal for the History of Science*, 19 (1986), pp. 147–76; John D. Cox, *Storm Watchers: The Turbulent History of Weather Prediction from Franklin's Kite to El Niño* (Hoboken, NJ: John Wiley & Sons, Inc., 2002), pp. 75–84; J. Malcolm Walker, *History of the Meteorological Office* (Cambridge: Cambridge University Press, 2012), pp. 23–70.

3 James Burton, 'The History of the British Meteorological Office to 1905'. The Open University, PhD diss., 1988), pp. 40–1; Pauline Halford, *Storm Warning. The Origins of the Weather Forecast* (Stroud: Sutton Publishing, 2004), pp. 145–51.

4 Sarah Dry, 'Safety Networks: Fishery Barometers and the Outsourcing of Judgement at the Early Meteorological Department', *The British Journal for the History of Science*, 42:1 (2009), pp. 35–56.

5 Pauline Halford, *Storm Warning. The Origins of the Weather Forecast* (Stroud: Sutton Publishing, 2004), p. 193.

6 F.H. Klein, *The Foretelling of the Weather in Connection with Meteorological Observations by F.H. Klein together with a Description of the Telegraphic Warning System Introduced in The Netherlands, June, 1860, as proposed by the Director of the Royal Netherlands Meteorological Institute Professor Dr. Buys-Ballot* (London: Benjamin Pardon, 1863), translated from the original Dutch by A. Adriani, p. 22.

7 Frank Henry Ludlam, *The Cyclone Problem: A History of Models of the Cyclonic Storm* (London: Imperial College of Science and Technology, 1967), p. 25.

15

NOTES ON METEOROLOGY

Robert FitzRoy

Source: Robert FitzRoy, *Notes on Meteorology* (London: Board of Trade, 1859), pp. 1, 11–12, 13–22, 23–25, 34–35.

ACQUAINTED as all are, more or less, with the causes and consequences of changes in the air we breathe, or in water, scarcely less important to existence, it may be interesting, and one may hope useful, to endeavour to illustrate some of their remarkable features. If they should be less philosophically treated than might be desired, I must hope that due allowance will be made; because it is as practical observers of nature that I now venture to ask for your attention to some few parts of meteorology which may be so far popularised as to be useful in every day life.

Meteorology is the science treating of air and water, wind and weather, heat, cold, and climate, in combination.

First, I shall endeavour to give a general view of the atmosphere, in order that the explanations subsequently offered may be properly understood, and then I will make very brief allusions to some of the most useful instruments, not attempting anything like a complete description to those who are more conversant than I am with most of them, but simply alluding to particular points in which certain instruments *now* adopted are supposed to be better than those which were used some years ago, and drawing attention to some of their peculiarities.

After having made this partial reference to instruments, I will try to show how they may be made most practically useful in every day life, by foretelling weather; and I will refer briefly to the effects of changes in the atmosphere, and particular winds, on climates generally. Cyclonology, or the investigation of circular storms, in connection with the general winds: and movements of the atmosphere, called sometimes (whether correctly or not) "atmospheric waves:" and moisture or vapour in the air, will be briefly considered.

[. . .]

I may here allude to the combined action of the earth's rotation with the circulation of the atmosphere. As any body of air which we suppose to move from the equator is making its way northward, whether stopped at the latitude 30°, or carried on to nearer the pole, the earth itself is turning; and the consequence of these combined actions is, that instead of that body of air going direct towards the pole, as the earth turns, the said mass of air, having the velocity of the equatorial

DOI: 10.4324/9781003650294-25

part of the earth (not of the polar part, which turns very slowly), moves eastward as it goes northward, and its actual course is towards the north-east (S.W. wind).

On the other hand, the cold air from the pole, which starts southerly, meeting parts of the earth that are turning faster and faster, comparatively, is left more and more behind, and becomes while going towards south and west, north-easterly and easterly in effect. On the authority of Herschel and Dové, such is the system of our winds, and if referred to when using instruments showing the pressure, direction, elasticity, and composition of the air, the facility with which nearly all that takes place in the atmosphere can be foretold or explained becomes satisfactory beyond expectation.

[. . .]

One may call a very simple diagram, a circle divided by a diameter from north-east to south-west, the thermometer compass. While the wind is shifting from south-west, by west, north-west, and north to north-east, the thermometer is falling, but while shifting from north-east, by east, south-east and south, towards south and south-west, the thermometer is rising. Now the barometric column does just the reverse. From north-east the barometer falls as the wind shifts through the east to south-east, south, and south-west, and from the south-west, as the wind shifts round northward to north-east, the barometer rises – it rises to west, north-west, north and north-east.

The effect of the wind thus shifting round when traced upon paper by a curve seems certainly wave-like to the eye; but I believe it to be simply consequent on the wind shifting round the compass, and indicating alteration in the barometric column.

If the wind remained north-east, say three weeks, there would be no wave at all – there would be almost a straight line along a diagram (varying only a little for *strength*). The atmospheric line, in such a case, remains at the same height, and the barometer remains at SO inches and (say) some three or four tenths, for weeks together. So likewise when the wind is south-westerly a long time, or near that point, the atmospheric line remains *low,* towards 29 inches. Thus such "atmospheric waves" may be an optical delusion.

The diagram alluded to above shows how the barometer and thermometer may be used in connection with each other in foretelling wind and consequently weather that is coming on, because as the one rises the other generally falls, and if you take the two together and confront with their indications the amount of moisture in the air, at any time, you will scarcely be mistaken in knowing what kind of weather you are likely to have for the next two or three days, – which for the gardener, the farmer, soldier, sailor, and traveller, must be frequently of considerable importance.

Among the objects to which meteorologists should chiefly devote themselves seems to be the selection of practically and immediately useful points, in the first instance, and purely scientific ones, for future objects, subsequently. Among those most useful, probably simultaneous observations, once or twice in the 24 hours, by general concert of a great number of observers, will be found the most immediately available, because they give a bird's-eye view of the earth's surface at a given time, and again synoptically for successive definite times, days or weeks, perhaps months, from which may be learned what is the regular or irregular succession of weather, what may be expected and what are the various causes and connected circumstances. Such a plan is now in progress, and we may hope in two or three years to know, at all events in the Atlantic, and the countries bordering on this nearest ocean, what is the usual course of changes and *sequences* of weather during certain cycles – to know beforehand what may be expected, and what changes certain occurrences will inevitably bring after them, so as to become enabled to turn such warning knowledge to practical general use.

I will endeavour now to explain, more in detail, some of the principal and general movements of our atmosphere, – looking at the world as a whole, and referring to generally prevalent movements, not to merely local peculiarities.

Pray suppose that you are looking at a section of a globe from pole to pole as if looking at our world from a position at some distance outside of it, and taking in the view of a whole hemisphere, and then remember that the effect of the sun's heat, the constant effect throughout the year, is to warm and elevate, or *continue* raising all the body of air in the neighbourhood of the equator from many degrees, say ten or twenty, on one side to many (perhaps more) on the other.

Around the whole circumference of the world it is that zone of the atmosphere upon which the sun exerts the greatest and most constant power, the effect being to keep the air in that particular region continually rising. But it cannot rise beyond a certain distance, because the counter effect of gravitation towards the earth's centre acts. It also gets chilled and cold, after rising some miles from the earth's surface; but being continually pressed upon from below, by air still more recently heated, the upper air must flow away *laterally* (like water seeking its level) in the higher regions of the atmosphere, where it may be said to overflow. Meanwhile, to the lowest spaces adjacent to the ocean or land whence such air has risen, other air will press in from either side to fill what would otherwise be a comparative vacancy; such other air being cooler, and drawn from afar, whether from south or north, or elsewhere. Now that air which has been first warmed, expanded, and raised, – then cooled in the upper regions – must go somewhere. It cannot go upwards beyond a certain limit. The consequence then is that it overflows, as has been shown, on each side, part going northward in a direction nearly corresponding with the shape of the earth's surface, or parallel with it; part going similarly towards the south – thus keeping up a continual movement of air towards the poles, *from all parts of the equator all round the world,* while there is another tendency constantly of the lower air, from the poles towards the equator, *near the earth.* A grand and immediate effect of these constant currents, or as one may say of this circulation, is the phenomenon of the trade and (so-called) anti-trade winds prevailing within certain limits, the former from near the equator to 20 or 30 degrees on each side of it, all round the world, and the latter in middle latitudes, – excepting where spaces of land intervene, and, from their heat or cold, occasion local or exceptional differences. Speaking *generally,* these trade and anti-trade winds around the world, upon which navigators depend so much, are caused by ascension of the equatorial air, and a supply of its place by other air pressing towards the equatorial places of deficiency.

Now that cold air (cooler air) flowing in, must of course come from some great source of supply! *Where* is that source of supply, one may ask, in sufficient quantity? If you compare the size of the world in its equatorial regions, the bulk of it around the equator, with the *small area* near each pole, and consider that the world is constantly turning, the central part being very large and moving rapidly in proportion to the polar, where it has very slow motion, it will appear evident that the mass of heated and expanded air near the equator must be very much larger than the body of cold air near the poles. It is not only very much larger in quantity, but it has a rotatory motion, very rapid, nearly equal to that of the equatorial earth, which air near the poles has not. The nearer the pole the less the rotatory movement of the earth, the nearer the equator the more rapid that movement; and the atmosphere, which goes nearly with the earth, moves similarly. The result of these combined movements, those of the earth and air, is that warm air which moves towards the pole from the central zone has another impetus, with the velocity of the equatorial regions, and as it goes towards the pole while the *earth turns eastward,* moves faster and faster than the earth does as the polar regions are approached. If you throw anything out of the window of a railway carriage in motion, it partakes of your motion, and instead of dropping behind, it drops abreast of the carriage. So with air that rises from near the equator in its movement towards the poles; and the consequence is, such a prevalence of westerly winds in middle latitudes that they are justly termed anti-trades. These prevalent winds are at first south-westerly, then westerly or north-westerly in the regions between the tropics and those of the arctic and antartic circles.

The converse action, which is the easier to understand at first sight, takes place from the pole towards the equator. The mass or body of air or wind, starting from near the pole, moves towards the equator with little or no circular or rotatory movement, because the body of air within the arctic regions is *comparatively* slow in rotation, that part of the earth moving very slowly compared with the equator. Any body of air from polar regions thence moves towards equatorial parts to fill the spaces being continually vacated there; but as it so moves the earth is turning to the eastward, and the result is a diagonal, a more or less diagonal movement of any such body of air across the world's surface, causing north-north-east or south-south-east, or north-easterly or south-easterly, or easterly winds; *more* easterly as you get nearer the equator; so that the wind which we feel here as a north-east wind actually left the arctic region as a northerly wind, and, probably, when it has passed further south, will become sensibly more easterly. The simple effect of the polar current becoming more easterly towards the equator may be shown easily by moving a pencil from either pole of a common globe, on a stand, towards the equator direct, while the globe is turned eastward. The line traced by hand makes a diagonal (or rather a spiral) line nearly between the two directions. (It is spiral on account of the globular form.) This effect is easy to see; but the other movement, from west to east, and the reason why a body of air moving from the equator towards the pole goes faster than the surface of the earth over which it moves, is not so evident *at first.* It is only after reflection one sees clearly that, the rotatory movement within some degrees of the pole being slow – compared with the rapid motion near the equator – if the air rising from the equator moves with the momentum of that part of the world, as it goes further towards the pole it must move faster and faster over the earth's surface in an *easterly* direction, and the result must be the general or prevalent westerly winds in middle latitudes, which have been called by Sir John Herschel and others, the anti-trades.

Referring again to a globe, its poles and equator, we see that extensive central masses of air rise from equatorial regions, move *above* laterally and towards the poles, and that polar currents move nearer the earth's surface towards the equator. Owing to the convergence of the meridians (that is, their approaching nearer and nearer), all the meridians meeting at the pole, from larger spaces at and near the equator, the polar supplies are deficient, and are made up for by parts of the equatorial current, as will be shown. All the body of air round the equatorial region, large and expanded as it is has a *tendency* to move towards the poles; but it *cannot do so,* there is not *space* enough for it; the law of gravitation acting through this fluid air, seeking its equilibrium, prevents its *rising* beyond a limit; when resisted and cooled it becomes compressed, and then struggles with the polar counter-current for room near the earth's surface. Suppose that a body of air at the equator which has the above described double tendency, (namely, to move eastward and towards the pole), starts from an equatorial place, and moves more and more westerly (from the westward) over the earth's surface as it goes towards the pole, because its motion in middle latitude is faster than the motion of that part of the earth from the west, and it goes, therefore, faster than the earth, – it becomes compressed into smaller space, and also meets with the polar current making its way towards the equator, (by the turning of the world making a diagonal (or spiral) course which is north-easterly, and afterwards easterly when it gets near the equator). There are thus two conflicting currents or bodies of air, one making its way northward and becoming compressed, having its vapour condensed and (as it were) squeezed out of it by the dry polar current which it meets, – and the other making its way towards the equator.

The effect of the antagonistic currents of air in raising the barometer about the "horse latitudes," and the return of part of the upper current, when cooled and condensed, to the equatorial regions, has been already described. We see the effects of these two conflicting currents in ordinary changes of weather in this country (nowhere more plainly) the wind continually going round by north, east, south, west, and north – round and round again in successive changes, comparatively seldom going round the other way, the retrograde motion being exceptional to the *general* rule, like the exceptional eddies which one may see in the currents of a river when streams are meeting one another, the eddies generally turning one way, but occasionally from some local circumstances, turning contrariwise, as exceptions to the general rule of direct *apparent* change, which, in north latitude, is as the hands of a watch move.

Perhaps these views may not be quite evident at first, more particularly that part relative to the equatorial or heated current going faster than the world, and, therefore, making westerly winds as it gets toward the pole; but it is the explanation given by Herschel and Dové. Dové bases upon it what he calls the "Law of Gyration," and to him is due the credit of having proved, if not detected, this law, by comparing most perseveringly, during the last thirty years, observations from all parts of the world; and he has shown that this law of gyration applies equally to all parts of the world; that it is not peculiar to the Atlantic or the Pacific, or north or south latitudes, but it is the general law throughout the world, exceptions occurring only over or near countries which are covered with snow or ice, or are localities where the sun's rays strike periodically and cause great heat. Such local circumstances interfere with the regular or normal movements of the winds.

From the "Law of Gyration," it is easy to trace the contests and effects of the two principal winds, which are quite different in their character, the one being warm and moist, having a great deal of watery vapour held in it as gas in an invisible state. Sometimes this vapour will be chilled or cooled by approach to the polar current near it (perhaps above, perhaps on one side), causing the vapour to become visible in the shape of cloud, or, when more cooled, precipitating it as rain; at other times being retained in the atmosphere, but invisibly.

Everybody may have noticed, when the wind has shifted from south-west to west or north-west, particularly during bad weather, that the first shift to the northward, or from the south even towards the west, is generally accompanied by a sharp squall, a good deal of rain and perhaps hail, after which the weather clears up. This is the usual course in our latitude – and, substituting south for north, in the southern hemisphere.[1]

The circulation of atmospherical currents need not be only horizontal – it may be at an angle with the earth's surface – or even vertical; but the first effect of any cold current impinging upon the warm equatorial, is to cause condensation of its vapours in the shape of cloud, if not rain, hail, or snow. After this first effect, being dry (comparatively) it acts like a sponge; it takes up the rest of the vapour that there may be in the air. There is no longer precipitation, but all the vapour that it can sponge up (as it were) goes away invisibly; and after a short time, there is a clear sky with a more or less polar wind.

Thus, the first effect of the cold air is condensation, and the subsequent effect of the dryness of the air is the sponging up and hiding the vapour (as it were) in itself: dry air from the polar regions being capable of containing a great deal more vapour in an invisible state than the warm air which comes loaded with aqueous vapour from the equatorial regions – loaded, indeed, with nearly as much as it can carry.

The struggle between these currents over various parts of the earth's surface, occasionally causes the circular storms, now usually known by the term Cyclones, or rotatory gales.

Whether so called, or named typhoons, tornadoes, hurricanes, or whirlwinds – circular or rotatory, more or less, they all are, not exactly, but in an elliptically curved form, more or less irregular rather than circular. The outline of such a circulation of wind in one locality, say a few hundred miles in area, may be represented by moving a piece of card-paper, with arrows to show directions of the wind, across a part of the earth's surface, as shown on a large map, – the North Atlantic, for example, between England and Newfoundland. The course of the arrows shows how the wind is circulating in that space; so that, while the wind is at one point westerly, at another it may be easterly, – here southerly, and there northerly: supposing such a piece of or card-board to represent a body of air in circular motion, and at the same time moving *bodily* onwards, as these great meteors (for so they may be called) do move, – that is to say, while circulating around their centres, they are also moving onwards across the earth or ocean's surface.

If such a circular card-paper be moved northward from near the tropic, or from the neighbourhood of the "horse-latitudes," across the Atlantic (as we know, from logs and other collected observations, such rotatory storms do move), the wind will be represented circulating round, while the meteor, or cyclone, travels on bodily northward.

When first touching Ireland the wind would be east, but, as it moves on, the wind there becomes south-east, south, and south-west (the air in such a meteor is supposed to be going round, so that there are different winds in different parts, which *may* be some hundred miles separated). At London such a cyclone might begin with a south-east wind – which, as it passed on, would become south, then south-west, west, and then north-west, and north: – or it might be east, north-east, north, north-west – and west, if the northern half of the cyclone only passed across London.

The turn or *apparent* sequence of the winds is, in these *first* cases, what sailors call *right-handed,* or with the hands of a watch; though the meteor itself, or the body of air in motion, moves round contrary to the hands of a watch, while the appearance to the eye seems to be the reverse: but, in the *second,* it is *left-handed.*

Take any place for example. Supposing the beginning of the movement there, is with an easterly or south-east wind, and that it shifts thence to the south, and south-west, and west, and then by north to north-east, – the effect to our eye is that of the wind shifting "right-handed," as we say it does, or "with the sun," but the wind in the cyclone itself has been going round actually in the opposite direction, or against the sun and contrary to the hands of a watch (the best reference for illustration).

In the southern hemisphere there has been some confusion, owing to our seamen always calling this direction right-handed or with the sun, and the opposite direction left-handed, whether in the southern or northern hemisphere, although when beyond the tropic in the southern hemisphere, and looking towards the sun, he appears to move "left-handed," as sailors call it. In dealing with circular winds of any kind, it is better to refer to the hands of a watch, rather than to nautical expressions, "right or left handed," "with the sun," or "against the sun."

There is a simple rule in connection with these rotatory winds, which is, that in the northern hemisphere, when facing the wind, the centre of its circulation is square to your right, and when in the southern, square to your left. All the terms, tornado (twisted), typhoon, cyclone, hurricane, and whirlwind, are but different words expressing similar rotatory movement or circulation, of vastly increased strength *at times,* but not *always* so excessive in force and duration.

We do not usually feel the total circular effect; we feel part of it; the wind shifting only from some one point to some others, because the whole meteor does not often go over any one place; a part of it goes across in general, and, therefore, places on one side of its centre are affected differently from those on the other side, where the wind will appear to shift differently. It is sometimes found that cyclones, instead of passing horizontally, or parallel to the earth's surface, are inclined at an angle (more or less acute) with the horizontal plane, and we, therefore, feel only parts of them; there may be several such circulations passing or following in the same direction, and revolving similarly, like eddies, with their lower portions, on one side only of the circulating meteor, touching the earth's surface, the other parts circling above our heads, so that we only feel a part of the change, and not the whole of the movement.

It should be remembered that there are several layers or strata of currents in the atmosphere, moving variously, and one above another.

When the centre of a cyclone passes over any place there is a sudden change from one direction completely to the other, with probably a lull between. If only an extreme or part passes, there is a gradual change through nearly half the compass – and very seldom are there instances of greater sensible change than this, through about half the compass, excepting when successive storms follow one another and cause complications.

Professor Espy of the United States has written much on this subject, and has published what has been supposed by some to be a different view; but in fact his theory is a corroboration of Dové's, and also supports such general views as are here offered. Professor Espy says that in a circular storm the wind blows from many directions towards the centre. It can be demonstrated mathematically that the consequence of the wind blowing from over large areas, some hundreds of miles in extent, towards a centre, will be a whirl near the centre, and a comparative lull in the centre, which is what actually takes place. You find the strongest parts of the whirlwind near the centre, but not in it; and the further you get from it, the more the wind diverges from a circular course. You may see a similar convergence in a trough of water with a hole in the bottom. If you open that hole the water begins to pour out, and the water near the opening which causes the movement moves fastest – the outer part moves more gradually. In air there is a somewhat similar effect, as there must be an elevation of the central part consequent on rarefaction or evaporation, which draws air from all other quarters: or there must be a *compression* consequent on two bodies of air passing each other and twisting air forcibly between: as you see in a river, when one body of water passes another, a succession of eddies caused by such a lateral pressure of currents passing each other. So it is in the atmosphere, only that there these eddies are much more complicated, extensive, and various, on account of the extreme *elasticity* of the air, which not only yields to pressure, but has always a tendency to expand again greatly afterwards – a resilient elasticity not found in water, which is almost incompressible and inexpansive.

With reference to this subject, I would refer to a passage published by myself as far back as 1839. I asked the question with reference to Reid's "Law of Storms." "Are not storms exceptions to the general winds or atmospheric currents, not the causes of them? Some persons may have laid too much stress upon such exceptions, and have rather overlooked the general principal features – those of the conditions which *prevail almost continuously.* Common winds occur throughout the year, except during short intervals; but hurricanes, or even ordinary storms are comparatively rare. May not opposing or passing currents cause eddies or whirls on an immense scale in the air, not only horizontal, but inclined to the horizon, or vertical?"

"In laying a ship to during a storm there are *other* points to be considered, besides the veering of the wind, such as the direction of the sea, perhaps against a current. I never myself witnessed a storm that blew from more than fifteen points of the compass, either successively or by sudden changes. In most, if not all of the storms to which I can bear testimony, currents of air arriving from different directions appeared to succeed each other or combine together. One usually brought the 'dirt,' to use a sailor's phrase, and another cleared it away, driving much of it back again, often with redoubled fury. One of these currents was warm and moist, another cold and dry, comparatively speaking. While one lasted the barometer fell or was stationary; with another it rose. At all places I have visited, or from which I have obtained notices on the subject, the barometer stands high with easterly, and comparatively low with westerly winds, on an average. Northerly winds in the northern hemisphere affect the barometer like southerly winds in the southern hemisphere."[2]

With regard to *foretelling* weather, I will only quote two or three very short passages from my own writings. "It should be remembered that the state of the air foretells rather than shows present weather (an invaluable fact too often overlooked)." I refer now to *foretelling* weather, because it is immediately connected with the general law of rotation, or Dové's law of gyration. "The longer the time between the signs and the change foretold by them, the longer such altered weather will last; and, on the contrary, the less the time between a warning and a change, the shorter will be the continuation of such foretold weather. If the barometer has been about its ordinary height, say near thirty inches at the sea-level, and is steady or rising while the thermometer falls and dampness becomes less, northwesterly, northerly, or north-easterly wind, or less wind, may be expected. On the contrary, if a fall takes place with a rising thermometer and increased dampness, wind and rain may be expected from the south-eastward, southward, or southwestward.[3] Exceptions to these rules occur when a north-easterly wind with wet (rain or snow) is impending, before which the barometer often rises (on account of the *direction* of the coming wind alone) and deceives persons who from that sign only expect fair weather." "Three things appear to affect the mercury in the barometer: – 1. The direction of the wind; the north-east wind tending to raise it the most, the south-west to lower it the most; and wind from points of the compass between them proportionally as they are nearer one or the other extreme point."

What it is that causes these peculiarities immediately connected with the direction of the wind, whether it is entirely owing to the air, the southerly or equatorial wind becoming lighter from its warmth, or from some electrical condition which is not traced out as yet, is not clearly proved. If connected with electricity, as is probable, we cannot now deal accurately with it. Here, only those facts are stated which have been distinctly verified.

"North-east and south-west may be called the wind's *poles,* or rather, *extreme bearings.*" If the wind hangs for two or three weeks in either of these quarters, the mercury will remain near its higher or lower point, the atmosphere (with a *north-east* wind) appearing steady. "The range of difference of height of the mercury due to change of direction *only* from one of these bearings to the other (supposing strength, or force, and moisture to remain the same) amounts in these latitudes to about half an inch, shown by the barometer as read off.

2. The amount taken by itself, of vapour, moisture, wet, rain, or snow in the wind or current of air (direction and strength remaining the same) seems to cause a change amounting, in an extreme case, to about half an inch: – and
3. The strength or force alone of wind from any quarter (moisture and direction being unchanged) is preceded or foretold by a fall or rise according as the strength will be greater or less, ranging in extreme cases to more than two inches."

The barometer is affected to the extent of two inches by the *strength* of the wind, it is affected about half an inch by the quantity of moisture in the air, and perhaps another half inch by the *direction* only of the wind. As these usually act more or less against each other, and in smaller proportions, the barometer is seldom so greatly affected; and its ordinary changes are not so marked.

Such rules have been contracted, on the scale of a large barometer for the use of fishermen and coasters, in the following manner: –

On one side being –	And on the other side –
Rise	Fall
For	for
N.E^{ly}.	S.W^{ly}.
N.W. – N. – E.	S.E. – S. – W.
Dry	Wet
Or	or
Less Wind.	More Wind.
Except*	Except*
Wet from	Wet from
N.E^{d}.	N.E$_{d}$.

* (Being the *only* exception.)[4]

[. . .]

Reference may *again* be made here to a useful rule already noticed. Supposing that a storm is rising, and that a seaman wishes to know where the greatest strength of wind is. If he *faces* the wind, in northern latitude, the centre of the circulation, or cyclone, will be square to his right; and in the southern hemisphere, if he *faces* the wind, the centre will be square to his left; therefore he knows in which direction to go to avoid that part where the greatest strength is, and must shape his course according to circumstances. He can hardly have a simpler rule. His position with reference to distance from the centre, whether it is near the ship or far off, can be ascertained by watching the falling or rising of the barometer, and the way in which the wind shifts. It requires only a sketch upon paper – a rough figure, with an hour or two's observation of the veering of the wind, to know exactly in which direction to steer, without studying any elaborate rules, which sometimes perplex more than they assist the seaman in the hour of need.

[. . .]

As so much has been said, during the last few years, about "atmospheric waves," I would again refer to them here. If wind veer round the compass, in the course of two or three days (more or less), or is many days in making a circuit – invariably, as it goes round, the barometer rises or falls according to the direction or strength of the wind. Supposing a diagram to represent 36 hours, and divided into spaces of three hours each along the upper horizontal line; while below, points of the compass are shown, from north around by east to north again, and continued to south; and at the side a scale of inches and decimals, from 28 to 31. Then suppose that the wind has gone round the compass once, or say once and a half, as happens occasionally; and that it has been an extreme case of depression, as in a storm. Then, if from (say) 30·3, with the wind at north, a shifting occurs, first towards the north-east, and then onward in the same direction around the compass – as the wind so shifts to the north-east, and is about to shift towards the east and south, the barometer foretells it, or falls beforehand. When the wind is north-east the mercury is lower, probably, that when it was at north. As it gets to the east the mercury falls, and gets lower still at south-east, and falls still more to south and south-west, where it is probably the lowest, because it feels the effects of the south-westerly or equatorial current most then, and may be down, let us suppose, to 28·2 inches. As the wind shifts round to south-west, west, north-west, the column in the tube rises, till, perhaps, the wind is north, or even north-east, when it may be as high as 30·8. It has been known in this country as high as 30·9. As the wind goes round again to the east, and south-east, and south, the barometer falls as before, and a line or curve traced upon paper, representing these falls and rises, or oscillations of the barometer, during a certain time, say these 36 hours, has an appearance like the outline of a wave of water; but as these apparent waves, or undulations, take place *exactly* as the wind shifts, and proportionally to its strength – and as, if the wind remains in one quarter for some days, or say two or three weeks together, the curve becomes almost a direct line, remaining at about the same elevation, it seems that there is an intimate and immediate connection between such a curve, or wave line, and the oscillation of the mercury, though not necessarily between the curve and any undulatory movement of the atmosphere above our heads. If a body of the atmosphere above us swelled upwards, like a wave, and fell again, as some suppose, as it were in "crests" and "troughs," how should we reconcile it with the fact of there being various currents passing over each other in the atmosphere in differing directions? Aeronauts who have been up in balloons know that from one stratum of air they passed into another, and another, and perhaps a fourth also, moving in different directions. There cannot be *vacancies* between the undulations of various strata of air. These different bodies of atmosphere could hardly be undulating like waves while having spaces between them, and interferences of cross movements. Waves of ocean have only elastic air above them, which does not impede their rise and fall materially; and they are only superficial, not reaching far down.

Were there a raising of any part of the mass of air, the lighter or equatorial portions, or winds, would rise the highest, and would expand; but, according to the "Wave theory" (here controverted), the reverse is the fact; you have the lowest part of the apparent trough of the wave, with the lowest barometer, that is, with the air, which is the *lightest* and most *expanded,* and ought (therefore) to rise up the *highest;* and you have, coincident with the heavy dry air, the highest part, or what is called the "crest," of the wave. Considering then these facts, and the exact correspondence of the movements of the mercury with the wind's direction, besides noticing the extreme variability traceable in such an atmospheric wave, (which can hardly be conceived to be motionless for *weeks,* as in the case of a *steady north-easterly* wind, and then going into extraordinary irregularity during *a day or two,*) we are led to the belief and assertion that what are commonly called "atmospheric waves" are rather delusive; and that there are waves in any line indicating oscillations of the barometer, but not such, in the atmosphere itself, as are usually adverted to by the expressions "trough" and "crest."

[. . .]

Another word may be here added by way of recapitulation respecting "atmospheric waves" (so-called).

Mr. Birt drew particular attention to these supposed undulations of atmosphere, by papers read at meetings of the British Association, and by a special Article in the Admiralty Manual of Science. Sir John Herschel, Le Verrier, and other great authorities then countenanced Mr. Birt's theory, and apparently sanctioned his opinion. Yet there is so much argument against those views, that even the highest names may scarcely warrant their implicit adoption. That there must be undulations in the atmosphere – constituted as it is – cannot be doubted, but that the curve traced on paper, representing the oscillations of a barometer, as the wind veers round the compass, corresponds to a mechanical, wavelike, undulation of the body of atmosphere – is not perhaps sufficiently proved.

One may demur to it on these grounds. First, the curve so traced, on paper, varies, not only with the barometer, but with the *direction* of the wind, which is invariably accompanied by change of pressure, consequent on the greater or less action of polar current.

Secondly, while the wind remains in one quarter – the curve, or line, taken as that of a wave, remains almost unvaried, except in consequence of altered *strength* of wind or *much rain,* which have each a *comparatively* small effect.

Thirdly, the lowest part of the curve (*called* the trough of the wave) always corresponds to the lowest barometer, or *lightest air;* whereas it is the lightest air that rises highest, as instanced at the equator; and, therefore, the crest of an atmospheric *wave* (so to speak) ought to be over the place of lowest barometer.

Fourthly, Aeronauts always find, and the upper cloud often show, currents above very different from those below. These superposed and successive strata, in rapid cross motion, must tend to destroy undulation. I am aware of what has been stated on this subject (atmospheric waves) in the invaluable Essay on Meteorology, in the last edition of the Encyclopædia Britannica, but with the utmost deference I submit that the experiment – of the undulations transmitted through successive (coloured) strata of fluids, *in a vessel* – did not meet the case of fluid strata of *air* moving horizontally, in various directions, across each other.

That there are tidal waves in the atmosphere, caused by the sun and moon, experiment has proved; but that they are *very* small has also been demonstrated. This subject, however, has yet to be further investigated, by means chiefly of the accurate barometrical measures instituted and carried on by Government during late years, in many parts of the world, and especially at sea. Such waves as these follow their causes, in periodic times – and *not* (in disregard of sun and moon) only corresponding to *direction* and strength of wind.

As electricity has been referred to frequently, it may be well to remark that with polar wind and fine weather there is much positive electric influence. With equatorial (*apparently*) little or none, as far as hitherto ascertained.

Source Endnotes

1 But liable to exceptions, when the wind retrogrades, or "backs."
2 Voyages of Adventure and Beagle. – Appendix. 1839.
3 Both glasses low, indicate snow.
4 This is the first systematic exhibition of the subject, in small space; and practically.

16

THE WEATHER BOOK

A Manual of Practical Meteorology

Robert FitzRoy

Source: Robert FitzRoy, *The Weather Book: A Manual of Practical Meteorology* (London: Longman and Green, 1863), pp. 102–105.

Simultaneous observations of our own atmosphere had been recommended by some of the most eminent men, who had turned a part of their attention to meteorology, even before Dr. Lloyd undertook and executed his valuable series, extending through ten years, in Ireland. From these, from Mr. Stevenson's three years in Berwickshire, and from many fragmentary notices, some of very old date, some in the last two centuries, and others of our own time, it was obvious that, over and around the British Islands, storms had an eastward course, or, in other words, came from the west *generally.* Exceptions, however, *seemed* to occur with easterly winds (from the east), which were not then understood.

In 1857 the Board of Trade invited general co-operation around and upon the North Atlantic, between the tropic of Cancer and the arctic circle. United States, Norwegian, Danish, Dutch, German, French, Spanish, Portuguese, and Italian observations were collated with those of our own coasts, ships, and observatories. The lighthouses (admirable points for such enquiries) were in the front rank, by Trinity House direction. Ships traversed the sea, provided with the few instruments indispensable, while, for central operations, the principal observatories afforded an unquestionably sound basis.

A series of charts was then commenced by Mr. Babington, which, in three years, increased to many hundreds. These outline charts have *no mark* on them *unnecessary* for expressing atmospheric condition or time, or locality by latitude and longitude. An outline only of land, and as few names as possible, are shown. Being intended to express consecutive simultaneous states of atmosphere – as if an eye in space looked down on the *whole* North Atlantic *at one time* and afterwards took similar views (much more extensive than '*bird's eye*') at regular intervals of hours or days, so as to obtain sequences of synoptic conditions – we called them 'synoptic charts.' It was subsequently suggested that 'synchronous' would be a better term, and it was adopted. Now, however, it appears to the writer that *this* word is less appropriate. In General Sabine's recent lecture at Cambridge, on

 DOI: 10.4324/9781003650294-26

Magnetism, the term synchronous is applied to observations made at the same local hour around the world: but these are not simultaneous. Truly indeed they are synchronous, but not in the sense that is required for occurrences happening simultaneously or synoptically, referred to time or view, at one place only, and one meridian. Therefore it is that we now abandon our erroneous appellation and revert to '*synoptic* charts.' Their principles of construction, and many other details respecting these very useful documents, are in the Appendix, and in accompanying diagrams.

Among the most instructive and permanently valuable, are those of the period which included two notable storms, that of October 25–26[1] and November 1, in 1859. This series was published as an atlas, with the Tenth number of the Board of Trade Meteorologic Papers.

First comparisons of such observations around our own coasts *seemed* decidedly to show consecutive '*waves*' of atmosphere, by the curves indicating *altitudes* of the *barometric* column.

Espy's great works, and his multitudinous curves closely covering many large sheets, *seemed to prove* the case indisputably.

Yet the present writer ventures to submit a very different view of the subject – one that, at all events, is in no degree contravened by any facts hitherto publicly recorded as having been anywhere reliably observed.

The two principal currents of air being different in their qualities, obviously cause states of pressure or tension, temperature, and other peculiarities, at any given place, corresponding to those of the air then present – a portion, however small, of a passing current. This current does not continue, in the temperate zones, for a long time; it may be a mere streamlet, as it were, of a few hours duration; it may continue during some days; it may persistently last through several weeks; but, while it is present, the statical conditions of air around an observer are of the same character, however different in quantity, or force – excepting when there is a *mixture* of air currents at the place of observation, in which very common, if not prevalent case, statical conditions vary with the relative proportions of elements intermingled.

(Such elements are now sufficiently known, and traceable, to admit of their being stated in a mathematical manner, and managed algebraically.)

It has been proved that great atmospheric currents circulate, extend to various distances horizontally in length and breadth, or either only, are sometimes side by side but moving in opposite directions, sometimes superposed or overlying. And it has also been proved that the mercurial column has an average height, in presence of one current, differing considerably from that which is the normal, average, or mean elevation under opposite influence, that of the contrary main current.

What must be the results of observation at any one place? As one of the principal air currents, polar or tropical, approaches, its peculiarities are gradually felt, and barometers alter (other instruments also.) The alteration continues in one direction as the current passes, until its greatest influence is past, and a contrary change begins, which increases as the opposite current advances, or extends its effects.

Thus the barometer rises and falls, or falls and rises, more or less regularly – as it is influenced by the passing currents of air; and as these movements are more or less directly with the surface *wind* which goes round the compass, 'gyrates,' or shifts with those currents (of which it is the effect and evidence), a line traced on paper, coordinating the barometric heights, will have a wave-like form, in which the crests will correspond to one current and the hollows (or troughs) to the opposite.

But this appearance *on paper* in no way demonstrates a real mechanical undulation of atmosphere, a body composed of many layers, or overlying strata, different in qualities, and moving in *various* directions simultaneously.

Source Endnote

1 Royal Charter.

17

WINDS AND THEIR COURSES; OR A PRACTICAL EXPOSITION OF THE LAWS WHICH GOVERN THE MOVEMENTS OF HURRICANES AND GALES. WITH AN EXAMINATION OF THE CIRCULAR THEORY OF STORMS, AS PROPOUNDED BY REDFIELD, SIR WILLIAM REID, PIDDINGTON, AND OTHERS

George Jinman

Source: George Jinman, *Winds and Their Courses; or a Practical Exposition of the Laws Which Govern the Movements of Hurricanes and Gales. With an Examination of the Circular Theory of Storms, as Propounded by Redfield, Sir William Reid, Piddington, and Others* (London: George Philip and Son, 1861), pp. 1–9, 92–96.

WINDS AND THEIR COURSES.

I assert, and more – I will prove it – that there never was such a thing as a really circular gale, or what is termed by Mr. Piddington and others, a TRUE HORIZONTAL CYCLONE.

Every gale[1] or hurricane has two distinct sides, formed by two distinct currents of air flowing in opposite directions, and crossing each other at two points, one on each side of the centre: thus –

Fig. 1.

DOI: 10.4324/9781003650294-27

A current of air cannot flow in a true circle – it is against reason and common sense to suppose so. A current of air in passing round a centre, cannot re-enter its original channel: it must either cross the first part, or flow round inside, as shown in Figs. 2 and 3: thus –

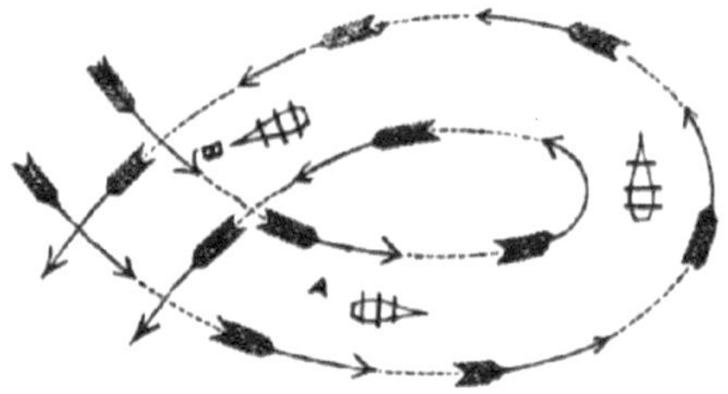

Fig. 2.

Fig. 3.

If it passed round inside, as in Fig. 3, then the whole current must ascend in the centre; but there is no evidence to show that such is the case, neither is it likely. But we have plenty of evidence to prove that currents of air cross each other: we need not refer to the published details of Hurricanes which have occurred for such proofs, for by casting our eyes aloft at the commencement or end of a gale, we may have occular demonstration of the fact: – for the clouds may be seen passing over from a direction very different to that from whence the wind is blowing; or if there be two strata of clouds, these may be seen crossing each other. It follows, then, that if a gale were formed as at Fig. 2, and a vessel should enter at *a*, as she ran on the wind would gradually veer through nearly every point of the compass, and her track would very nearly form a circle; but on arriving at *b,* where one branch of the current crosses the other, the wind would be found to shift several points rather quickly – it might be suddenly. A short calm might even take place here: for that part of the current of air which passes over the other must leave the earth's surface and ascend in order to do so. Here, then, we have the cause of the sudden shift of wind, or short calm, so often reported. It is perfectly ridiculous to suppose that all the vessels have passed through the centre of a storm, which are reported to have done so: nine-tenths of them never were near it.

Figure 2, then, may be taken to represent the course of a gale formed by one current of air: but every gale or hurricane, and even every whirlwind, is formed by the joint action of *two currents of air,* and there are two points of confluence or crossing in each, as is shown in the diagram (Fig. 1.)

This being the case, it follows that whenever a ship is put off before the wind in a hurricane or gale, she is almost sure to meet with a sudden shift, unless she is in the current which passes under the other. Should she be in the latter position, she may have no sudden shift, but she will have the full force of the storm, for where one current passes under the other, the wind in it is pressed down and concentrated on the surface.

It is a common remark, that the wind blew as though it had blown through an archway or funnel, and such is literally the case: hence the unequal force of the wind at points equidistant from the centre. I shall explain this more folly when treating of the storms of each hemisphere separately.

It may be asked, How can you prove this? How can you prove the fallacy of "*the Circular Theory*?" My answer is – By the analysis of the very logs which have been put forward in its support.

I will now proceed to give a general description of what I conceive to be the form, &c., of a hurricane or gale.

Every hurricane or gale, as I have before said, has two distinct sides, formed by two currents of air, flowing in opposite directions and crossing each other at two points (Fig. 1.) The two sides are not always equally developed at the earth's surface – that is, the one often blows harder than the other; it is but seldom, in fact, that the winds in the two sides are equal in force. The greatest force will be found at either *confluence,* but rarely of equal force at both. It may blow with hurricane violence at one *confluence,* and only a moderate gale at the other, or it may blow with hurricane violence at *both confluences,* and only a moderate gale at the sides.[2]

The Centre of a Storm is not the most dangerous part. There are times in which it would be possible for a steam-ship to pass from one side of the gale to the other, across the centre, and yet not experience more than a moderate gale, whilst vessels on either side of her, under the confluences, may be exposed to violent hurricanes. A vessel may become embayed, as it were, in a storm, in such a position that it may be almost impossible for her to escape altogether uninjured. There are many positions, indeed, in which a ship, especially a dull sailing one, may be overtaken, and be unable to escape. One *confluence* is generally more distinctly marked than the other, especially in extra-tropical latitudes; so much so indeed, that for a long time I imagined that there was but *one*: which at that time I named the *Centre-line,* or *core* of the gale. The one on the west side of the centre, in either hemispheres, is always more marked than the one on the east side. When I say *more marked,* I mean that the *meeting* of the two winds may be seen more distinctly: as, for instance, in the North Atlantic, when the wind *flies* from S.W. to N.W. This takes place at the confluence on the west side of the centre. On the east side, where the winds forming a confluence are from the S.E. and S.W., their meeting is not so distinctly marked. The wind seldom *flies* from S.E. to S.W. – it more usually *veers,* but rather more quickly than at any other part of the storm. The reason is simply this: the S.E. and S.W. winds, flowing out from the equator, are more highly charged with vapour, and are warmer than a westerly or northerly wind, and consequently have a greater tendency to ascend and to mix with each other. A N.W. wind, on the contrary, is a dense, cold wind – a descending wind – and on coming in contact with a S.W. wind, it condenses the vapour which the latter has taken up, thus producing torrents of rain at their junction. Here we have one material proof that the winds do not blow in circles: for, on passing from one wind into another, as from a S.W. into a N.W., a long bank of *nimbus* (rain-cloud) may often be seen stretching away from S.W. to N.E., or with the wind just left, and at right angles to the wind entered, before which it is being rapidly driven; showing distinctly that the two winds are in contact, and blowing at right angles to each other. This could not be the case if the wind blew in circles – if it was one continuous current of air blowing round a centre.

Form or Shape of a Gale: – In extra-tropical latitudes they become extremely elongated: so much so, that the length is often equal to twice or three times the breadth. Within the tropics, they are no doubt very nearly circular at times: but the trade-wind, increased in force, often forms a part of the hurricane itself. Hence it is difficult to define the exact shape. For all practical purposes, the diagrams are sufficiently near the truth.

According to the *Circular Theory,* the air forming a hurricane is distinct from, and forces [its] way through, the surrounding air. This I deny. A hurricane is formed by the local atmosphere, set in motion by the effect of electricity, and perhaps magnetism.

The Direction of the Wind cannot be taken as a guide to the bearing of the *centre.* The bearing of the *centre* may vary from one to sixteen points from the direction of the wind. If the centre always bore eight points from the direction of the wind, and if the *centre* was the only dangerous part of a storm, then the Law of Storms would really be what many persons, whose avocations do not require them to apply it practically, appear to consider it: – a simple science, requiring no great depth of thought or mathematical knowledge – such as may be acquired by the most ordinary capacity, or applied by the merest tyro, when necessary. Such, however, is not the case. The centre is not the only dangerous point; the direction of the wind is no guide as to the bearing of the centre; the barometer does not always fall when we are advancing on the centre, neither does it always rise when we are receding from it.

THE BAROMETER is no guide as to the distance of the centre. It is generally lower at one confluence than the other. Its rise or fall in a storm depends upon the way in which the atmosphere is charged, and whether the storm-currents are ascending or descending. The greatest *rise* will take place near the confluence on the north or west side of a storm in the northern, and on the south or west side in the southern hemisphere. The greatest *fall* will take place at the confluence on the opposite side of the storm, in both hemispheres. The cause is easily explained. Suppose a storm to be travelling to the westward in the northern hemisphere: the two currents of air forming the western confluence will be, one from the north and west, the other from the north and east. Both are cool, dense winds, neither has any tendency to ascend; consequently one prevents the other from passing freely; one dams the other up; the particles are pressed closer together; the whole presses more heavily on the earth's surface; and the natural consequence of this is, "*the Barometer rises above a mean height.*" The winds forming the opposite confluence are from the south-east and south-west; both are highly charged with moisture; both readily ascend. The barometer *falls* accordingly, and will seldom, if ever, rise above a mean height at this part of the storm. The same rule applies to gales travelling to the eastward; it is also applicable to the southern hemisphere, by changing north to south and vice versa; east and west remain the same.

In the North Atlantic I have often seen the barometer rise above a mean height when near the verge in the rear, but never when in front of a storm, except when in one travelling to the westward.

Within the tropics, a considerable fall may take place at both confluences, since both currents are highly charged with vapour. Its amount, however, is no sure guide as to the force of the wind likely to be encountered. When it is great, a strong current of air usually flows in to restore the equilibrium, and a ship situated some distance from the point of least atmospheric pressure, may feel the force of this current more than one immediately under it. Consequently a ship, having only a slight fall of the barometer, may have a heavier gale than one on board of which the barometer has fallen considerably. It follows that the least change in the height of the mercury, and any disturbance in its regular ebb and flow should be carefully noted, and its cause ascertained if possible. A rise or fall of $^{5}/_{100}$ of an inch will often indicate the approach of a storm as clearly as a fluctuation of one inch will at other times. It is a common complaint that "the barometer gave no warning of the approach of a hurricane." I doubt much whether such a case ever really occurred, and I believe such complaints have arisen, not from any real fault of the barometer, but from the erroneous opinions entertained respecting the influence of the atmospheric currents on that instrument. All such childish marks as *Fair, Changeable, Stormy,* &c., should be effaced from all barometers intended for nautical purposes, since they are only calculated to mislead, and give an erroneous idea of its real movements.

PRECURSOR WAVE. – So common is it to meet with *a heavy swell* in the trade-winds, and also in extra-tropical latitudes, that little dependence should be placed on this phenomenon, as an indication of a coming storm. Heavy swells and rollers are often produced without the aid of strong winds. Mr. Piddington tells us (*Horn Book, p.* 205) that Colonel Reid gives an interesting instance of the excessive rise of the barometer and precursor wave before a cyclone – "being that of the *Medway* steamer, in which he (Col. Reid) was a passenger, which vessel experienced fine weather on the eastern side of the Atlantic, but a long, low swell coming from the N.W. prevailed, being at first westerly, then increasing and coming nearly from the north, on the 1st October, in Lat. 46° N.; Long. 14° W. This swell was proceeding from a cyclone raging on the western side of the Atlantic."

Is it possible to conceive anything more absurd, than to suppose that, *because* there was a swell from the N.W., in Lat. 46° N., and Long. 14° W., it *must* have proceeded from a cyclone raging on the western side of the Atlantic? In the first place, if the swell proceeded from a cyclone raging in the portion named, ought it to have come from the N.W. and northward at all? Ought it not to have come from the westward? The fact is, the westerly, north-westerly, and northerly swell, together with the excessive rise of the barometer, plainly indicated that the *Medway* was crossing the tail-end of a cyclone, or gale, the body of which had already passed to the N.E. Thus, whilst the Colonel was looking for the enemy in front, it had out-flanked him, and was nearly in his rear. Now, what would be the fate of an army placed in such a predicament? I imagine it would run the risk of being severely handled, if not destroyed. And have not numbers of ships been exposed in the same manner, through their commanders trusting to the absurd *Circular Theory*; through trusting to the absurd idea that the *centre* was the *only* dangerous point; that that centre *always* bore about eight points from the direction of the wind; and that the barometer *always* gave ample warning of its approach? I venture to say that scores have been so exposed. Numbers of ships have been taken aback by the sudden shifting of the wind, thrown on their beam-ends, dismasted, and perhaps destroyed, at a time when their commanders least expected it; when they imagined (as was really the case) that they were miles away from the centre. Ask any *Circular Theory* gentleman to explain how it happened, and his answer would be, "Oh! it was from the incurving of the wind;" or "The gale had recurved;" or in fact, anything but the truth.

A sudden shift of wind may occur, without either the ship or the gale altering its position, from the effects of the celestial bodies on different parts of the gale. The moon *repels* aqueous vapour, and, consequently, strengthens or increases the force of a wind flowing from either pole towards the equator. This is more especially the case when she is in the same hemisphere – thus: suppose a ship to be at A, (Fig. 1., N. H.,) with the wind at west or west-south-west; the moon's rising (especially from first quarter to foil moon) would strengthen the opposite current B and weaken A; consequently B may force its way under A, compelling it to ascend, thus bringing a sudden shift of wind to the ship at A; and it may blow furiously for a few hours, gradually dying out as the moon draws towards the western horizon, become nearly calm, and then the original current A will spring up again. Here we have the cause of the "*flying in*" and "*backing*" of the wind so often met with in the North Atlantic, and even in the English Channel. The effect of the sun is the reverse of that of the moon, with certain exceptions, to which I shall refer hereafter. The sun *attracts;* the moon *repels.* Veerings of the wind, from the causes above mentioned, take place at both confluences of the storm, but the wind seldom flies so suddenly on the eastern, as it does on the western side of a gale. It veers and hauls more gradually.

I have said that every gale has two sides, formed by two distinct currents of air. Now each of these MAIN-STREAMS is formed by a number of belts or veins *(minor streams)* running nearly parallel to each other. These belts or veins are often very distinctly marked by long bands of cloud *(cumulus)* stretching *with* the wind *at the sides* of a storm, but often *at right angles* to it *near the confluences.* The *force* of wind varies considerably at times in the different veins. The *direction* of the wind also varies considerably in passing from one belt to another, often as much as two points. Sometimes the wind between the belts of cloud is drawn towards, at other times repelled from, them. These belts often appear to converge in the distance, sometimes at one end only, sometimes at both. This however is only apparent, for, cross them where we will, they present the same aspect: they appear to converge in consequence of their being curved. Particular attention should be paid to these belts of cloud, since they often indicate pretty nearly the position of a ship in a gale. At the *sides* of a gale they stretch in long, well-defined bands *with* the wind, and *without any other clouds above them;* at either *confluence,* they either *cross* the wind, or we may observe *another stratum of cloud, at a greater elevation, crossing them at right angles.* Every attention should be paid, at all times, to the form and direction of the clouds, since they often indicate the approach of a storm long before the barometer, or anything else. No matter in what form they may appear, whether cumulus, cumulo-stratus, cirrus, cirro-cumulus, or stratus, the course they are pursuing should be carefully noted. If there be two strata, note if they cross each other – and if so, in what direction. Or if there be one stratum only, and the clouds forming it are coming from a different direction to the wind, note accordingly – project them, and compare with a diagram of the same hemisphere – and in this way a pretty good idea will be obtained of the position of the ship or place of observation with reference to the position of the gale.

SUDDEN GUSTS OF WIND IN A GALE OR HURRICANE. – These are produced by one current passing over another and descending, in the same manner as when the wind is blowing "*off shore*" where the land is high. When this is the case, sudden gusts strike the water between the shore and the place where the true wind is blowing, which is at some distance from the shore. The intervening space remains in a state of calm, interrupted only by the gusts above mentioned and by eddies. Here we have a fac-simile of what takes place in a gale or hurricane. Let us suppose a ship to be at C, (Figure 2, N. H.) she may have a heavy gale interrupted by a sudden gust in consequence of its passing down from over the opposite current. As the gale advances over her, on her arriving at A it will become nearly or quite calm, and remain so until the opposite current or side D strikes her in all its fury; the wind being, perhaps, from a point nearly opposite to that from which it was blowing at C. On the other hand, when the wind blows "*dead on shore,*" where the land is high, it often falls light near the shore – often it is nearly calm – or it is deflected from its original course and blows along the land: thus eddies and whirlwinds are often formed. The same effects are produced by one current of wind blowing dead on to the other; (as at B, Fig. 1, N. H.) the upper portion of the current passes over the other, whilst the under portion is either repelled along or carried round with the opposite current, as shown by the small arrows in the diagram.

[. . .]

THE CIRCULATION OF THE ATMOSPHERE.

A treatise like the present would, of course, be extremely imperfect if it took no notice of the general system of Atmospheric circulation, on which all connected with Storms must depend; yet in touching upon this subject, I do so with considerable diffidence. The theory I am about to state, I by no means wish to state dogmatically; for far more time and data would be required to establish it in all its details than I have now at my command. At present I only suggest it as a matter of discussion, and as a theory which appears to me probable in itself, free from the gross inconsistencies which may be pointed out in those of Captain Maury and others, and capable of affording a satisfactory explanation of a greater number of atmospheric phenomena than any other system with which I am acquainted.

I believe then that the Circulation of the Atmosphere depends entirely on the electric or magnetic currents produced by the influence of the celestial bodies, and especially by that of the sun. The rays of that luminary generate currents of electricity, and these – currents of air. The laws of equilibrium require the existence of a counter current for every stream thus generated; for the air in flowing off not only leaves a partial vacuum at the point of departure, but displaces a portion of the atmosphere in its progress, which, flowing outwards, has a natural tendency to recurve and fill the vacuum caused by the outflowing current.

Whether the above be the cause or not, the following general Laws will be found to obtain: – A current of air, generated in the warmer, and passing off towards the cooler regions, in a north and westerly direction, (in the Northern Hemisphere) becomes cooled, and is gradually repelled, so that it assumes first a westerly and then a south-westerly direction, thereby returning to the warmer regions. A current flowing from the cooler towards the warmer regions, in a south-easterly direction, as it becomes warmer, is repelled to the E. and N.E., and it thus returns to about the same parallel from which it started. A current flowing out from the warmer towards the cooler regions, in a north-easterly direction, is not repelled but rather attracted, its course thus becoming more northerly as it advances; whilst in the same way, a current flowing from the cooler towards the warmer regions, in a south-westerly direction, becomes more southerly.[3] The above laws apply to the Southern Hemisphere if we interchange the words South and North; East and West remaining the same; and they apply to all systems of winds whether stationary or progressive.

Trade-winds and monsoons are nothing more than different portions of stationary systems or gales. The S.W. monsoon and S.E. trades of the Indian Ocean are two distinct currents. I at one time thought otherwise, but ocular demonstration has convinced me that this is the case. The crossing of the winds and points of contact are clearly marked, and it is by these crossings, &c., that the trades are at times rendered so squally and uncertain. The westerly stream near the equator is common to both Hemispheres, and it is between this and the counter-current (the trades) that the hurricanes are formed, whilst most common gales are formed at the trade confluences, the currents generating them taking the general direction of the main streams. Gales, however, may be formed wherever decomposition of water is taking place to any extent; for the gases by ascending will generate an electric current, and this again an air current as before described.

The changes which take place from the influence of the sun and moon, cause the same current at one time to ascend and at another to descend, and it is during these changes that most of the gales and hurricanes are formed. The *sun's crossing the line* is often attended by violent storms known as *the equinoctial gales.* These are produced by the influence of the sun and moon becoming suddenly reversed as it were, whilst at the same time those bodies act more powerfully in opposition to, or in conjunction with each other, when near the equator, than at any other time.

Captain Maury, in his *Physical Geography of the Sea,* attributes the N.E. trades and S.W. monsoons to the effect of the earth's diurnal rotation. But how a N.E. and S.W. wind can both be produced in this manner, is more than I can comprehend. He says that a current of air in flowing from the tropics towards the equator is carried to the westward, thus producing the N.E. and S.E. trades, but that in flowing out from the latter to the former it is carried to the eastward, and thus produces the S.W. monsoon in the Northern, and N.W. monsoon in the Southern Ocean. I can of course easily understand how, IF *the earth slips from under the atmosphere,* a portion of air flowing from the Tropic of Cancer may appear to take a S.W. direction, and thus produce a N.E. wind. But how the same portion of air, in passing from the equator to the Tropic of Capricorn, can be impelled to the S.E. by the same influence, namely, diurnal rotation, I cannot conceive. For the fluid would not only have to flow to the eastward faster than the parts of the earth's surface over which it is passing, but also faster than the point at the equator from which it set out, so as to bring that point to bear N.W., otherwise in flowing off, it could not produce a N.W. wind.

Speaking of the *red dust* which falls near the Cape Verde Islands, he says: "They (the Cape Verde Islands) are situated in the direction which theory gives to the upper current flowing from the Oronoco and Amazon with its *rain-dust,* and they are in the region of the most frequent showers of *rain-dust,* all of which, though they do not absolutely prove, are yet strikingly in conformity with this theory as to the Circulation of the Atmosphere." Now, how is it possible that dust, or air, or anything in fact can be carried from the Oronoco or Amazon to the Cape de Verdes by the effect of diurnal rotation? The course is in direct opposition to it. If the earth, in rotating, slips from under the atmosphere (which I much doubt) then the air, and, consequently, anything floating in it, should appear to move from east to west. If the *actual* direction of a portion of air, set in motion as the Captain supposes, be northerly, as from the equator towards the pole, then its *apparent* direction should be to the N.W., and it should advance upon the pole in a spiral from the S.E., and if the *actual* direction be from N. to S., *i.e.* from the pole towards the equator, then its *apparent* direction should be to the S.W. (We are of course speaking of the Northern Hemisphere.)

The Captain says, "The better to explain, let us take a terrestial globe for the illustration. Bring Madeira, or any other place about the same parallel, under the brazen meridian: put a finger of the left hand on the place, then moving the finger down along the meridian to the south to represent the particle of air, turn the globe on its axis from west to east, to represent the diurnal rotation of the earth, and when the finger reaches the equator, stop. It will now be seen that the place on the globe under the finger is to the southward and westward of the place from which the finger started; in other words, the track of the finger over the surface of the globe, like the track of the particle of air upon the earth, has been from the northward and eastward." This is undoubtedly correct; but if the finger be continued along the brazen meridian towards the Tropic of Capricorn to represent the air flowing from the equator towards the pole, turning the globe as before, and stopping on the finger arriving on the tropic, will the place on the globe under the finger be to the S.E. of the place on the equator from which the finger set out? Undoubtedly not. It will be S.W. of it, as the place on the equator was S.W. of Madeira.

Thus far Captain Maury, in endeavouring to assign a cause for the N.E. and S.E. trades, has considered the atmosphere as a spherical shell in a state of rest, and not partaking of the earth's motion of rotation. How are the westerly winds to be accounted for? He supposes his particle of air, in passing out from the equator, to have an initial velocity of rotation; or in other words, the earth is no longer slipping from under it. Instead of pausing now to discuss the effect of such an initial velocity (which be it remarked is an *angular* velocity) on the path of a particle *over the surface of a sphere,* we will only ask the following preliminary questions: – If the atmosphere be in a state of rest, and not revolving with the earth, *how* is this initial velocity acquired? *Where* is it first acquired? Why should it he rather acquired at the equator than at any other point? And if acquired gradually and previous to arriving at the equator, what becomes of the apparent westerly motion and the formation of the easterly winds? The atmosphere cannot at the same time be rotating and not rotating, or dependent on the earth's motion at the equator, and free everywhere else. Yet Captain Maury's theory would demand this. We would urge the above questions on his careful consideration, and until they are satisfactorily cleared up we would recommend him, *preux chevalier* though he may be, not to throw down the gauntlet, and challenge all comers in defence of his theory of the Circulation of the Atmosphere, ingenious as it undoubtedly is, and only needing one thing to recommend it – namely, to be supported by Possibility, and by Facts.

FIG. 1.

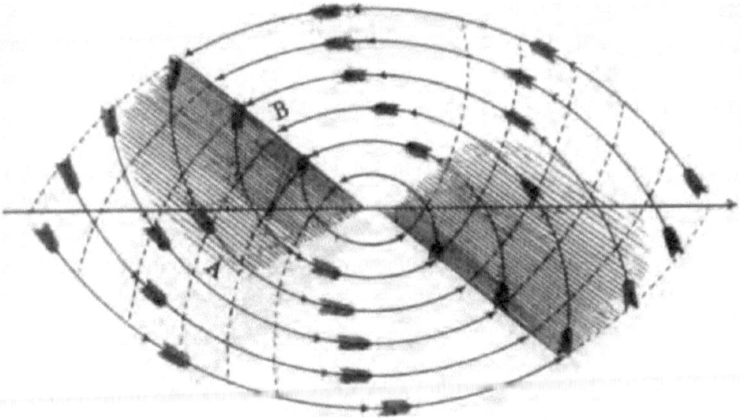

FIG. 2.

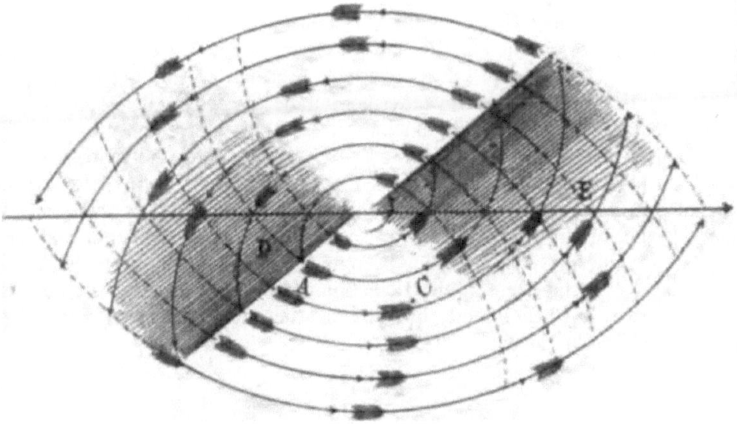

NORTHERN HEMISPHERE.

The *dust* which falls near the Cape Verdes most likely comes from the neighbourhood of the Oronoco or Amazon, but the earth's diurnal motion has nothing to do with its passage. It is carried across by the current forming the right-hand side (looking to the eastward) of an immense stationary gale, the counter current to that forming the so-called N.E. trades. It is in the electricity which forms the life-spring of this and all other currents that we must seek the force by which the dust is carried up and borne onwards. As this electric current becomes checked in its course to the E. and N.E. the *dust* gradually descends, and a portion of it is carried far out into the Atlantic by the displaced air which is flowing out to the westward.

It is deposited by a different current from that by which it was carried up, in the same way that rain is deposited by a different current from that which carried up the vapour. A vapour-bearing current is not always a rain-bearing current, for it must meet with a condensing medium in the form of a counter-current or cold surface, before it can deposit its vapour in the shape of rain, snow, &c. And as it often happens that the vapour-bearing flows over the condensing stream, the vapour is condensed between the two, so that the rain, &c., falls into and is borne onwards by the latter, which is the counter-current to the one that carried up the vapour. Thus, in England, an easterly wind is often a rain-bearing wind, whilst the moisture has been taken up by one from the south or south-west.

Source Endnotes

1 The strictest observations of natural phenomena, leads us to the conviction that a current of air is *never* set in motion without the formation of what we may term a *Complementary Current,* opposite in its direction to the one just originated, and crossing it in two points, though not necessarily in the same plane with it. The object of this Complementary Current is to restore the equilibrium of the atmosphere. As every wind forms one side of such a system [illegible] we have just described, we designate such a pair of currents A GALE. The author, however, wishes it to be distinctly understood, that in using this term, he has no wish to convey any idea of the relative force or velocity of the wind in either current. These may be either faint airs, or may blow with hurricane violence.

2 I am satisfied that many of those which have been described by some writers, as double Hurricanes or Cyclones, were nothing more than the two confluences of the same storm.

3 I have said "the currents are repelled or attracted as they become cooler or warmer." Perhaps I should rather have said, as they become positively or negatively charged; since I believe heat to be the effect of electric action. I do not believe, for instance, that the sun's rays convey any heat directly, but only that they render the electricity of the earth active, and this produces heat.

2.4

SEMI-LAGRANGIAN

Editorial Headnote

2.4 Semi-Lagrangian

Although Robert FitzRoy and George Jinman adopted the Lagrangian approach to weather forecasting and cyclone modelling, moving away from the Eulerian approach, there was another weather scientist, Francis Galton, for whom the appropriate method was a middle way between the two: the semi-Lagrangian approach. His method was based on the representation of the flow field by streamlines and the elaboration of glyph and iconic maps and schematic micromaps. The three Galton texts reproduced here illustrate the advantages of combining the two approaches and obtaining an instantaneous picture of the overall flow field and the other meteorological elements.[1]

Francis Galton (1822–1911), a British polymath and pioneer in fields as diverse as statistics and eugenics, as well as a botanist and leading geographer, became interested in weather mapping after his African expeditions in the early 1850s, during which he practised geographical cartography.[2] His first foray into meteorology was a retrospective chart of the weather in England in 1861, which was privately printed by his friend, mathematician William Spottiswoode.[3] His 1863 book, *Meteorographica or methods of mapping the weather,* which he published at his own expense, presented an original system of symbols that advocated for ensuring that what he conceived to be "iconic weather maps" gave a three-dimensional picture of the weather's progress. Galton was a promoter of initiatives and projects who, two years earlier in July 1861, had sent a circular to many European observatories asking them to telegraph him the data of their observations so that he could collect them and use them to produce weather maps for December 1861. He received invaluable help from the Netherlands, Belgium, Austria and Germany, and much less help from France and Italy (precisely the two countries where the Eulerian approach would gain the most strength).[4] The introduction to the work and the maps reproduced here describe a new system of figures and symbols that more fully express the power of the visual sense: while barometric and thermometric readings are shown in figures, wind direction and strength and cloud cover are represented by circular and hexagonal symbols (arrows, marks and shaded types, respectively). Galton is interested in the representation of iconic barometric pressure maps – rather than isobars or traditional contour lines – because while wind direction is easy to represent (since it moves in the plane of the earth), the other elements need to be plotted in 'a third dimension' (p. 5). With the presentation of barometric maps or grids, which contain symbols with shaded types, he aims to apply 'a similar pictorial treatment to that employed for geographical elevations', i.e. barometric elevation and depression zones and other features should be 'pictured by contour lines and shadings, on the same principles as mountains and valleys' (p. 5). In doing so, Galton seeks to introduce the visual and geometric language developed by Edmund Halley, Léon Lalanne and other cartographers and engineers, finding that its applicability (and its vivid pictorial treatment) results in iconic, three-dimensional images of barometric pressure distribution.[5]

In the article 'Recent Weather', published in *The Reader* (a weekly magazine on literature, science and the arts) on 19 December 1863, the micromap representation usurps isobaric cartography. Galton expresses great perplexity at the 'audacious dogmatism' of many meteorologists using the isobaric line technique (p. 730). Through observation of the westward movement of storms, Galton perceives that 'any prevision of the weather founded on' such a westerly translation must have been 'wholly in error' (p. 730). Galton explains this by the fact that there is an undoubted tendency for storms to move westward, so that their character is so modified that the barometric lines of one day have no resemblance to those of the previous day. The inadequacy of isobaric lines to timely represent storm motion, a trope identified with views critical of the Eulerian approach, is central to this text. By opting for symbols and figures instead of isobaric lines, Galton seeks not so much to discredit the value of the latter as to convey more information through schematic representation.

Galton's study on 'A Development of the Theory of Cyclones', which was published in the *Proceedings of the Royal Society of London* in 1863 (although it was received on 25 December 1862), is a more theoretical articulation of the same semi-Lagrangian perspective. Analysing his high-dimensional graphs and streamline micromaps, Galton notices a completely unexpected pattern: whereas in low-pressure areas the winds rotate counterclockwise in inward spirals (as expected in cyclones), in high-pressure areas the winds rotate clockwise in outward spirals. Galton calls these calm areas *anti-cyclones*, as opposed to cyclone patterns. An area of high pressure is thus defined as

> a locus of dense descending currents, and therefore of a dispersion of a cold dry atmosphere, plunging from the higher regions upon the surface of the earth, which, flowing away radially on all sides, becomes at length imbued with lateral motion.
>
> (p. 385)

These three texts by Galton consider the foundations of the adoption of the semi-Lagrangian approach in different conceptual respects. The first, *Meteorographica*, takes the most ambitious cartographic view, showing iconic and glyph maps and schematic micromaps representing streamlines or the flow field. The second text questions the sufficiency of isobaric lines to adequately represent storm motion, thereby justifying the introduction of new symbols and figures. While the third proposes the new concept of an anticyclone based on the observation of schematic micromaps. Each considers a different facet of the semi-Lagrangian perspective, and in certain aspects, such as the anticyclonic concept and streamlines, they are decades ahead of twentieth-century dynamic meteorology.

Notes

1 Aitor Anduaga, *Politics, Statistics, and Weather Forecasting, 1840–1910. Taming the Weather* (London and New York: Routledge, 2020), pp. 186–92.
2 Derek Forrest, *Francis Galton: The Life of a Victorian Genius* (London: Paul Elek, 1974); Karl Pearson, *Life, Letters and Labours of Francis Galton* (Cambridge: Cambridge University Press, 1940–1930), 3 vols.; John C. Waller, 'Gentlemanly Men of Science: Sir Francis Galton and the Professionalization of the British Life-Sciences', *Journal of the History of Biology*, 34 (2001), pp. 83–114.
3 Nicholas Wright Gillham, *A Life of Sir Francis Galton: From African Exploration to the Birth of Eugenics* (Oxford: Oxford University Press, 2001), p. 142.
4 Crispin Tickell, 'Meteorographica and Weather', in Milo Keynes (ed.), *Sir Francis Galton, FRS. The Legacy of His Ideas* (New York: Palgrave Macmillan, 1993), pp. 140–51, on pp. 54–61.
5 Michael Friendly, 'The Golden Age of Statistical Graphics', *Statistical Science*, 29:4 (2008), pp. 502–35, on pp. 514–16; Katharine Anderson, *Predicting the Weather: Victorians and the Science of Meteorology* (Chicago: The University of Chicago Press, 2005), pp. 200–04.

18

METEOROGRAPHICA, OR, METHODS OF MAPPING THE WEATHER

Francis Galton

Source: Francis Galton, *Meteorographica, or, Methods of Mapping the Weather: Illustrated by Upwards of 600 Printed and Lithographed Diagrams Referring to the Weather of a Large Part of Europe, During the Month of December 1861*, (Cambridge: Macmillan, 1863), pp. 3–6.

METEOROGRAPHICA.

THE accompanying Charts and Maps are contributions to a branch of Meteorology which is theoretically divisible into two separate portions, though they may be more or less united in practice, and it is convenient to class them under a single term, – Meteorography.

I mean by that phrase, 1st, the art and practice of tabulating Observations which have been made simultaneously at numerous stations, each record being inscribed in the geographical position of the place where it was made; and, 2ndly, the subsequent step of delineating the General Results of the Observations in a pictorial form.

When lists of observations are printed in line and column, they are in too crude a state for employment in weather investigations; after their contents have been sorted into Charts, it becomes possible to comprehend them; but it requires meteorographic Maps to make their meaning apparent at a glance.

The necessity and character of the double process I have described, may be illustrated by the well-known methods employed in Geography. If lists of latitudes, longitudes, altitudes, angles, and itineraries are laid before a Map Maker, he is able to elaborate the shapes of mountain ranges, the courses of rivers, and the irregularities of coast; but to arrive at these desired results he must *protract* his materials, by placing each datum upon a chart in its proper geographical position. His next labour is critical and artistic; he has to weigh conflicting observations, to make the best of omissions and obvious irregularities, and, finally, after deducing his results, to delineate them in a map, in order to give permanence to his ideas, and to make them intelligible to others.

DOI: 10.4324/9781003650294-29

A scientific study of the weather on a worthy scale, seems to me an impossibility at the present time from want of accessible data. We need meteorographic representations of large areas, as facts to reason upon, as urgently as experimental data are required by students of physical philosophy.

Meteorologists are strangely behindhand in the practice of combining the materials they possess. There are more than 300 skilled observers, using excellent instruments, scattered over Britain and the Continent, who transmit observations taken thrice daily, to Meteorological Societies or Government Institutes. Besides these, are the same number of Lighthouse keepers, who transmit good returns, of at least wind and cloud, to one or other of the three great Lighthouse Boards of Britain. Lastly, many observers publish independently. Yet throughout this mass of labour that practice of general combination is absent, which is required to utilize it as it deserves. No means exist of obtaining access to any considerable portion of these observations, without great cost, delay, and uncertainty; much less are they to be obtained in a "reduced" and never in a meteorographic form. The labour of a meteorologist who studies the changes of the weather is enormous before he can even get his materials into hand and arrive at the starting point of his investigations. In the ordinary course he has to apply, with doubtful chance of success, to upwards of 10 Meteorological Institutes in Britain and Europe, for the favour of access to the original documents received by them, and to fully 30 individuals besides. He has next to procure copies, then to reduce the barometer and thermometer readings to a common measure, and, finally, to protract them on a map.

I feel that all this dry, laborious, and costly work, which has to be undergone independently by every real student before he can venture a step into the scientific part of his work, is precisely that which should be undertaken by Institutes established for the advance of Meteorology. A debt of gratitude is due to Admiral Fitz Roy for his methodical daily reports of observations made at 20 English and 5 Foreign stations, but I believe them to be insufficiently numerous, extended, or frequent to afford a just knowledge of the winds even of a general kind; and I think it would have been a most valuable addition to the octavo Blue Book (I trust it will be annual) published by the Board of Trade if it had included an extended list of observations collected at leisure, instead of a mere reproduction of the telegraphed weather reports. The materials existed in readiness for use, and the expense of printing them would certainly not have been more considerable than that of the copious extracts from logbooks, which were inserted, and which they would have advantageously superseded. Records of this description would be many times increased in value if, instead of being printed in lines, they were printed compactly in squares so as to fall in some degree under my first division of meteorographic charts.

Thus, London 29.27 NNW 2. VI. 34 1. to stand for	Bar. 29.27.	Wind N.N.W.	Force 2.	Clouds VI.
	Therm. 34°.	Excess above Wet Bulb 1°.		At London.

Or better still, if an arrow head replaced the letters N.N.W. I am supposing the tables to be printed in line and column, each square containing the observations made at a station whose position is *somewhere or other* within the square, and more accurately determinable by a standard chart of reference, or by the inscription of latitudes and longitudes, instead of a name, in the case of sea observations. Still more desirable would it be if, in addition to numerical tables, the Board of Trade were to publish a series of small meteorographic Maps, like those I append, and which I will shortly describe.

A sustained series of publications of this kind, extending over two or three years, would give an extraordinary impetus to the scientific study of Meteorology. They would supply the necessary materials in a manageable form, for arriving at a general knowledge of the distribution of the various elements of the weather; they would afford means of testing the extant theory of "forecasts" with a rigour impossible at the present time, and they would necessarily improve it.

If extensive tables of reduced observations were issued in England, we might look for the co-operation of Meteorological Institutes on the continent (who already publish voluminously) in following our example. The conflict of different Measures would create no serious difficulty. All Germany uses Reaumur and Paris Feet, as France, Belgium, Holland, and Switzerland use Centigrade and Metres. Europe is thus divided into three great sections, each maintaining a common system of Measures throughout its whole area, and so far complete in itself. Each section might therefore be represented by separate tables, and it would be no serious trouble to collate the three together, if they were printed on the same scales and projections. A Meteorological, like a Geographical, Map is independent of Units of Measurement.

Entertaining the views I have expressed on the necessity of meteorological Charts and Maps, and feeling confident that no representation of what *might* be done would influence Meteorologists to execute what I have described, so strongly as a practical proof that it *could* be done, I determined to make a trial by myself, and to chart the entire area of Europe, so far as meteorological stations extend, during one entire month, and I now publish my results. They consist, 1st, of a printed series of morning, afternoon, and evening weather charts, of each of the 31 days of December 1861, making 93 maps in all, each of which contains about 80 stations. 2d. A printed series of 14 maps on a fourfold scale, one to each open page of the book, with the object of illustrating in greater detail some portion of the weather changes there represented. 3d. A printed summary of the barometric heights corresponding to each of the 93 maps, also included in the body of the book. These are followed by – 4th, A lithographed map in red and black of the barometer, thermometer, rain, and wind, corresponding to each of the 93 maps. 5th. A photo-lithographic reduction of a tentative MS. diagram, showing the wind currents during each of the 93 epochs of observation. This is based on the printed materials alone. 6th. A photo-lithographic reduction of seven maps, showing the nature of the changes of the wind in Britain during a gale that occurred in the course of the month, which happen to be imperfectly represented in the ordinary charts. 7th. A photo-lithographic reduction of four of my ordinary MS. wind charts, to illustrate what I shall hereafter describe as "anti-cyclones." With the exception of the 2nd and 6th of the above series, *every* map includes the *same* area. The maps and charts have been drawn on various scales, but each separate set has a map of Europe annexed to it, on the scale used throughout the set.

Sources of Information. – After communication with most of the leading meteorologists of the continent (to many of whom, especially Professors Quetelet, and Buys Ballot, and the late Professor Kreil, I am under great obligations), I forwarded circulars and blank schedules to their separate stations, asking for reduced observations during the then coming month of December, and promising in return, to those who fulfilled the conditions of my circular, a copy of the resultant charts. Observers are scattered unequally over Europe, and in the case of those who promised me returns, the inequality was still more conspicuous. I had therefore to supplement deficiencies as I best could, partly by the special kindness of men of science and partly by copies of documents transmitted, not to me, but to the different meteorological Institutes.

The result was, that I could not find enough stations to fill every part of a chart of Europe, if it were printed on a larger scale than that which I have adopted, but I had materials for a four-fold (superficial) scale in the case of Germany and about a six-fold in that of Britain. I therefore compromised the difficulty of disposing my numerical materials to the best advantage, by issuing a complete series of small general Charts and occasional illustrations of portions of their area on a larger scale, instead of a few large comprehensive sheets, as I had originally intended. I also issued the numerous lithographs which I have already described, in which the whole of my materials, whether published or unpublished, has been more or less utilised. I think the stations in my main series of printed charts are close enough, if the vacant areas were filled up, to give a just general idea of the barometer and thermometer; and that the scale of the 14 side maps is nearly, if not quite sufficient for wind and rain.

It may interest those who are not acquainted with the present activity of Meteorologists, to learn the number and distribution of regular stations in Europe, where accurate observations are made thrice a day. In addition to those mentioned in the following list, are a large number of individuals who observe regularly and publish occasionally, but do not transmit their observations to any Institute. This list is approximative only.

Number of Observers.		
62	British Meteorological Society includes England and Wales, but neither Ireland nor Scotland.	
65	Scotch Meteorological Society.	
Lighthouses	Trinity House, for English lighthouses, Northern Lights, Edinburgh, for the Scotch, Ballast Board, Dublin, for the Irish,	No returns of the weather are permitted to be issued by the light-keepers to private persons, except under the especial authority of the several Boards.
6	Royal Observatory, Brussels. Professor Quetelet.	
6	Royal Observatory of Utrecht. Professor Buys Ballot.	
80	Meteorological Society (The Government Statistical Office), Berlin. Prof. Dove.	
120	Meteorological Institute of Vienna. (The late Professor Kreil.)	
None within the area of my Charts	Meteorological Institute of St. Petersburgh. Professor Kuppfer.	35 Stations in Europe and Asia.

The 25 returns issued by the Board of Trade, and the 50 others, printed for private circulation by M. le Verrier, are only made once a day, and are therefore excluded from this list; neither do I mention the continental lighthouses, for I do not know how far the weather is observed with regularity in any of them. I might have added the logs of ships, of which many are forwarded to the Board of Trade.

There was no central institute in Switzerland, though there are many excellent observers, and there were schemes for instituting a wide system of Swiss simultaneous observations; neither was there any recognized Institute in Denmark or Norway. Whether by accident or misunderstanding, several promised communications from Denmark have never reached me, to my great regret, for its weather was closely linked with our own. From Sweden I could obtain nothing, from France next to nothing, from Bavaria only the valuable observations made at Munich. From Italy I had considerable hopes held out to me, but little fruit. The interior of Ireland is wretchedly represented, and would have presented a gap, like France, were it not for two eminent astronomers and some chance assistance besides. This is a serious deficiency, for Ireland holds a very important meteorological position in respect to England. I was also unfortunate in the accidental absence of reports from lighthouses on its eastern coast.

Every schedule of returns bears, as Meteorologists are aware, some intrinsic evidence as to its own value, which is strengthened almost to certainty by collating it with returns from adjacent stations. Now, I can most gladly testify, that of the numerous light-keepers whose returns, by the especial kindness of the Irish and Scotch Boards, were transmitted to me, and to which, by the favour of the Trinity House, I was permitted access, the great majority were distinguished by a most painstaking exactness. I believe they would make valuable observers of *all* the elements if their barometers were put in order, their thermometers properly exposed, and instructions given them to reduce their barometric readings to the nearest hundredth of an inch. I have used several of their barometric returns in compiling my symbolic tables, though they were not sufficiently accurate to print in the charts.

The lighthouses on the French coast (whose returns failed me at the last), and those of the North Sea and Baltic, ought to supply a valuable cordon of meteorological observers.

Charts of Numerical Record.-My 93 printed charts contain a selection from my materials, printed in their proper geographical positions; and they do a little more than this, because I have represented pictorially two of the weather data which admit of being accurately expressed by symbols. It is evidently quite as precise to use the symbols 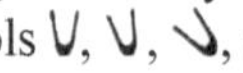, as the letters N, NNW, and NW, while the former have the advantage of telling their tale directly to the eye. With similar precision do the series  express five states of the sky,-entirely clouded, mostly, half, a few clouds, and none at all; and these are as many varieties as I cared to insert. The difficulty is greater with the other records. We want to depict the barometer to the nearest hundredth of an inch, while its range exceeds 2½ inches; in other words, we have upwards of 250 possible variations to register. As these are far too numerous to be attempted by depths of shading or differences of colour, we have no other option than to make use of numerals. As with the barometer, so in a less degree with the thermometer; the distribution of moisture is perhaps to be managed graphically. Again, the size of the chart increases in direct proportion to the share occupied by the individual symbols, and a volume of large maps is a serious inconvenience, it is too embarrassing for frequent use, it requires space to be spread out upon, trouble and even strength to handle it, and the maps cannot be examined without standing and bending over them. I concluded, from these considerations, that my problem was to fix upon cloud and wind symbols, and to print the rest in figures with the utmost compactness. After many experiments I adopted the present system, and had types cut on purpose. I gratefully acknowledge the assistance of my friend Mr. William Spottiswoode, Her Majesty's Printer, in bringing these typographical attempts to a successful conclusion.

Meteorographical Maps. – We will now leave the subject of charts of mere record, and consider the method of embodying their results. It takes a vast number of co-ordinates to determine the course of any irregular line, and a still greater number of pieces of mosaic or stitches of tapestry to express a very simple picture; so, conversely, a few judicious sweeps and shadings of a draughtsman's pen, may embody the simultaneous observations of hundreds of meteorologists. The ultimate condensation of bulky reports into small intelligible maps is the object we must keep stedfastly in view.

Our first difficulty is to map five elements which refer to the same area and mutually affect each other; viz., wind, cloud, barometer, thermometer, and moisture. I should fear it was hopeless to expect to combine them all without confusion in one or even into two maps; but, let that be as it may, we have first to consider good plans of delineating them separately. The course of the wind (on the earth's surface) affords no difficulty, and the principles on which I have treated it are pretty satisfactory. First, there are the photo-lithographic maps of two gales, and of four selected epochs, reduced from my own manuscript; they contain, in a small though legible form, the original observations, as a justification for the bold lines drawn through them, which represent the course of the wind currents, such as I interpret them to be. Secondly, there are the red and black maps, reduced coarsely by hand, from my own manuscript charts, which contain about 200 observations. Thirdly, there are the photo-lithographs of all the 93 days in a separate map. I drew these last, from my printed materials alone; they are of interest as showing the class of errors into which one is liable to be led by the employment of too limited a number of observations, and are intended to be compared with the far more trustworthy red and black maps. These are accompanied by arrows that refer to *groups* of observations, in justification of my lines of wind currents.

Considering the remarkable interest now testified by the English public in the law of gales, the lengthened and repeated notices in the Journals during the progress of each of them, and the energy which the Meteorological Department of the Board of Trade shows on those occasions, I should think that miniature diagrams like those I have given, would be very proper subjects for publication; they would be more intelligible and precise than columns of verbal explanation, and the justificatory charts of observations might be published at leisure. As an example of what I mean, the following miniature woodcut, which was reduced from my MS. chart by Photography –

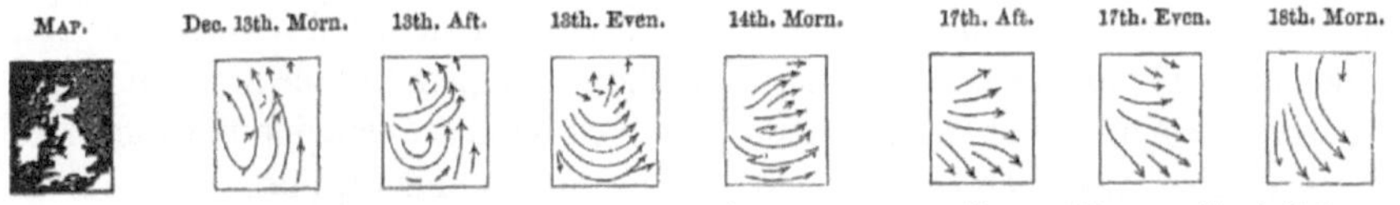

Change from a South to a West Gale. From a West to a North Gale.

embodies all I have been able to conclude from the observations of about 150 stations concerning the change of the wind during the December gale of 1861. It would only require some previous organization, to secure the necessary returns by post long before the interest of the event had passed by.

It requires practice before the lines on the diagrams can be drawn in an effective manner. I proceed by laying short heavy marks on my manuscript chart, wherever the general direction of the wind is unmistakeable; next I push about the marks, as freely as the conditions admit, until they stand at convenient distances apart, and tell their tale clearly, then I mark the places where they stand, and afterwards brush them away. This, however, is not all; for attention has to be paid to the convenience of comparing one map with another, and the lines in each of them should have a distinct relation, wherever it is possible, with those of the map that preceded it. Where the wind is wholly unchanged, the lines employed to represent it should be the same in number, and occupy the same position as in the previous map. Where, as is more usually the case, the direction is somewhat altered, the lines should be made to differ no more than the alteration requires.

In studying the wind in my 93 photo-lithographs, it should be recollected that as the weather is often slow in change, the printed charts which precede and follow the one under examination will throw fresh light upon it; secondly, the side charts will be found very valuable; and lastly, the wind in the mountainous parts of Europe (which is to be made out from the diagram A. 1 on the first page,) is so complicated that it had better be altogether ignored.

I must warn persons who may hastily compare my deductions with the original charts, that it requires patience and some training of the eye to interpret the wind currents satisfactorily. Had the stations been more numerous, local deviations would have balanced each other, and been lost in the mass; but here, with only 80 stations, they confuse the eye. With the still fewer, and less frequent observations that meteorologists commonly employ, and the more limited areas they deal with, there is wide room for fancy, especially when the air is much disturbed and thrown into eddies. The direction given by one station will often be so flatly contradicted by an adjacent one, that obviously both cannot be indications of the same great system. Then arises the question, which of the two is to be ignored? Exercising the right of occasional suppression and slight modification, it is truly absurd to see how plastic a limited number of observations become, in the hands of men with preconceived ideas. If they wish to see cyclones, they will find them everywhere; so with winds blowing to centres, or winds blowing from centres. The advantage of numerous stations in the case of wind, is well seen by comparing the photo-lithographs of the 200 stations with the Charts of only 80 stations that refer to the same days. Not only is there a delicacy of curve to be deduced from the former, which is wholly invisible in the Charts, but the very essence of the mode in which the wind has made its changes, is hidden or even misrepresented in the latter.

The original MS. charts of the winds will be deposited in the library of the British Meteorological Society, where any member of that Society, who might care to do so, would be able to satisfy himself of the data upon which I have founded my representation of the currents.

We have now seen that the direction of the wind (neglecting the depth of its current and other winds moving in various directions at higher altitudes), can be easily represented. I also see that I could have indicated its force by varying the thickness of the lines, and its irregularity by drawing them with an unsteady stroke. The movements of the wind are on the plane of the earth, and can, therefore, be described by directive marks on the plane of our paper; but the other meteorological elements are of a different character, and introduce a "third dimension." They require a similar pictorial treatment to that employed for geographical elevations; in other words, areas of elevation and depression of the barometer, &c., must be pictured by contour lines and shadings, on the same principles as mountains and valleys. I have used shaded types, •, ∗, ⊙, ◯;, in the printed Barometrical tables, the intensity of shade representing the divergence from an average; and two colours, the one for excess, and the other for deficiency. They seem to me to be clear so far as they go, but the curves of nature are represented at a serious disadvantage by the mosaic work of types. Far better are the red and black lithographed maps of barometer and thermometer, drawn upon this principle 

though with different shading. They were compiled by placing marks like small chessmen, on the printed charts and drawing contour lines round the areas on which I found them grouped. In doubtful cases, I referred to my unpublished materials, and in the outlying corners of Ireland and England was able to make considerable though critical use of lighthouse returns. I have taken the total barometric pressure, *not subtracting the vapour tension,* reduced to the level of the sea and to 32° Fahr.; and the thermometer, as reduced to the level of the sea, allowing 1° for each 300 ft.

I imagine the new art of photo-zincography (or -lithography) as practised with such remarkable success by Sir Henry James in the Ordnance map office, and more recently by others, is likely to be of important service to meteorology. It is not to be expected that the meteorologist should work delicately, like an artist, nor is it advisable he should waste his time in making fine strokes. He naturally, writes with a free hand, draws his lines boldly, and ultimately produces a large chart, from which a clean tracing is readily made, that looks well enough at a distance, but would be too coarse and far too bulky to lithograph in fac-simile. At this stage the new process comes in most helpfully; for it throws the rude outlines and thick touches of the manuscript into a delicate miniature on a sheet of zinc, or on a lithographic stone, ready for the printer to cover with ink, and throw off his impressions, exactly as from any other lithograph. One matter is, however, essential, which I am sorry I did not attend to sufficiently in the maps I annex, namely, to draw or stamp the marks with a sharp and densely black stroke on perfectly clean paper. Therefore, though the results are perfectly effective, they are coarse and do not exemplify the process in the way it deserves.

I may state that on enquiry of the artist (Mr. Ramage, 8, West Brighton Crescent, Portobello, near Edinburgh,) who made two of my photo-lithographs, I found the cost of producing 100 copies of a reduction from a clear chart, was at a less rate than 3*d.* a square inch, every expense included. But while praising photo-lithography, I know to my cost that this art is by no means so matured, as to be universally practised with success. The black and red sheet was to have been made by a London photo-lithographer of considerable pretension, who, after detaining the MSS. for half a year sent me his results so ill executed as to be wholly unfit for publication. I was ultimately obliged to have the maps lithographed in the ordinary way, as they now appear. I regret the absence of delicacy of outline, which photographic reductions from my original MS. would have shared with the little woodcuts, I have inserted above.

Errata. – I had intended drawing up a table of *errata,* and am sorry to acknowledge they would have been numerous; but reflecting that the observations I received are all liable to error, I thought it best to let the charts tell their own tale. Whoever works at them will find obvious irregularities to the extent say of one in ten records. These, whether due to errors of the press, or of record, or to local peculiarities, must be thrown out of consideration. Some barometric readings have a constant error, but the errata are not numerous enough to create real difficulty.

METEOROLOGICAL RESULTS.

Those which I have to offer on the present occasion lie comparatively on the surface; for my aim thus far has been mainly meteorographic. I may claim credit for myself in having issued, at considerable cost and labour, as many observations relative to this one month of December, 1861, as are published in the Weather Reports of the Meteorological Department of the Board of Trade, during an entire year, to say nothing of the numerous forms under which these, and still more abundant materials, are illustrated.

The surface that is mapped in these Charts, is inconsiderable when compared to the area of the whole earth, but is far from inconsiderable when compared to a strip reaching from the Equator to the Pole, and therefore containing every climate and nearly every meteorological phenomenon which the earth affords. The space they include, extends from the 42d to the 62d degree of north latitude; consequently a similar area added to the north would include the remotest Arctic regions, while one to the south would dip far into the trade winds, and another still further to the south would almost include the Equator.

Bearing in mind the vastness of these areas, it may well astonish us to find frequent wind currents sweeping with an unbroken flow across our Charts, being obviously portions of enormous systems which far overpass their limits, and probably connect the winds of the tropics on the one hand with those of the Arctic circle on the other. Still more surprising is it to remark the simultaneity of the wind changes. We may see systems of currents converting themselves into entirely different combinations, with perfect unanimity throughout the whole area over which our observations extend, showing the rapidity and the distance to which every aerial impulse is communicated, or, in other words, testifying to the remarkable mobility of the air. It is difficult not to conclude, even from these limited series of weather maps, that ordinary changes of wind and sky have their sources in far more numerous and distant regions than is commonly supposed.

These maps bear testimony to the shallowness of that lowest stratum of the atmosphere, the movement of which causes our winds. We observe that the Alps form a barrier which the winds rarely overleap without change of direction, and that the mountains of Eastern Germany usually divert them. No ordinary liquid can give an idea of that combination of mobility and thinness that characterizes the aerial strata, unless it be a film of oil lying on a pool of water.

Many meteorologists will refer with eagerness to these wind Charts, to see how far they may confirm or oppose the theory of cyclones. I deduced from them, as was explained in a short memoir read before the Royal Society, (*see* Proceedings 1863, p. 385) that they testified to the existence, not only of cyclones, but of what I ventured to call "anti-cyclones." If the lines of wind currents, in the black and red lithographs, are compared with the barometrical charts immediately above them, one universal fact will be found throughout the entire month. It is that on a line being drawn from the locus of highest to the locus of lowest barometer; it will invariably be cut more or less at right angles, by the wind; and, especially, that the wind will be found to strike the *left* side of the line, as drawn *from* the locus of highest barometer. In short, as by the ordinary well-known theory, the wind (in our hemisphere) when indraughted *to* an area of light ascending currents, whirls round in a *contrary* direction to the movements of the hand of a watch, so, conversely, when the wind disperses itself *from* a central area of dense descending currents, or of heaped up atmosphere, it whirls round in the *same* direction as the hand of a watch. I confidently appeal to these maps, and especially to the original MSS. whence these charts have been reduced, to confirm the theory. I annex four woodcuts that I had made before the idea of anti-cyclones had occurred to me. They were intended to illustrate the different systems of winds that blew during the month. They are not so accurate as might be now deduced from the red and black charts, for they were only based on the printed observations. Nevertheless they will serve our purpose.

Dec. 1. Cyclone.

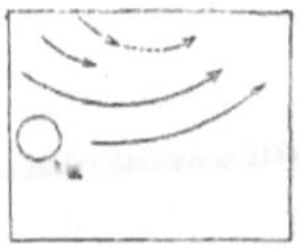

Dec. 14, 15. Cyclone.

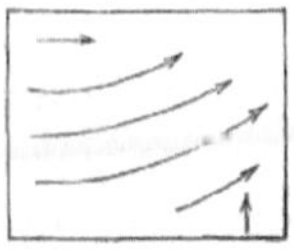

Dec. 2, 3. Anti-cyclone.

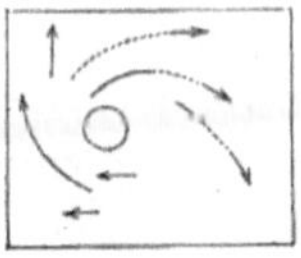

Dec. 19. Cyclone and Anti-cyclone.

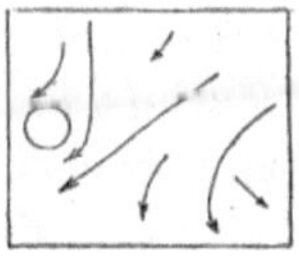

Many of the charts during the continuance of the gales, and especially the side maps of December 10th and 11th, show gentle winds and clear sky in close proximity with violent storms. They testify to the value of warnings telegraphed to a whole line of coast from a central office, whenever gales are known to be blowing at particular stations.

The areas of barometric elevation and depression are enormous, and in their main features very regular. They are easily recognized by the lithographed maps, in black and red. There is no case in which the Charts include the whole breadth or length of any one of these areas, and there are cases where clearly not one-half of them is included, yet the map is about 1,200 geographical miles in height, and 1,500 in breadth. They do not move with regularity, ridge behind ridge, like waves of the sea, but they are ever changing their contours and their sections. They also vary in the speed and directions of their movement of translation.

In addition to these great fluctuations, there are numerous smaller ones, and it occasionally happens that the barometric readings have a widely-spread tendency to show marked though minor irregularities in adjacent places; or, to use an intelligible though inaccurate metaphor, there are violent ripples on the great barometric waves. My data are wholly insufficient to do justice to these minutiæ. In Switzerland, where the mountains retard aerial equilibrium, the barometer seems very irregular in respect to those few stations which I have had space to insert in my maps. I fear, however, there has been some misunderstanding here, as to my request to have the barometric observations reduced to the sea level, and to the temperature of 32° Fahrenheit. No doubt when the Swiss meteorologists have set in action their purposed 80 stations, such irregularities as may really exist, will become peculiarly instructive.

I cannot see in the maps before us that surface winds have always had much influence in causing or remedying the great barometric variations, and conclude they were frequently effected in the regions above our ken. The areas of calm are generally those of highest barometer, never those of very low barometer. They are invariably the centres of partial whorls of wind, or are situated between conflicting currents. The winds adjacent to calms are nearly always gentle and capricious.

The westing of what was originally a South wind and the easting of North winds, due to the "convergence of meridians," is apparent, but only in a slight degree. It is masked by the cyclonic and anti-cyclonic curves.

I do not find the temperature of the same wind to be so uniform as I should have expected. There is, however, one marked condition of temperature and cloud in connection with the wind, which is persistent, beautifully marked, and full of interest. During the latter third of the month a phase of weather prevails with much variety of detail, but steady in its main features, and is well illustrated by the side maps of December 21st and 25th. A westerly wind sweeps across the North Sea over Belgium, Denmark, and Germany east of the Elbe; it is accompanied with a sky that is invariably overcast, and with a warm temperature. This is sharply bounded by an area of confused wind, of perhaps 100 miles wide and 500 miles long, that shifts its position in the land between the Rhine and the Elbe. Here the sky clears considerably, and the air is much colder. West of this again is another area also sharply bounded, where the sky is absolutely pure, the cold intense, and an east or north-east wind has plunged down upon the earth.

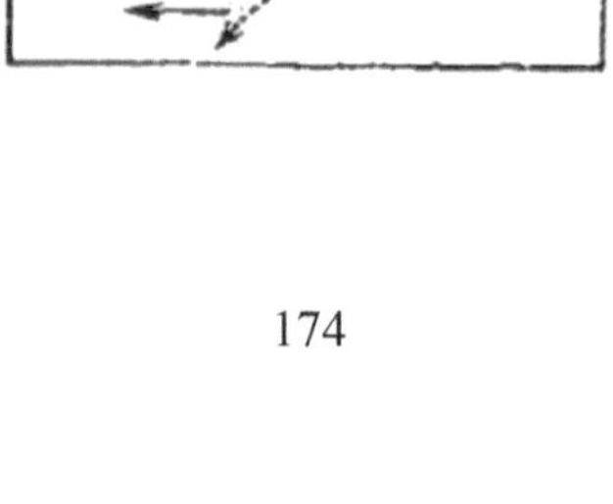

June 1863.

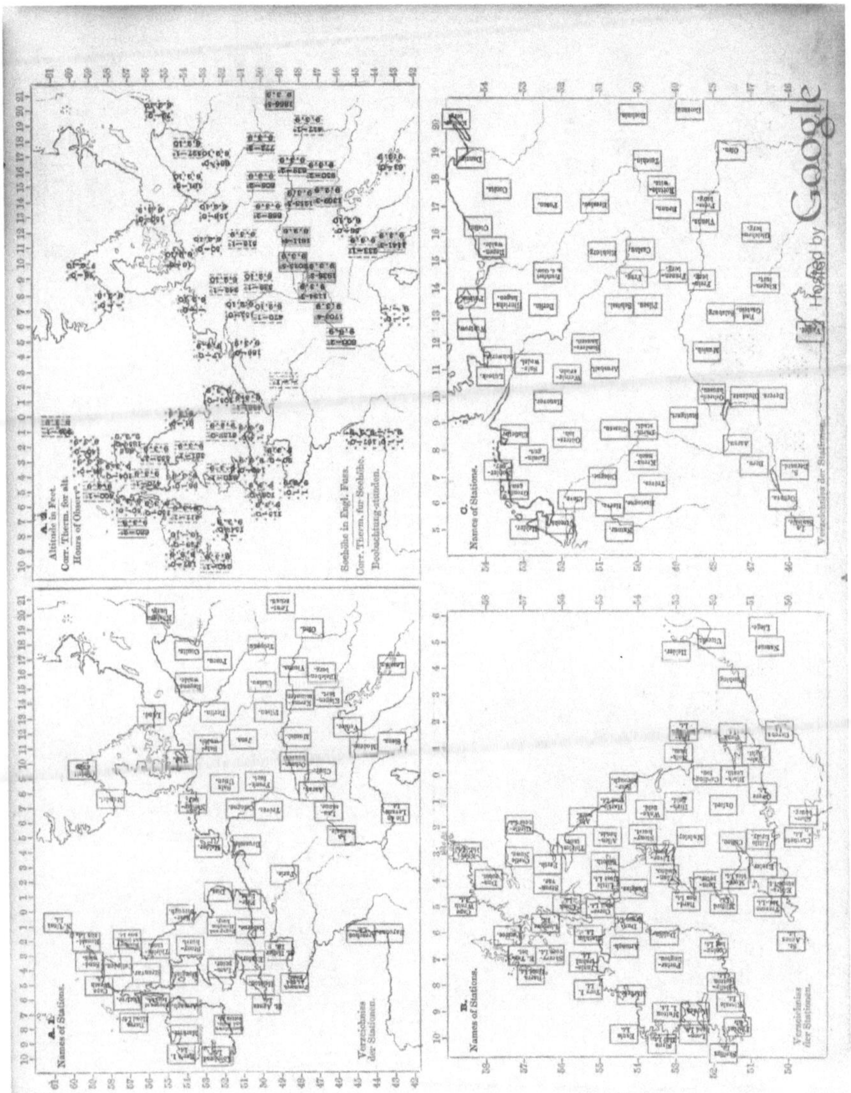

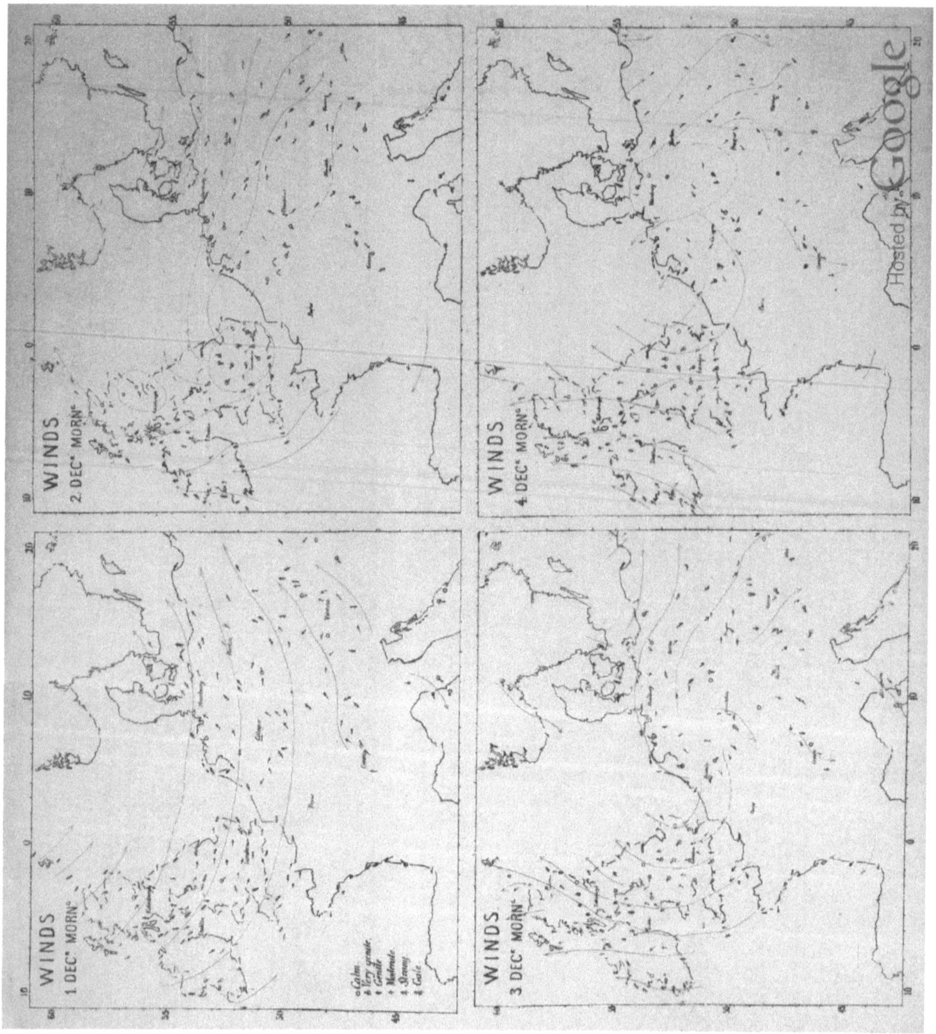
WINDS
1 DEC^r MORN^g
WINDS
2 DEC^r MORN^g
WINDS
3 DEC^r MORN^g
WINDS
4 DEC^r MORN^g

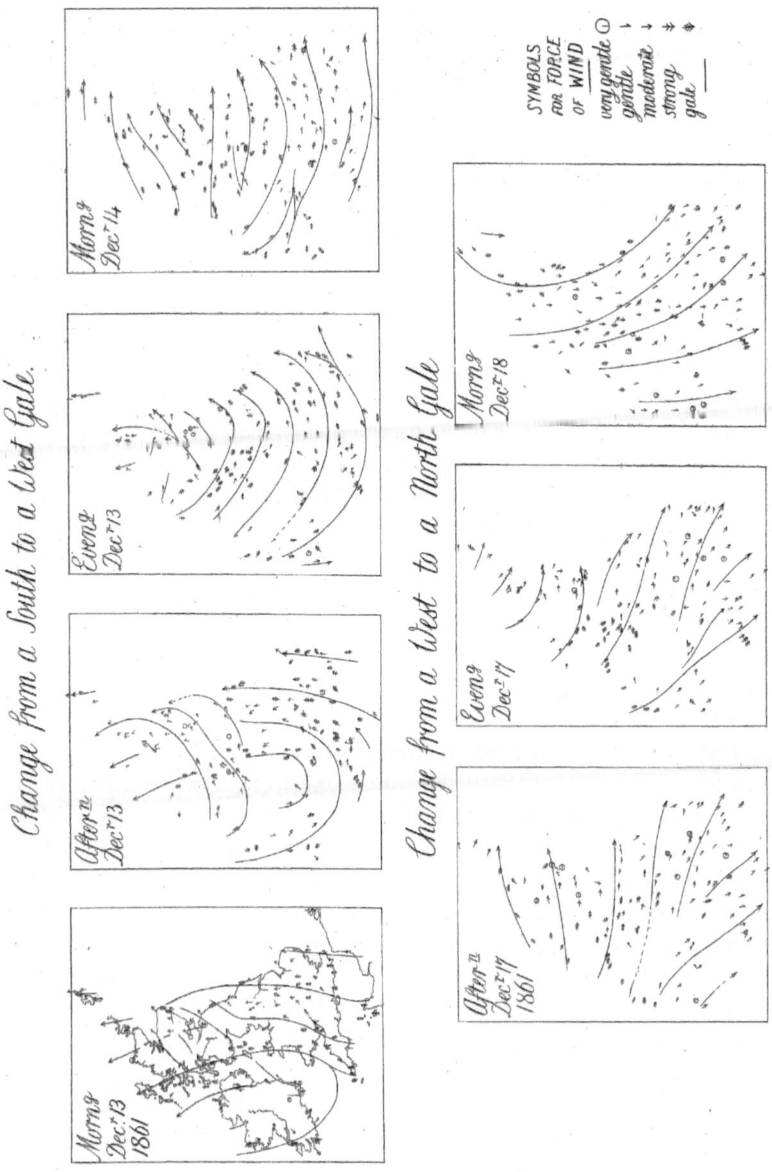
Change from a South to a West Gale.
Morn^g Decr 13 1861
After^n Decr 13
Even^g Decr 13
Morn^g Decr 14
Change from a West to a North Gale
After^n Decr 17 1861
Even^g Decr 17
Morn^g Decr 18
SYMBOLS FOR FORCE OF WIND
very gentle
gentle
moderate
strong
gale

19

RECENT WEATHER

Francis Galton

Source: Francis Galton, 'Recent Weather', *The Reader*, 2 (19 December, 1863), p. 730.

SCIENCE.

THERE is great, good news for meteorologists. The Director of the French Imperial Observatory has stepped forward to supply a crying want. He has commenced a daily issue of weather-charts, embracing the greater part of Europe, to which any one may subscribe as for a newspaper, and receive his copy in London by the forenoon post. This *Bulletin* is plainly but effectively lithographed. The map is indicated by a white space in a faint blue ground. Only enough colour is used to make it visible; strong contours are avoided, lest they should interfere with the meteorographical lines that are traced on the chart. Arrows are drawn over the several stations to show the direction of the wind; they are more or less feathered, according to its force. The barometric heights are written in minute figures, and iso-barometric lines are boldly drawn at intervals of five millimètres (= two-tenths of an inch).

The observations telegraphed each morning from some sixty stations, scattered between Scotland, Norway, and St. Petersburg on the north, down to Gibraltar and Sicily on the south, are reduced, lithographed on the map, and printed in time for circulation by the evening mail from Paris. The *Bulletin* is an institution worthy of our most grateful recognition. It even contains more than the map. The observations themselves are given at length; those referring to previous days which happened to have been delayed too late for use are appended, and a page is devoted to astronomical and meteorological correspondence. The cost in Paris of a yearly subscription is 36 francs; in England, postage included, it is 58 francs, or about £2. 7s. It dates as a publication from November 14th, though tentative charts from the 1st of October were made and circulated privately.

Our first impression on examining these charts is surprise at the narrow limits of what is really known compared to the audacious dogmatism too common among meteorologists. The area of Europe, vast as it is, is too small to exhibit the whole of any one weather phase. For instance, we have no representation of an entire cyclone in any one of these numerous maps. The lower half of what may have been cyclones lay over England on no less than five distinct occasions; but

 DOI: 10.4324/9781003650294-30

we know nothing of their upper halves. We have not the slightest data to justify us in saying that the portion of which we have no direct knowledge is symmetrical with the rest. That which appears to have been a part of a cyclone may have been in reality a narrow protruding gulf of barometric depression. When, in our further remarks, we speak of cyclones, we must not be understood as implying that they were more than partially developed.

Another obvious fact is that a limited area of very high barometer is just as dangerous a neighbour as a similar area of very low barometer. The greatest danger of all is when these two are separated by a comparatively narrow limit. The most violent of all the storms before us had this peculiarity. The weather of Dec. 2nd and 3rd had marked cyclonic curves, with England as a centre, but on the 4th an area of very high barometer appeared suddenly in Spain, whilst the area of lowest barometer was in Denmark. The storm of Nov. 4th and 5th was distinctly anticyclonic; from Nov. 16th to the end of the month, and especially on Nov. 20th and Nov. 29th. the Continent was under anticyclonic influence.

The cyclonic and anticyclonic movements of the wind are universal. In every case where the areas, either of very high or of very low barometric pressure, are small, the direction of the wind is compounded of two influences – the one of indraught towards the area of lowest barometer, or of outflow from that of the highest, as the case may be; and the other an eddying movement, to the left-hand in the first case, and to the right-hand in the second. We need not dwell on the well-known explanation of these influences.

If the iso-barometric lines were simply traced from these charts, and were presented to a meteorologist, he would be able immediately to supply the direction of the wind in those well-marked cases of which we are speaking. There is but one apparent exception, and this was on Nov. 8. when a storm of a very confused character raged over France. The iso-barometric lines of the chart indicate no noticeable tension on that occasion; but, when we examine the figures on which the lines are based, we see a vast amount of local perturbations on a small scale. Thus there was a difference of 3–10ths of an inch between London and Belgium, though only 5–10ths between London and Spain; and yet this, taking it as a whole, was the line along which the tensions varied most rapidly.

There have been in all five well-marked cyclonic storms between the 1st of October and the 12th of December – namely, on the 1st, 12th, and 25th of October, on the 2nd of November, and on the 2nd of December. In every case the centre of the storm included the north of Ireland, and in every case the condition of the atmosphere had quite altered its condition by the morrow or the next day. The lines of equal barometric pressure which had formed a series of concentric curvatures of small radius round the area of lowest pressure, almost always deployed themselves into nearly straight lines that stretched across Europe, sloping slightly towards the north or east. It seems evident to ordinary observers that any prevision of the weather founded on the westerly translation of an aërial eddy must have been wholly in error in these cases. There is an undoubted tendency to a westerly movement in the storms and in the weather generally that is pictured by these charts, but its character is so modified during the proces of translation that the barometric lines of any one day have grown out of resemblance to those of the day previous. We are sure, in the charts had their dates removed and were shuffled out of their series, that no living meteorologist who did not know them by heart could place them back again into their proper order.

F. G.

20

A DEVELOPMENT OF THE THEORY OF CYCLONES

Francis Galton

Source: Francis Galton, 'A Development of the Theory of Cyclones', *Proceedings of the Royal Society of London*, 12, 1863, pp. 385–386.

IV. "A Development of the Theory of Cyclones." By FRANCIS GALTON, F.R.S. Received December 25, 1862.

Most meteorologists are agreed that a circumscribed area of barometric depression is usually a locus of light ascending currents, and therefore of an indraught of surface winds which create a retrograde whirl (in our hemisphere), because they bring to their destination a lateral impulse, partly due to the greater easterly speed of the earth's surface whence the southern portion of the indraught took its departure, and partly due to the less easterly, or we may say greater westerly, speed of its northern portion.

Conversely, we ought to admit that a similar area of barometric elevation is usually a locus of dense descending currents, and therefore of a dispersion of a cold dry atmosphere, plunging from the higher regions upon the surface of the earth, which, flowing away radially on all sides, becomes at length imbued with a lateral motion due to the above-mentioned cause, though acting in a different manner and in opposite directions. The currents necessarily travel with diminished radial speed as they widen out from their central area of dispersion, and the eastward tendency of the northern portion of the system and the westward tendency of the southern become more overpowering. It may be presumed, on consideration of the extreme mobility of the air, that a continuous dispersion of currents would result in the yielding of the east and west winds, which had no tangential movement of their own, to the curvature of the others, and that we should witness a disposition of currents like those in the annexed diagram, which is copied from an actual occurrence on December 2, 1861. The appearance is that of a centre of calms whence currents flow in radial lines, rapidly curving to the right and forming a sort of "anticyclone."

DOI: 10.4324/9781003650294-31

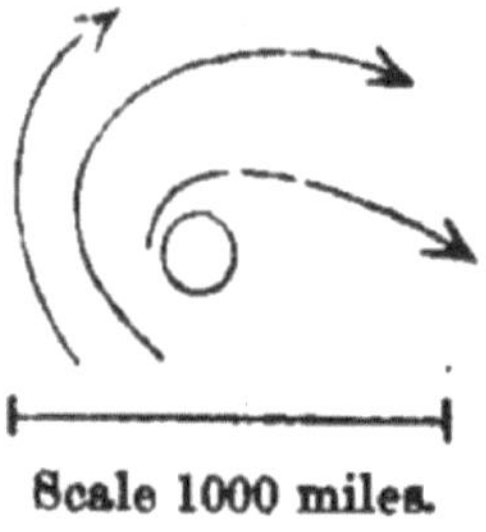

Fig. 1.

Dove's law of gyration is so fertile in result, that it accounts for the same direct rotation of a cold wind by a wholly different process. As an antithesis to his theory of cyclones being due to an equatorial current pressing against quiescent air, he adds (Law of Storms), with a view of illustrating his position, and not of meeting cases that practically occur, polar cyclones, "if they exist," would have a direct rotation. It is not necessary to allude further to his well-known theory – it is sufficient to show that two separate causes cooperate in producing a rotation or curvature of currents such as I have described. I have not the slightest doubt that a strong curvature of atmospheric currents to the right does frequently exist, owing to the descent of cold air from above; for in lately charting the weather of Europe thrice daily during a month, I found it more or less present on from fifty to sixty occasions. Its existence is consonant to what we should expect. It is hardly possible to conceive masses of air rotating in a retrograde sense in close proximity, as cyclonogists suppose, without an intermediate area of direct rotation, which would, to use a mechanical simile, be in gear with both of them, and make the movements of the entire system correlative and harmonious.

The result I have thus far arrived at, and which I should look for hereafter, is that whenever the barometer shows circumscribed areas of marked elevation and depression at distances not exceeding 1500 miles apart, a line drawn from the locus of highest to that of lowest barometer would be cut by parallel wind-currents at an angle of about 45°, in the way shown in the diagram.

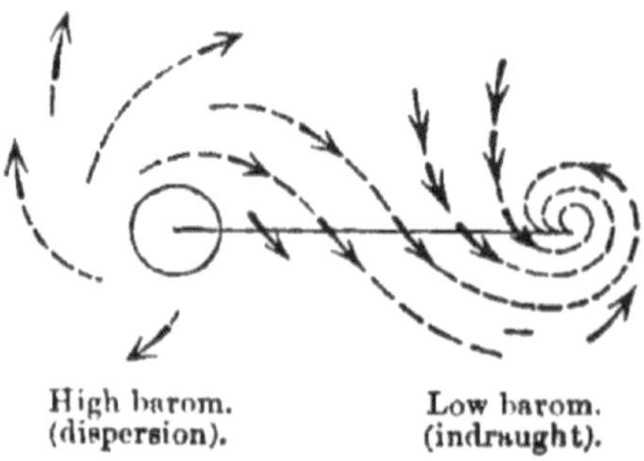

Fig. 2.

I doubt if it be of advantage to investigate the changes of wind produced by a system of indraught and dispersion passing over any locality, because the barometrical sections vary so rapidly as to make the incoming portion unsymmetrical with that which has already passed over.

Part 3

STATISTICS AND THE HEGEMONY OF THE EULERIAN APPROACH

3.1

DUTCH STATISTICAL METHOD

Editorial Headnote

3.1 Dutch Statistical Method

The science of statistics flourished in Europe and America in the mid-nineteenth century, and with it came numerous societies, congresses and publications exploring its potential in science, administration and society. The development of, and theories and approaches surrounding, statistical science were intrinsically linked to nation-state building, population growth, industrialisation and public administration and health. Its efflorescence also had an impact on meteorology. At the sixth International Statistical Congress in Florence in 1867, a distinction was made between meteorological statistics (a space-based "meteorology of simultaneous events") and climatology and daily meteorology (a time-based "meteorology of successive events").[1] For weather scientists, a fundamental concern was understanding the relationship of statistics to meteorology, which some approached from the perspective of deviations and others probed by demanding statistical rigour. Christoph Hendrik Diederik Buys Ballot (1817–1890) argues that deviations from mean values – rather than direct measurements – reveal physical causes, while Alexander Buchan and James Stark – who will be discussed in the next section – seek to reconcile simultaneous observations with direct measurements of atmospheric pressure (reduced to sea level).

Buys Ballot was a professor of mathematics at the University of Utrecht who became world-renowned in science in the 1860s for his rule of thumb describing the relationship between wind direction and barometric distribution in the northern hemisphere, a rule known as Buys Ballot's Law (see Vol. II, 3.1 'Forecasts and warnings'). His article 'On the Great Importance of Deviations from the Mean State of the Atmosphere for the Science of Meteorology' (1850), reproduced here, sets out a method of enquiry for reconciling the science of meteorology with statistical techniques. Buys Ballot was one of the pioneers of synoptic meteorology and, as director of the Royal Dutch Meteorological Institute (Koninklijk Nederlands Meteorologisch Instituut, KNMI), his statistical method helped to establish systematic weather observation and issue weather forecasts.[2] His abiding interest in statistical meteorology and mathematical statistics led him to found the KNMI in 1854, which he directed from that year until his death in 1873. The aim was to protect a country highly vulnerable to phenomena such as floods, tides and storms.[3] Buys Ballot was a firm follower of John F.W. Herschel and his idea of a worldwide network (see Vol IV, 1.1 'Imperial network'); indeed, by 1850 he had published an appeal 'to all friends of meteorology' to participate in a join project in his country. However, unlike the British astronomer in 1836, Buys Ballot argues that the standards of preciseness and correctness required in astronomical measurements were neither claimable nor applicable to meteorology.

In his 1850 article, published in the *Philosophical Magazine* (then a highly respected journal and a platform for the dissemination of physical theories and findings), he holds that deviations from mean values illuminate the truth of physical causes. Buys Ballot dismisses the idea that direct measurements are trustworthy values. His section II on temperature states: 'it is from [deviations] that we must derive the exhibition of the state of temperature, not from the absolute observed temperature' (p. 45). What he states about temperature applies equally to all meteorological elements. His thesis is that the method of means is appropriate to astronomy, not to meteorology. This method is only efficient if the scientist has many observations made under similar conditions – which is not the case in meteorology. Buys Ballot therefore seeks to convert unreliable measurements into reliable ones by proposing a new calculation of observations based on the method of residuals, a method developed by William Whewell in 1847. In subsequent publications, Buys Ballot provides insights into the method of residuals (and the calculation of the so-called "normals") and its application to the study of weather changes.[4]

The second text selected here concerns the dissemination of Buys Ballot's forecasting method. F.H. Klein (dates unknown) was an experienced Dutch seaman and captain (then retired) of the East-Indiaman who contributed to the dissemination of Buys Ballot's method in publications on both sides of the English Channel and in 1863 published *The Foretelling of the Weather in Connection with Meteorological Observations*, accompanied by the addendum *A description of the telegraphic warning system introduced in the Netherlands, June, 1860*, as proposed by Buys Ballot. The original text was written in Dutch and translated into English by A. Adriani, a professional chemist, who added some explanatory footnotes (pp. iii-iv). Based on observations made since 1857, Buys Ballot had devised a system that used deviations from normals for weather forecasting purposes. The rules of this system were set out in the document "Eenige Regelen voor aanstaande Weersveranderingen in Nederland, voornamelijk in verband met de dagelijksche Telegraphische, Utrecht, 1 July, 1860" (Some Rules of Weather Change in the Netherlands in Connection with the Telegraphic Reports Utrecht, July 1, 1860), which served as the basis for Klein's work. Almost simultaneously, Buys Ballot made his method known to the British academic world in a paper 'On the System of Forecasting the Weather pursued in Holland' read at the 1863 annual meeting of the British Association for the Advancement of Science.[5] Buys Ballot revised the English manuscript of Klein's work and added a table that was not in the original (p. iv). In his 1863 work, Klein sets out the main points of the rules laid down by Admiral Robert FitzRoy and compares them with Buys Ballot's system (see 3.1 'Forecasts and warnings). He adduces that, while FitzRoy's rules are useful for isolated observers, the deviation system will be more useful in furthering the interests of meteorology. The reason lies in the significance of system, as opposed to that of rule. Whereas in the system adopted in the Netherlands each observer can consider his observations as a part of a larger whole, in the British system the observer perceives that the many valuable observations are inevitably isolated and unconnected. FitzRoy's approach is not deemed holistic enough to the overall observational framework in which Buys Ballot and Dutch meteorologists place weather forecasting.

Notes

1 Aitor Anduaga, *Politics, Statistics, and Weather Forecasting, 1840–1910. Taming the Weather* (London and New York: Routledge, 2020), pp. 201–02.
2 Ewoud van Everdingen, *C.H.D. Buys Ballot 1817–1890* ('s-Gravenhage: D.A. Daamen, 1953); L. Harold Burstyn, 'Buys Ballot, Christoph Hendrik Diederik', in Charles Coulston Gillispie (ed.), *Dictionary of Scientific Biography* (New York: Charles Scribner's Sons, 1970), vol. 2, p. 628.
3 David Baneke, 'Who Predicts? Scientific Authority and User Expertise in Dutch Storm Warnings 1860–1920', *Science, Technology, & Human Values*, 2024, pp. 1–29.
4 Marcel Boumans, *Science Outside the Laboratory: Measurement in Field Science and Economics* (Oxford: Oxford University Press, 2015), pp. 75–6.
5 Christoph Hendrik Diederik Buys Ballot, 'On the System of Forecasting the Weather Pursued in Holland', in *Report of the Thirty-Third Meeting of the British Association for the Advancement of Science* (London: John Murray, 1864), pp. 20–1.

21

ON THE GREAT IMPORTANCE OF *DEVIATIONS* FROM THE MEAN STATE OF THE ATMOSPHERE FOR THE SCIENCE OF METEOROLOGY[1]

Christoph Hendrik Diederik Buys Ballot

Source: Christoph Hendrik Diederik Buys Ballot, 'On the Great Importance of *Deviations* from the Mean State of the Atmosphere for the Science of Meteorology', *London, Edinburgh and Dublin Philosophical Magazine*, 37 (247), 1850, pp. 42–49.

THERE are some points in meteorology which certainly are no longer disputed, but which nevertheless are not always sufficiently present to the minds of meteorologists, nor so publicly pronounced and admitted as to compel them to action, and serve as a torch to illumine the path leading to the penetralia of science. Though these facts may appear simple, particularly in England, where they have been treated upon, still even there it cannot be considered superfluous to view them in another light, and to introduce them in connexion with a Dutch prize question, which, in my opinion, is the most important meteorological question that can be brought forward, and which fully characterizes the new meteorological period developed by Dove.

On the other hand, England has such boundless merits as to meteorological observations, and has such great facilities, in consequence of her extensive influence and wide-spread observatories, for the application of those facts in all parts of the earth, that she has a right to receive information, through her own periodicals, of what is taking place on the continent. I therefore decided (on being solicited by the Society of Physics of Berlin to send in annually a report of, and opinions on, what might occur in meteorology) to adopt also for insertion in an English periodical, by modifying some points of less interest, a paper drawn up for the Journal of the Society, *Die Fortschritte der Physik.* The truths to which I allude were particularly remarked by me in 1846, when I wrote my work *Les Changements Périodiques de Température dépendants de la nature du Soleil et de la Lune déduits d'observations Neerlandaises de* 1729–1846, in which those particulars are to be found, chapter vii. p. 104 *et seq.* But let us proceed to those propositions.

DOI: 10.4324/9781003650294-34

I. The average temperature which prevails at any certain place is not that which is generated there by the action of the sun, &c., and which would depend simply on the latitude and the elevation of the ground, but is remarkably changed by the influences of other regions, particularly by the action of the winds.

II. That average temperature, such as it is obtained anywhere from observations during a series of years, for the different months or days of the year, will by no means always prevail at those places for the determined month or day of each single year. On the contrary, observations give generally great variations; and it is precisely the magnitude of these variations which it is of the utmost importance to learn.

III. What we asserted regarding the temperature in proposition II. applies equally to all meteorological indications; it is of great importance to become acquainted with those variations of the barometer, and of the force and direction of the wind.

IV. The most efficient means for prognosticating the weather are, the employment of the electric telegraphs and of self-registering instruments, because they facilitate and make possible a tabular union of the variations mentioned in II. and III.

In the following remarks these propositions are more fully developed.

Art. I. No one can call in question our first proposition, we shall not therefore demonstrate it; we will only observe, that it is the winds that modify this temperature; we will endeavour to ascertain how much is to be ascribed to each wind.

Let Θ represent for longer, θ for shorter spaces of time, the *mean theoretical temperature* of a place, and let MT and *mt* represent the *mean temperature deduced from long series of observations,* OT and *ot* the *observed mean temperature* for that longer or shorter space of time at the same place in a given year; then we need only annex some distinctive symbols to these three signs in order to show of which space of time we are speaking. Therefore, when we speak of years or seasons, we shall annex the first letters of the words *year, y*; *winter, w*; *springs sp.*; *summer, su.*; *autumn, a,* at the foot of the great letters; and when we speak of months, we shall annex the first letter of the name of the month after the small letters. Thus Θy will signify the mean temperature of the year, θf the mean temperature of February, which ought to prevail at a certain place, if that place did not receive warmth from and impart warmth to the surrounding parts: it will readily be perceived that these values must ever be the same for each place situated in the same latitude and elevation above the sea. On what does it depend that MT*y* and *mtf* are not found alike in all the places in the same parallel, but always different in every different place; lower in this, higher in that, than Θy and θf? If there was a parallel circle which extended wholly over a continent, or if we passed regularly over a parallel circle in a ship, and if we determined the MT*y* in *n* different selected places, then would $\frac{\sum \mathrm{MT}y}{n}$ be nearly expressed by Θy; the value so obtained would, however, be somewhat greater than Θy, because there is more air drawing towards the north than towards the south over a whole parallel circle, that southern air at the same time being warmer; and also because, near the equator, the latent heat which is employed in the vaporizing of water is greater than that which is freed by rain; and, on the contrary, in higher latitudes there is more freed than expended on the formation of vapour. This is also a cause why on the land Θy must be something less than at sea. If we would nicely calculate this influence, we ought equally to distinguish between rain that is formed in different altitudes. Properly we ought to infer theoretically Θy, and equally so θf, from the warmth which emanates from the sun to us every day; from the warmth which every day and night issues from beneath the surface of earth; and from the warmth that is produced by animals, consumed by plants, lost by radiation, given by condensation of vapour. The difference between Θy so obtained, and MTy, (MT$y - \Theta y$), at a certain place, is to be ascribed to the influence of the wind during the year; similarly, $mtf - \theta f$ is the influence of the wind which prevails in February at that place. How far we are as yet from such a determination may be seen, for example, from the meteorology of the late illustrious Daniell. But even if it could be obtained, it would, however, not be fit for a great space of time; for certain it is, that the resultant of the winds of a whole year would not have the same influence on the temperature as the different components. This will appear more evident to any one who directs his attention to the differences $mty - \theta y$, $mtf - \theta f$, $mtjl - \theta jl$, &c. He would find, not only for every place, but also for every month of the year, different influences for the same wind; the same wind producing a different cooling or heating power in different seasons, a circumstance of which account is not taken in computing the resultant.

Art. II. We will suppose that not only Θ and θ, but also MT and *mt* were obtained at a certain place from a long series of observations, that MT therefore could be considered as the equilibrial state of temperature at a determined season of the year at that place; then we should find that the differences OT – MT, *ot* – *mt* are to be ascribed to these circumstances: – 1st, that the winds had not in that space of time the same direction as usual; and 2nd, that the distribution of temperature at the surrounding places was quite different from the usual distribution during that space of time, or shortly before. Thus it is necessary that we know the variations, not only for the place itself for which we desire to explain the temperature, but also for the surrounding places, since the variations at the first place must be explained partly from the variations at the latter. The most important causes are always to be sought in the variations (deviations); it is from those that we must derive the exhibition of the state of temperature, not from the absolute observed temperature. Even now, when we give absolute temperatures, we do in fact give deviations; namely, deviations from the arbitrary zero of temperature. Positive prognostications will always be given under the form of deviations, and best by deviations from the mean state of the atmosphere. Certainly it were to be wished that we better understood the theoretical temperature (Θ), which must take place for certain soils and elevations above the sea, and for every latitude, in order also to be able to give the deviations from those theoretical values; but for want of that knowledge the deviations from the mean value are the main points for consideration.

Art. III. Temperature is that state which most attracts the attention of the public, in consequence of the immediate influence it has on the relations of human life; but scientifically, all states are equally important. The pressure of vapour is perhaps the most uncertain element, and of which we can give little explanation, from the manner in which it is measured; electricity is not anywhere (with exception of Brussels and Kew) sufficiently known either as to its origin or as to its quantity; it is not therefore expected that the deviations of these elements can be given. It is the winds that bring us the warm air and cold waves; they must be accurately noted as to direction and strength, but it will avail little to give their deviations from the mean direction; the winds must therefore be noted with their real directions. On the contrary, as to the barometer, the deviations again are of the greatest importance, especially as here the theoretical state is known for every latitude. Everywhere, where attention has been paid to the deviations of the barometer, which at the same time have been observed at different places or successively at the same place, it has led, in connexion with the observation of the wind, to practical discoveries. Who knows not the fruits of the labours of Redfield, Reid, Piddington, Thom? In later times also, the investigations of Mr. Birt on the atmospheric waves, which in November pass over England and Europe, are only deduced from the knowledge of the deviations.

As we must give so much importance to the deviations of the barometer and thermometer as to assert that they are applicable in most meteorological investigations; as we further consider how much trouble it causes for a calculator to subtract the value obtained from every observation which he uses from the mean value at that place for that day, whilst the observer can so very easily, whilst he is noting his observations, note in a column next to it the difference from the mean, we dare propose and urgently invite the observer everywhere, where the mean of each day is sufficiently known, or even where this is known with some approximation, to take this little trouble. We should then, after a lapse of five or ten years, be able to modify the mean of the temperature, and then to determine it with greater certainty. A book that contains such means for every day of the year for many places would certainly be a most useful book; and if one once possessed that, then it would be sufficient for the observers to communicate the deviations; from time to time the corrections could be published in the form of supplements.

Art. IV. England has given the example of communicating the state of the weather[2] at a certain instant in all parts of the country, at least to Glasgow, by means of telegraph, as I saw to my surprise in the Greenock *Daily News*. My proposition made in the *Changements de Température* was actually in operation. Later I have also learned that that plan is likewise proposed to the British Association: thus the communication of the weather becomes more scientific than it hitherto had been. Telegraphs can, better than self-registering instruments, answer the purpose of approaching to a foretelling of the weather at all places before it exists there; they give us the opportunity of being informed of the weather before it has passed from one place to another. We can be on our guard and arrange our occupations (our observations in the first place) accordingly; remarkable phænomena will be noticed by different observers in corresponding manners at different places, and many other advantages will result. It is true, we shall be able likewise to see from the graphic self-registering instruments the state of the weather at every instant at all places where they exist (and this is the main point of agreement of both), but only *long after* those circumstances are past; from the telegraphs we are acquainted with it *instantaneously.* It will yet be *some time,* we avow it, before we become sufficiently acquainted with the laws of the transit of the states of the weather, to be able to foretell with any certainty even from the known state of the weather at the neighbouring places (particularly, also, because to the class of "surrounding places" belong the places above us, and the winds and temperature reigning there); but we shall *never* arrive at this point except by means of exhibition of simultaneous observation. When we draw a series of theoretical deductions from a given distribution of meteorological indications (wind, heat, pressure, &c.), and compare them with the subsequent distribution which is really found to follow as an effect of the first, we cannot fail to penetrate better the origin and influence of the winds, and moreover the variations of temperature produced by them.

Therefore it will be of importance to improve our graphic methods of representing simultaneous observations, and either by that means or by numbers to exhibit those states in a tabular form. As to the graphic method, it is to be seen in the articles of Mr. Birt, Report of the Meeting of the British Association, 1844, 1845, &c., that there is a want of a uniform method for noting three or more variations at the same time. A good graphic method would have enabled Mr. Birt to communicate more in fewer pages, and to put every reader in a state to try his conclusions and to extend them.

Martins, in his annotations to the translation of Kämtz, has endeavoured to exhibit more than three variations at the same time; but that method, in our opinion, is not very clear.

Since the investigation of the great November atmospheric wave is probably to be continued, it is perhaps not inconsistent to propose another manner of exhibiting it. We can form a map of all the places where observations are made, and give to each place a sign, for example (*a*), (*b*), &c.; then the position of those signs in the other part of the drawing must express the height of the barometer at those so indicated places in the following manner. Let us draw round that map a circle, which we consider as the section of a vertical cylinder, in the plane of which those places are situated; let us further suppose that at the places (*a*) (*b*) the barometers are placed with the lower surface of the mercury in the same plane of that circular section; so will the top of the highest standing barometer lie in a plane parallel to the first, and give a circular section of the cylinder; the distance between the two circles will be proportional to the height of the barometer at that place. This will be true for all places; and the height of the letters or signs (*a*) (*b*) noted at the respective sections in the surface of the cylinder would indicate the height of the barometer at those places. If we imagine our eye in the axis of the cylinder, and we project those supposed points (*a*) (*b*) on the horizontal plane, or what is the same thing, if we apply the method of descriptive geometry to this, the greater or lower height of the barometers at (*a*) (*b*) will be represented by points, *α*, *β*, in circles concentric with the first, but more or less removed, whose distance from the first (the difference of the radius) will be proportional to that height. So the first circle may represent a very low barometrical height, as 730 millimetres, and each millimetre that the barometer in (*a*) or (*b*) is higher may be represented by one or two millimetres' distance of the circle (*α*) or (*β*) from the first. But we can go a step further; we can divide those circles from the common centre point into a number of sectors, thirty sectors I suppose: in one sector the letters may signify the height of the barometer at the 20th of October, in another at the 21st, and so on; every sector may be applied to one day. So we shall have the barometrical range at every one place, and the different states at the different places simultaneously, equally well and clearly represented, while the map within the circle indicates the distribution with respect to the surface of the earth. If, by chance, the points (*a*) in the different sectors lie in a circle, it indicates that the barometer at *a* has not varied during the month; if they lie in an ellipse, then there have been two highest and two lowest states; we see immediately those heights and the dates. So, as to the simultaneous comparison of the barometer heights, the eye catches immediately the letters that are higher or lower, and we see on the map in the middle the line of places where the barometer was at the highest or the lowest. If there actually is a wave, then it is clearly exhibited thus to the eye; if the eye has difficulty in discerning the wave, then probably there is no wave, or perhaps that wave must have been marked out by the foregoing or following days.

It is even possible to go a step further, and, for example, to represent at the same time the thermometer state by the distances at which we set the letters (*a*), (*b*) (the signs of the places) from the radii which divide the circle; the more on the right, for example, the higher the state of the thermometer. It is evident, however, that the more variations we exhibit, the more is lost in simplicity. We could have represented also latitude, longitude, time, barometer, thermometer. How much soever we improve the graphic representation, we must however always, where there is not merely a view required but an accurate calculation, have recourse to a tabular exhibition. To promote this, the Utrecht Provincial Society for Arts and Sciences has issued the following prize question: –

> Since the usual indications of thermometer, barometer, and anemometer throughout the year, that is, the periodical functions by which those indications are expressed, are now known for a great number of places, at least with approximation, and thus a sufficient basis is laid for the investigation of non-periodical changes; the Society wishes –
>
> After having calculated the mean indications of thermometer, barometer and anemometer, over as great a number of places as possible in Europe and Asiatic Russia for periods of five days for two successive years, –

1. The deviations of thermometer and barometer from their above-described usual state in each of those periods, should be united in tables;
2. The manner of transit of those deviations in time and space should be investigated; and
3. Those deviations should be compared and brought in connexion with the winds that prevailed during each of the said periods, and with the deviation from their usual direction which they exhibited.

The importance of this prize proposition in connexion with the various researches which are required for its investigation, has induced the author of this question to augment the offered prize with one hundred and fifty guilders. It is wished that the answer be received before October 1, 1851, by M. C. von Marle, Secretary of the Utrecht Provincial Society for Arts and Sciences. The name of the author, as usual, in a separate letter.

Utrecht.

Postscript. – I have not mentioned in my memoir that the observations instituted by the Royal Society have lately been computed, and that the means of each month are in fact subtracted from the mean of the months of the same name. I had not seen the last part of the Philosophical Transactions in which these computations are placed.

Source Endnotes

1 Communicated by the Astronomer Royal.

2 This publication was principally planned and is carried into execution by the proprietors of the London *Daily News.* The report of the state of the weather at 9^h in the morning at numerous places is published every day in that journal. – G. B. Airy.

22

THE FORETELLING OF THE WEATHER IN CONNECTION WITH METEOROLOGICAL OBSERVATIONS

F. H. Klein

Source: F.H. Klein, *The Foretelling of the Weather in Connection with Meteorological Observations*, (London, 1863), pp. 5–6, 20–23.

THE FORETELLING OF THE WEATHER, &c. &c.

It is one of the most interesting facts that, in our century of rapidly-growing civilization, the exact sciences have been brought to bear upon the daily pursuits not only of the professional manufacturers and the general public, but also that within the last thirty years an entirely new branch of science has sprung up, which, under the name of Meteorology, is not as yet so well appreciated by the public at large, and the seafaring public especially, as it really deserves. This may be partly due to a misunderstanding of the teachings and the non-acknowledgment of the real value of the warnings given by the science alluded to; partly also to the deeply-rooted popular prejudices and erroneous belief in and explanation of some phenomena which have become impressed upon the mind from early infancy. The following pages contain the ideas on the value of Meteorology, expressed by a former seafaring man, who takes a great deal of interest in what is done in his native country in Meteorological pursuits.[1] The small pamphlet which ensued, owes its origin in Dutch to the desire expressed by the editor of a daily newspaper to inform its readers of the results of the Meteorological observations daily made at various places of the Kingdom of the Netherlands; which results, if merely published without any explanation as to the value and tendency of such figures, might not be well understood at all by the public at large, neither also by those who are more interested, – viz., that large portion of our community who spend the greater and best part of their life at sea.

It would be erroneous to suppose that Meteorological observations could inform us *daily* quite correctly with the state of the weather for every respective day, or even extend our information for some days to come. There is, no doubt, a great likelihood that such will be the progress of the united efforts of the many eminent men who devote their time to the study of the laws which

 DOI: 10.4324/9781003650294-35

regulate the state of the weather; but for the present we must content ourselves with what is.

[. . .]

This laborious and far from easily accomplished work has been successfully brought about by the Royal Netherlands Meteorological Institute, and hence the inhabitants of the Kingdom of the Netherlands are put in possession of valuable means to pronounce with a high degree of accuracy about the state of the weather.

It may not be quite out of place to mention, that a most eminent maker of physical instruments, M. Olland, at Utrecht, has contrived to construct barometers in such a manner that it is possible to read off on these instruments the *deviations* by immediate observation, so that the slight calculations necessary to find the *deviations* are avoided.

It is evident from what has been said, that the use of such instruments can only be available where the normal reading is well ascertained.

The following rules may be drawn up respecting the state of the weather, as taken from the observations according to the system of deviations: –

1. Pretty fair weather may be expected when the deviations in all our places of observation are +; it is less favourable when the sign – is everywhere exhibited.
2. When the deviations on the northerly stations obtain +, while, on the contrary, – is put to the deviations southward, one need not be apprehensive.
3. When the deviation on the southerly stations is +, and on the northerly –, one may expect a gale of wind.
4. It is an indication of great danger when it so happens that while the deviations of places situated to the north and southward from our place of observation exhibit at the very same moment the sign +, the deviation at the place where we are should exhibit –, for it indicates that at our place of observation a gale of wind will be most violent.
5. When the range of our observations extends further north or southward, it will be sometimes seen that some of these places of observation exhibit the sign +, others again the sign – to their deviations, while other places of observation again further distant towards the north or south will exhibit the sign +. Such a state of affairs is indicative of irregular disturbances in the atmospheric equilibrium, whereby the southerly or northerly currents alternately become the masters. It is very possible that the deviations will exhibit the next day a change of signs, and still no other inference can be drawn from such occurrence than *unsteady weather,* changeable. This occurrence is very remarkable when at one of the places of observation a rather large difference is observed between the reading of the previous day and the next.

   It may temporarily happen that Groningen reads higher than Maastricht; and things may be reversed the next day, and a gale may blow at the same time.
6. The more regularly the reading of the barometer decreases in height from the north towards the south, – in other words, when there is to be observed

a certain regularity in the deviations proceeding from the north towards the south, or if there be observed a decrease from the sign + to a less positive, – pretty settled weather and an easterly direction of the wind may be expected.
7. When, on the contrary, the deviations regularly decrease in the direction from south towards north, pretty continuous bad weather and westerly winds may be reckoned upon.
8. When places situated to the westward of our place of observation have relatively to ourselves a lower reading, – *i.e.,* if their deviation, as compared with that of our place of observation, is *minus*, – it may be counted a favourable indication.
9. A falling of the barometer of about two millimetres per hour, especially if such happens when the barometer reads already low, indicates that a rapidly growing gale of wind must be looked for.
10. A rough or bad sea occurring at Brest, or nigh the English coasts, is to be considered as a bad sign.

Although the observation of the above rules may lead to the almost certain escape of dangerous gales of wind, it must not be lost out of sight that Meteorology is only of very recent origin, and that in many parts of the world, and even Europe, observations are either not made at all, or only over a limited extent; and it hence follows that our knowledge is yet rather limited, so that, however well our observations are conducted, it may ultimately be found that influences now unknown modify to a more or less extent our present rules. It is in that aspect highly desirable that communications of observations made regularly should be rendered more rapidly available by increased telegraphic communications.

The rules laid down by Admiral Fitz Roy, inferred from the *reading* and *motion* of the barometer, may be reduced to the following main points: – Even the public at large is pretty well aware of the fact that a rise of the barometer is pretty freely accompanied by a decrease of wind and rain; whereas on the reverse taking place, *i.e.,* fall of the barometer, it is pretty freely accompanied by increase of rain or wind. It is also pretty generally known that a high reading of the barometer indicates fine weather; a low reading, on the other hand, bad weather. What has to be observed is only the *motion* of the mercurial column, and no regard should be paid to what often occurs on the dials or scales of barometers; viz., the words *fair, rain, wind*, &c. It may happen that what we would call *bad* weather occurs during a high reading of the barometer, and very fair weather may be experienced during a rather low reading of the barometer. The following rules, however, are, generally speaking, reliable: –

1. When the barometer reads at its usual height and shows a tendency to rise, while at the same time the thermometer falls and the atmospheric moisture decreases, the amount of which is best ascertained by the psychrometer, either north-westerly, northerly, north-easterly winds – or if it rained and blew previously, a decrease of wind and rain, may be looked for.

2. If, on the contrary, the barometer falls, while at the same time the thermometer rises and the moisture of the air increases, south-easterly, southerly, or south-westerly wind may be looked for.
3. When during winter time the barometer falls and the thermometer reads relatively low, a fall of snow may be expected.
4. When a north-east wind, accompanied with showers of rain, hail, or snow, is about to set in, the barometer will rise, being thus affected by the current of air about to begin blowing; but though such rise of the barometer proceeds, the general state of the weather does not become more favourable.
5. When it happens that the barometer reads below its usual height, a slight rise of the mercurial column predicts a decrease of wind, or a change of the direction of the wind to the northward. If, however, the reading of the barometer might happen to be extraordinarily low, *the first* rise of the mercury ordinarily indicates either an increase of wind, or heavy squalls of wind in the direction from north-west to north-east. A continued rise of the mercury of the barometer will, only *after* such increase of wind, indicate improvement of weather, provided, however, the thermometer falls. If, however, the thermometer remains stationary, it is probable that the wind will turn to the south or south-west, especially if the rise of the barometer be rather sudden.
6. The most dangerous squalls of wind to the north or northwesterly quarters occur shortly after the mercury, having been very low, begins to show a slight rise.
7. The mercury of the barometer has a tendency to rise when the surface of the metal in the tube shows a convexity; whereas when there is a tendency to fall, the surface of the metal assumes a concave shape.
8. A rapid rising of the mercury indicates changeable weather, a slow rise the contrary. A rapid fall of the barometer shows heavy wind, accompanied by rain or snow; alternate rise and fall is a sign of unsteady, showery weather.
9. Although, as a rule, the barometer falls with southerly winds and rises with northerly winds, it is no rule without exception, for the reverse is now and then observed to take place. In such case, it is fine and fair weather, with a southerly wind; while if the wind blows in a northerly quarter, it is blowing hard, and accompanied by rain, hail, or snow.
10. When during the blowing of a westerly wind the barometer suddenly falls, often a heavy gale from the north-west to the north-east may be expected.
11. When a heavy gale of wind begins in the direction of the easterly or south-easterly quarter and changes towards south, the barometer will continue falling until the wind is about to shift quarter. It often so happens that a calm sets in for a while, but on a sudden the gale of wind blows again, sometimes even very violently, while the mercury rises and the thermometer falls as soon as the wind is about to shift to the north-west or north.

The above are the main points of the rules laid down by Admiral Fitz Roy. These rules, it cannot be denied, are useful to isolated observers; but it is my opinion that the system of deviations, if it becomes more generally known and appreciated as it deserves, will be found more serviceable to promote the interests of Meteorology, since the basis of the system adopted in Holland is such that every observer who adopts it may consider his observations as being a portion of one great entire whole, whereas in the other case a large number of valuable and good observations remain necessarily isolated and unconnected.

Source Endnote

1 His native country is what in Great Britain is commonly called Holland, more correctly Kingdom of the Netherlands, which throughout centuries past has been, and is yet in every respect, along with other nations, at the head of civilization, and where all sciences, and particularly Meteorology, are taken up by men second to none elsewhere, while the most is made of the means at their command. For all who take any interest at all in science, and in Meteorological pursuits more especially, we strongly recommend a journey to Utrecht, the city where the Central Meteorological Observatory is placed, a celebrated university and many other institutions. – A.

3.2

SCOTTISH ANALYTICAL METHOD

Editorial Headnote

3.2 Scottish Analytical Method

The middle decades of the nineteenth century saw an increasing influence of statistical science on meteorology in Britain. Although this movement had much to do with the growing role of vital statistics in the study of British demography – reinforced by the cholera epidemic of 1854 – it took different forms and developments in England and Scotland and bore little resemblance to statistical meteorology in the Netherlands. The sources in this section illustrate different ways in which these new approaches to statistics responded to the perceived needs and demands of weather forecasting in Britain. While Christoph H.D. Buys Ballot devised an empirical-statistical method for storm warnings (see Vol. I, 3.1 'Dutch statistical method'), Alexander Buchan (1829–1907) – who partly followed in the wake of James Stark (1811–1890) – developed a synoptic-statistical method for weather forecasting purposes. However, each in their own way contributed to the convergence of statistics and meteorology that would illuminate weather forecasting from the late 1860s onwards.[1]

As Scotland's first Superintendent of Statistics, a post created following the introduction of the Registration Act of 1854 for the registration of births, deaths and marriages in Scotland, James Stark proposed a new method of estimating average temperatures that differed radically from that used by James Glaisher in England. As a Fellow of the Royal College of Physicians of Edinburgh, he had long worked in the field of vital statistics and public health. His method of producing statistical tables differed from the English model introduced by William Farr (the Registrar-General's statistician), which was geared towards predictive social science.[2] Yet, his disagreements also extended into the field of meteorology. His method of estimation differed from Glaisher's in privileging exact means over faulty deductions of means and holding that the calculation of strict means, rather than the deduction of mean values, was the only safe rule for the estimation of mean temperatures. In the Registrar-General's Report of 1847, Glaisher, who was the director of the Greenwich Observatory, had been firmly convinced that the method of reducing observations and determining mean values 'would some day lead to empirical laws'.[3] In the 1862 article reproduced here, Stark denounces Glaisher's practice of 'withholding the strict means of the different series of dry and wet bulb readings, and only publishing deductions' (p. 265). He criticises Glaisher for modifying the temperature exact means by using certain tables to correct them for diurnal and monthly range. Consequently, the mean temperature given is not the 'true mean, but an estimated mean' derived from the combined observations made with the self-registering and dry-bulb thermometers (p. 264). Stark attaches great importance to the statistical rigour that is the proper way to discover atmospheric laws which is central to meteorology, according to him, and which had been systematically neglected by Glaisher and his followers.

As Secretary of the Scottish Meteorological Society and a schoolmaster forced to give up his profession because of a laryngeal disease, Alexander Buchan imbibed Stark's statistical rigour as the most appropriate method for discovering not only the general laws of the atmosphere, but also the interdependence of meteorological phenomena.[4] Buchan's first text in this section introduces his analytical method of synoptic weather charts and shows how, in its final form, it shared essential features with Edme Hippolyte Marié-Davy's method (see 2.2 'Eulerian approach'). His 1863 study of storms in Europe, published in the *Transactions* of the legendary Royal Society of Edinburgh (founded in 1783), discusses the implications of synoptic maps and derives relationships from the distribution of atmospheric pressure. Unlike Buys Ballot, who had access to real-time observations, Buchan had to rely on data from the Paris *Bulletin International*, which provided him with observations from 135 sites across Europe. His study is based on two main sources. On the one hand, he reintroduces the argument previously formulated in part by Stark, according to which deviations from the mean give a distorted picture of the pressure distribution. On the other hand, following Marié-Davy, he draws isobarometric lines on the map at all the points where the pressure is the same, but only after reducing the observations of the barometric pressure at sea level, which is essential in his analytical method. The first and only map reproduced here – of the eighteen maps collected in the article – includes observations of surface pressure and wind speed and direction; it also shows how the barometric pressure around a storm area is contoured. Careful analysis of the data leads him to assert that 'storms most commonly assumed' circular or oval shapes as prefigured by isobarometric lines (p. 200), and that these types of lines – not isotherms – were appropriate for representing storms.

Buchan compiled his meteorological knowledge in his magnum opus, *A Handy Book of Meteorology* (Edinburgh and London, 1867), which saw an enlarged edition in 1868. The second edition, of which extracts are reproduced here, consisted of 371 pages with eight plates and included an enlarged chapter on storms.[5] In 1871 he published an expanded version, *Introductory Textbook in Meteorology*, which was well received as a university textbook in British academia. It soon became the standard text on meteorology in English.[6] Buchan's *Handy Book* had great resonance outside Britain; in their classic work on the foundations of dynamical meteorology (1898), the meteorologists Hugo H. Hildebrandsson and Léon P. Teisserenc de Bort described it as the work that most accurately reflected the new ideas of modern meteorology, so much so that 'the first investigations conducted in different countries were at first nothing more than a verification of those ideas'.[7] Buchan's treatise has a threefold structure. The first part is the most important for weather forecasting purposes, and presents the barometric distribution as the field that determines the meteorological conditions during the passage of a storm. The second part provides the general rules for the relationship between pressure and winds which Buchan derived from the synoptic analysis of maps and which were in fact an extension of Buys Ballot's baric wind law. The third discusses the formation and maintenance of storm systems, the mechanism of which is described by an empiricist explanation of the thermal convection theory of James Espy and Elias Loomis, based on surface data and cloud observations.[8] The extract reproduced here from the first section, 'History and scope of meteorology', provides insights into the meteorological societies established in the Western world and their key role in storm forecasting, while the extract from the final section, 'Storm-warning', provides an overview of the various storm warning services in place at points across the geography of the British Empire.

Together, Buys Ballot, Buchan and Stark illustrate the attempts at convergence of statistics with meteorology in the Netherlands and Scotland. In both cases, it resulted in a drift away from the Lagrangian approach of Robert FitzRoy and others in England. However, by distancing himself from Buys Ballot, Buchan embraced the Eulerian specification of the baric field: the field of barometric distribution determined the patterns of temperature, wind and precipitation. In his later writings, he provided rules describing the relationship between pressure and winds, which meant an extension of Buys Ballot's baric wind law.

Notes

1 Aitor Anduaga, *Politics, Statistics, and Weather Forecasting, 1840–1910. Taming the Weather* (London and New York: Routledge, 2020), pp. 201–47.

2 John M. Eyler, *Victorian Social Medicine. The Ideas and Methods of William Farr* (Baltimore, MD: Johns Hopkins Press, 1979).

3 William Marriott, 'Some Account of the Meteorological Work of the Late James Glaisher, F.R.S.', *Quarterly Journal of the Royal Meteorological Society*, 30:129 (1904), pp. 1–28, on pp. 3–4.

4 J. Malcolm Walker, *History of the Meteorological Office* (Cambridge: Cambridge University Press, 2012), pp. 32–4; Marjory G. Roy (ed.), *The Life, Achievement and Legacy of Alexander Buchan MA LLD FRSE FRS (1829–1907): Secretary of the Scottish Meteorological Society from 1860 to 1907* (Edinburgh: Royal Meteorological Society, 2007); Arthur Mitchell et al., 'Contributions towards a Memorial Notice of Alexander Buchan, M.A., L.L.D., F.R.S.' *Journal of the Scottish Meteorological Society*, 14 (1908), 101–18; Amy Ackerberg-Hastings, 'Buchan, Alexander (1829–1907)', in Bernard Lightman (ed.), *The Dictionary of Nineteenth-Century British Scientists* (Bristol: Thoemes Continuum, 2004), vol. 1, pp. 322–323.

5 Gisela Kutzbach, *The Thermal Theory of Cyclones: A History of Meteorological Thought in the Nineteenth Century* (Boston, MA: American Meteorological Society, 1979), pp. 71–5.

6 J.P., 'Scrap Album of Alexander Buchan', *Weather*, March 1950, pp. 119–22, on p. 120.

7 Hugo H. Hildebrandsson and Léon Philippe Teisserenc de Bort, *Les bases de la météorologie dynamique* (Paris: Gauthier-Villars, 1898), p. 105.

8 Gisela Kutzbach, *The Thermal Theory of Cyclones: A History of Meteorological Thought in the Nineteenth Century* (Boston, MA: American Meteorological Society, 1979), p. 74.

23

ON THE FALLACY OF THE PRESENT MODE OF ESTIMATING THE MEAN TEMPERATURE IN ENGLAND

James Stark

Source: James Stark, 'On the Fallacy of the Present Mode of Estimating the Mean Temperature in England'. *Proceedings of the Royal Society of Edinburgh*, 4, 1862, pp. 264–265.

In the Meteorological Tables for England, published at the public expense, viz., those of Greenwich Observatory, and those of the fifty-five meteorological stations in England, appended to the Registrar-General's Quarterly Reports, the mean temperature as given is not the true mean, but is an estimated mean, deduced from the united observations made with the self-registering and dry-bulb thermometers, the exact means of these instruments being first altered by certain tables, with the view of correcting them for diurnal and monthly range. The author objected to this mode of estimating the mean temperature, because the tables used for the purpose of correction were drawn up from insufficient data, and because the principle was bad of having observations made by one instrument, whose readings were liable to constant errors, made the basis for the correction of observations made by a more trustworthy instrument.

The tables for the correction of diurnal range used in England were drawn up from a five years' series of observations; but by a reference to the Scottish observations, it was attempted to be shown that a five years' period was far too short to elicit even a true mean annual temperature; while, by a reference to the mean temperature of corresponding months during five consecutive years, it was found that such a period gave still more incorrect results for the months.

It was then demonstrated, by a reference to the Greenwich monthly and daily observations, that the readings of the dry-bulb thermometer, at every two-hourly period, failed to give the true mean temperature either of the day or of the month, as they always missed the period of the day when the maximum temperature was attained, which, as indicated by the self-registering thermometer, was from 1° to 5° higher than any of the readings taken at the two-hourly periods. The consequence was, that the mean temperature deduced from the dry-bulb two-hourly readings was always *below* the truth, and it was impossible from them to ascertain

DOI: 10.4324/9781003650294-37

at what period of the day the highest temperature was attained, so that the chief elements for the calculation of the amount of diurnal range were wanting.

The data, however, which were used for the formation of tables of correction for the self-registering thermometric observations were shown to be even more faulty than those used for that of the dry-bulb; *firstly,* from the circumstance that the faulty dry-bulb mean was held to be the only true mean, and made the basis for the correction of the self-registering mean; and, from the circumstance that the whole of the self-registering observations were made with Six's registering thermometer, an instrument now considered inaccurate. By a reference to the Makerstoun Observations for a period of three years, and to those of Scotland for the last two years, it was demonstrated that, *when the self-registering thermometers are of proper construction,* the strict mean of the maximum and minimum readings appear to require no correction whatever, but indicate the true mean temperature. Particular attention was directed to the Makerstoun Observations for 1844, when hourly readings were taken with the dry-bulb thermometer, their strict mean giving the mean annual temperature as 44°9, while the strict mean of Rutherford's self-registering thermometer gave the annual temperature as 45°0, than which no results could be closer.

In the present state of the science, the author contended that the only safe rule for the estimation of mean temperature was to take the strict mean of the self-registering thermometers, provided these were of proper construction; and he strongly denounced the practice followed in England, of withholding the strict means of the different series of dry and wet bulb readings, and only publishing deductions, which the facts brought forward in his paper proved must, in most instances, be erroneous.

24

EXAMINATION OF THE STORMS OF WIND WHICH OCCURRED IN EUROPE DURING OCTOBER, NOVEMBER AND DECEMBER, 1863

Alexander Buchan

Source: Alexander Buchan, 'Examination of the Storms of Wind which Occurred in Europe during October, November and December, 1863', *Transactions of the Royal Society of Edinburgh*, 24, 1865, pp. 191–194, 196–198, 200–203.

XVI. – *Examination of the Storms of Wind which occurred in Europe during October, November, and December* 1863. By ALEXANDER BUCHAN, M.A., Secretary to the Scottish Meteorological Society. (Plates XIII. to XXI.)

(Read 3d April 1865.)

A brief account of the weather of this period as regards temperature was read before the Royal Society last year. It was drawn up at the request of Professor BALFOUR, to accompany his paper "On the Remarkable State of Vegetation in the Edinburgh Botanic Garden in December 1863."

From the 26th of October to the end of December the weather was in every way remarkable. Though frost occurred in the end of October and beginning of November it was not severe, and the temperature continued on the whole seasonable till the 12th of November. From this date till the end of the month it ranged unprecedently high, being 9° above the average temperature of the season. It then fell for the next ten days, but on no occasion below the average; and again rose considerably above the average during the week ending with the 18th of December. Under this genial weather vegetation in the open air advanced rapidly to a state of forwardness not usually seen till the month of March. In December 245 plants were in flower in the Gardens in the open air, and of these 35 were spring flowers. The frost which had occurred was insufficient to damage, to any material extent, 210 autumn-flowering plants; and the high temperature of November,

DOI: 10.4324/9781003650294-38

which was as high as what ordinarily occurs in the beginning of May, brought the spring flowers prematurely into bloom, so that there was to be seen the rare spectacle of sweet peas and hepaticas flowering together.

Thus, then, the atmosphere during this time was in a most abnormal condition in respect of temperature, which, of all the elements concerned, plays the most conspicuous part in destroying its equilibrium. It is not surprising, therefore, that the weather was equally remarkable, or even more so, for storms. From the 27th October to the 18th December, eleven well-marked storms passed over Europe in succession.

Since the space embraced by storms frequently includes the greater part of Europe, it is only recently, owing to the extension and growing popularity of meteorology, and the countenance now happily given to it by most European governments, that sufficient data could be obtained for a satisfactory treatment of the subject. For the observations of the principal observatories of Europe are too few in number, and at too great distances apart, to enable any one to lay down the isobarometric lines and general course of the winds, without drawing largely on conjecture and imagination. The recent multiplication of meteorological observatories is a great step in advance toward the discovery of the law of storms.

Observations have been received from 135 places scattered over Europe, from the Mediterranean to Archangel in the north of Russia, and from the extreme west to the Ural Mountains. All parts of Europe are pretty well represented except Central Russia, the south-east of Austria, and Turkey. The following are the sources from which the observations have been obtained: – The places in Scotland have been selected from the stations of the Scottish Meteorological Society; and most of the places in England and Ireland from Admiral FITZROY's Tables, published daily in "The Times," – the omitted observations on Sundays having been, to some extent, supplied by the observers themselves. Most of the continental stations have been taken from the lists given in LE VERRIER's bulletins of the weather, published daily in Paris, and from the "Meteorologische Jaarboek" of Dr BUYS BALLOT of Utrecht. I am further indebted to Dr BALLOT for his valued assistance in supplying me with additional observations to those printed in the Jaarboek. The observations from Russia were kindly furnished by M. FERDINAND MÜLLER, assistant in the Physical Central Observatory of Russia; those from Sweden, by M. BONNIER, Stockholm; those from Norway, by M. C. FIARNLEY, director of the Observatory, Christiania; and those from Denmark and Greenland, by Professor HOLTEN, Copenhagen. Rev. FRANCIS REDFORD supplied the observations from Silloth; Mr E. J. LOWE, those from Nottingham; Mr W. C. BURDER, those from Clifton; Mr HENRY DENNY, those from Leeds; Mr WILLIAM JOHNSTON, those from Banbury; Captain WILLIAMSON, those from Dublin; and Mr A. DICKEY, Queen's College, those from Belfast. The importance of observations from Iceland and Faroe was not overlooked, but we regret to say that no observations were made in those places during the period. I beg also to return my most grateful thanks to the MARQUIS OF TWEEDDALE and BARON BRUNOW for the

interest they took in this inquiry in procuring some of the most valuable of the observations, especially those from the north of Europe.

CONSTRUCTION OF THE TABLES AND MAPS.

The observations at the different places were made at 8 A.M. At the few places where they were made at a different hour, such as 9½ A.M. at Dublin, a slight correction was adopted to bring them into accordance with the others. The amount of this correction was deduced from the preceding and succeeding observations at the place, modified by the apparent course and rate of motion of the storm, as suggested by the observations of neighbouring stations. It not being necessary for this inquiry to descend to the thousandth of an inch of barometric pressure, or parts of a degree of temperature, the tables and maps may be accepted as representing the pressure and temperature of the air over Europe, with scarcely any deviation from the truth.

The *barometric* observations (Table I.) were brought to English inches, and then reduced to 32° and sea-level. Each observation, so reduced, was entered in its place on the map, and lines were then drawn through all those places where the pressure was equal. These isobarometric lines are given for every two-tenths in the difference of the pressure, – for 30·5, 30·3, 30·1, 29·9, &c., inches.

The lines of *temperature* have been laid down on a different principle. For lines exhibiting the actual temperature would fail to show, in a sufficiently clear manner, the real bearing of this important element, since the isothermals of October, November, and December run in a very irregular manner over the continent of Europe. Hence not the actual temperature, but the difference between the actual temperature and the mean temperature of each clay at the several stations is traced on the maps.

Dr BUYS BALLOT has calculated the mean temperature of many places in Europe for every alternate day of the year, and for a few other places ten-day means. The results were published in 1861, in "*La Marche Annuelle du Thermomètre et du Baromètre en Neerlande et en Divers Lieux de l'Europe.*" In the observations of temperature in the "Jaarboek," 1863, not the actual temperature, but the deviations from the mean temperature of each day, are alone given. These I have adopted *simpliciter.* Of the other stations, I have calculated the mean temperature of each day, using for this purpose Dr BALLOT's tables, Professor DOVE's mean temperatures, as given in "*Darstellung der Wärmeerscheinungen durch Fünftägige Mittel,*" 1863; the same author's "*Monats-und-Jahresisothermen,*" 1864; and the data in the Scottish Meteorological Society's "Proceedings" bearing on the subject. The differences between these daily means and the daily observed temperatures are entered in Table III., in which the minus sign shows that the temperature was under the mean; and if no sign is used it was above the mean. The stations are pretty well distributed over Europe, and are sufficiently numerous to show the changes of temperature which occurred near the earth's surface before, during, and after the successive storms. They were entered on the maps, and then, as in the case of the barometer, lines were drawn through those places where the differences were equal. They show where the temperature was the average 0°, and then in succession where it was 4°, 8°, 12°, &c., above the average or below it.

[. . .]

OBSERVATIONS OF THE BAROMETER.

The observations of the barometer are the most important of all the observations, since it is within the area where the barometer falls to some extent below the average that storms occur. Speaking roughly, the mean atmospheric pressure for these months is 29·9 inches. Therefore, the space comprehended within the isobarometric line 29·7, and the other lines showing a less pressure, may be called, for convenience' sake, *the area of low barometer.* Hence, while we trace the progress of these low pressures over Europe from day to day, we trace at the same time the progress of the storms.

[. . .]

FORM OF STORM AREAS.

The forms of forty-two different areas circumscribed by the isobarometric lines admitted of examination. Of these, thirty were either circular or slightly elliptical. In ten cases, the major axis of the ellipse was nearly double the length of the minor axis, and in one case it was three times the length. In two instances, 29·5 on the 11th November, and 29·3 on the 3d December, the outline of the areas was very irregular, owing to the occurrence of two central depressions in one case, and three in another, within it. It follows from this, that the storms most commonly assumed a circular or oval form, and that the ellipses were seldom much elongated.

The area over which the storms spread themselves was very variable in size, being seldom less than 600 miles across, but often two or three times that amount. This area was not constant, even as regarded the same storm from day to day, but varied in size, sometimes contracting and sometimes expanding. If it contracted, the central depression at the same time gave signs of filling up, and the storm of dying out. On the other hand, if the area widened, the central depression generally became deeper, and occasionally was broken up into two or more separate depressions, which appeared to become separate storms with the wind circling round each, as shown in the maps for 11th November and 3d and 17th December. These different depressions, however, soon reunited, and the storm proceeded as before.

[. . .]

DIRECTION IN WHICH THE STORMS ADVANCED OVER EUROPE.

The direction in which the storms advanced from the position they occupied on one day to the position they occupied on the next day, could be ascertained in twenty-four cases. In eleven of these, the progressive movement was to the N.E.; in four to the E.; in four to the S.E.; in two to the E.S.E.; and one to the E.N.E., S.S.E., and S.W. Thus, twenty-two travelled towards some point in the quadrant from N.E. to S.E., and only one took a westerly direction. Hence, these storms travelled as often toward the N.E. as toward all other points of the compass put together, and almost every one toward some point between N.E. and S.E.

The storms seldom proceeded in the same uniform direction from day to day. Though generally the change was not great, yet occasionally it was so. Thus of the many interesting features which marked the storm of the beginning of December, none were more remarkable than the sudden changes of its progressive movement. I have added in the Appendix observations relative to this storm at shorter intervals of time than 24 hours. From these it appears that the centre of the storm on the 2d was near Liverpool at 9 A.M. (Plate XX.); Worcester at noon; Oxford at 3 P.M.; Cherbourg at 6 P.M.; and Oxford at 9 P.M. The greater number of observations at 9 P.M. show three depressions: – 1*st,* at Shetland, 28·88; 2*d,* at Oxford, 28·89; and, 3*d,* in Holland, 29·22. It is very probable that the storm had separated into two parts near Liverpool, one of which took a north-easterly course toward Shetland, and the other a south-easterly course toward Cherbourg. At 9 A.M. of the following morning (Plate XX.), the first had advanced to the N.E. to Christiansund; the second had advanced northward to Shields; and the third had advanced eastward to Denmark. At 9 P.M. of the 3d, the first was no longer visible, having probably left Europe by the North Cape, while the second had advanced to the north of Holland. At 9 A.M. of the 4th (Plate XXI.), it had advanced to Copenhagen, had greatly diminished in area, and the central depression was five-tenths less than on the previous morning; and at 9 P.M., the observations at Christiania, Kiel, and Konigsberg show that the atmospheric equilibrium was restored, and the storm consequently had died out.

Most of the storms left Europe by the North Cape or the north-east of Russia; but two of them (Storms III. and X.) wasted away and died out before reaching Russia.

[. . .]

COMPARISON OF THE BAROMETRIC AND THERMOMETRIC LINES.

The observations of the thermometer do not equal in importance those of the barometer, for this among other reasons, that while the barometer measures the weight of the whole atmosphere pressing on it, the thermometer gives only the temperature of that portion of the air which is in immediate contact with the earth. There appears to be little apparent connection between these lines at first sight; for while the barometric lines approach more or less closely the curves of the circle or the ellipse, the lines of equal thermometric disturbance present the greatest possible irregularity of form. When, however, the attention is confined to the region of greatest barometric disturbance, a remarkable connection is at once observed.

It will be seen that in all cases the temperature rose a few degrees over the space toward which and over which the front part of the storm was advancing, and fell at those places over which the front part of the storm had already passed. In other words, the temperature rose as the barometer fell, and fell as the barometer rose. Generally, the temperature in advance of the storm was above the average, and in the rear of the storm below it. But if it was considerably above the average in advance of the storm, it was still above the average when the storm had passed, though lower than it was before.

In one or two cases the temperature, after falling a little, rose in what appeared to be the wake of a storm; but in these cases the observations of the following day showed that another storm was advancing close upon the one already past. The high temperature thus indicated the approach of the second storm, and properly belonged to it.

OBSERVATIONS OF RAIN AND CLOUD.

We learn from the observations that as long as the barometer did not fall below the mean, there was no continuous rain anywhere, but blue sky prevailed, varied with partially clouded sky or with fog. But when the barometer fell, the sky began to be obscured, and rain to fall at intervals; and as the central depression advanced, the rain became more general, heavy, and continuous. After the centre of the storm had passed, or when the barometer had begun to rise, the rain generally became less heavy, falling more in showers than continuously; the clouds began to break up, and fine weather, ushered in with cold breezes, ultimately prevailed.

In order to show where the greatest amount of rain fell, Table IV. has been prepared, giving the rainfall at those places over which, or near which, the storms passed. It is necessary to explain that, as far as known, the rain given in the table fell during the twenty-four hours preceding the date of each entry. Thus, as the centre of Storm II. was west of Ireland on the 1st November, and in the centre of England on the 2d, the rainfall of the 2d took place as the storm travelled between these places. The rainfall at the Irish, English, and French stations, over which the front part of the storm had passed, was excessive – two inches having fallen at Brest; one inch at Liverpool and Dover; and about three-quarters of an inch at L'Orient, Galway, &c. On the same day the rainfall in Scotland was everywhere small, none falling at many places, and the largest fall being about one-sixth of an inch at Portree, in Skye. In Scotland, where the fall of rain was small, the wind was feeble, in no case blowing a gale; and the barometer, though low, had varied little during the twenty-four hours, and a pressure almost equally low prevailed for a considerable distance round. On the other hand, where the rainfall was in excess, violent gales prevailed, the fluctuations of the barometer had been great, and the isobarometric lines were much crowded together in the vicinity. On the following day the storm had passed eastward to Denmark. The rain over the west of England and of France had diminished, but increased over the *east* of England, and in Belgium and the Netherlands, over which the storm had travelled on its way to Denmark. The Irish rainfall was small, but not so small as would have been, but for the advance of Storm III., whose rainfall swelled the amount; and the same cause increased the Scottish rainfall.

[. . .]

OBSERVATIONS OF THE WIND.

Every one of the storms on each day presented the winds under the same conditions, viz., whirling round the area of low barometer in a circular manner, in a direction contrary to the motion of the hands of a watch, with a constant tendency to turn inwards towards the centre of lowest barometer. The wind in storms neither blows round the centre of least pressure in circles (or as tangents to the concentric barometric curves), nor does it blow directly towards that centre. It takes a direction nearly intermediate, approaching, however, more nearly the direction and course of the circular curves than of the radii to the centre. To this general rule none of the eleven storms any day offered an exception. When the centre of the storm was in a situation where observations were made all round it, the following were the general directions as observed: –

At places	S.	of the centre of least	pressure, the	wind was generally	S.W.
,,	S.E.	,,	,,	,,	S.
,,	E.	,,	,,	,,	S.E.
,,	N.E.	,,	,,	,,	E.
,,	N.	,,	,,	,,	N.E.
,,	N.W.	,,	,,	,,	N.
,,	W.	,,	,,	,,	N.W.
,,	S.W.	,,	,,	,,	W.

The greater the force of the wind at any place, the more nearly did it approximate to the directions here indicated. On those occasions when no observations were obtained from one or more of the sides of the storm, such observations as were obtained followed the same rule. Hence the atmosphere on every occasion rotated round the centre of the storm; and it should be kept in mind that this is no theoretical statement, but the result of observations faithfully put down on maps.

It will follow from this, that as the storms advanced to the eastward the general veering of the wind at places lying north of the central path of the storm would be from the N.E. by N. to W.; and at places to the south of the centre, from the N.E. by E. and S. to N.W.

[. . .]

STORMS OF THE MEDITERRANEAN.

The observations from Austria, Turkey, Greece, Russia, and Syria are too scanty to enable us to trace satisfactorily any of the storms which occurred there during the period. There is enough, however, to show that the conclusions which may be drawn from those storms which passed over northern and western Europe cannot safely be applied to the storms of the Mediterranean, as regards their form, the direction from which they come, and the course generally pursued by them.

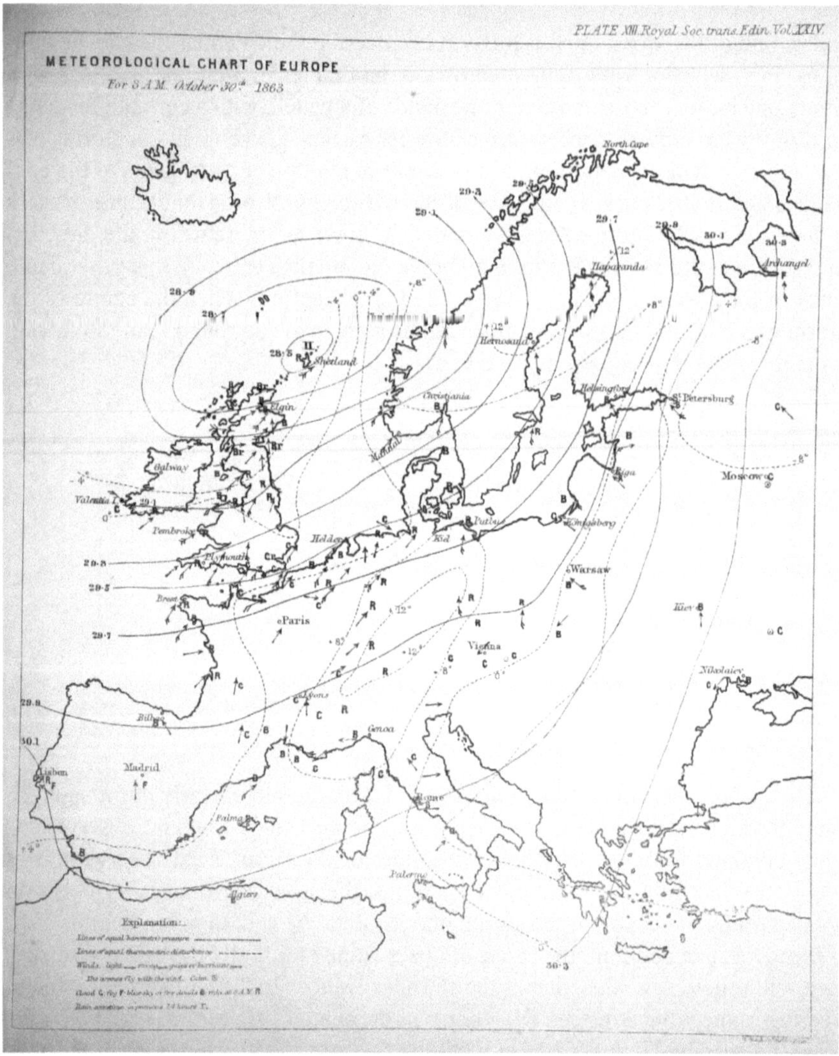

METEOROLOGICAL CHART OF EUROPE

For 8 A.M. October 30th 1863

25

A HANDY BOOK OF METEOROLOGY

Alexander Buchan

Source: Alexander Buchan, *A Handy Book of Meteorology* (Edinburgh and London: William Blackwood and Sons, 1867; 2nd ed. in 1868), pp. 9–14, 339–348, 350–355.

METEOROLOGICAL SOCIETIES AND STORM-WARNINGS.

19. The establishment of Meteorological Societies during the last twenty years must also be commemorated as contributing in a high degree to the solid advancement of the science, which, more than any other, depends on extensive and carefully-conducted observations. In this respect the United States stand pre-eminent, the observers there numbering about 800. Great Britain is also well represented in the English and Scottish Societies, which together number about 150 observers. In Austria, Switzerland, France, Prussia, Italy, Russia, Norway, the Netherlands, and other European countries, meteorology is being widely cultivated. In Austria alone the number of stations is 118, and in Switzerland 83. Considerable attention has also been given to the rainfall in Great Britain and Ireland; and, chiefly through the self-denying exertions pf Mr G. J. Symons, London, about 1500 rain-gauges are now registering the rainfall of the British Islands. An inquiry has been carried on during the last six years, having for its object the determination of the causes which affect the rainfall in the basins of the Rhône and Saône. Observers in Germany and Great Britain have been secured to co-operate with the French observers, and under the able management of M. Fournet, important results respecting the rainfall and the progress of storms will be obtained, from which it is hoped measures may be taken to avert the calamity occasioned by those great floods which periodically carry devastation and ruin over that part of France.

20. A special object of meteorological inquiry is to ascertain the degrees of heat, cold, and moisture, peculiar to different localities, and the usual periods of their occurrence, with a view to discover their effects on the health of the people, and on different agricultural products. With regard to questions of such general interest affecting the health and food of the people, these societies have already collected much valuable information, which but for their aid could not have been obtained.

DOI: 10.4324/9781003650294-39

21. But perhaps none of the arts have benefited to so large an extent by the labours of meteorologists as navigation. The knowledge thereby acquired of the prevailing winds in different parts of the earth during the different seasons of the year, the region of calms, the parts of the oceans swept by devastating storms at particular seasons, and the laws of storms, has saved innumerable lives and much property; and by pointing out the quickest routes to be followed, has shortened voyages between distant countries. In connection with this department of the science, the name of Captain Maury will always be remembered with gratitude for the signal service he has rendered to navigation. The good work thus begun by him has rendered intercourse among nations safer and more expeditious; and when future observation has supplied the materials requisite to enable us to correct the inevitable mistakes and fill up the blanks of his ocean charts, the benefits this celebrated meteorologist has conferred on the human race will more conspicuously appear.

22. *Prediction of Storms.* – Another fruit of the multiplication of meteorological stations is the prediction of storms and the foretelling of the weather. It is impossible to overestimate the value of storm-warnings to the shipping interests. In the north temperate zone, observation shows that storms almost invariably come from some westerly point, and thence follow an easterly course. In the United States of America, it is easy to warn seaports of the approach of storms; for as soon as a storm appears in the western States bordering on the Rocky Mountains, it is intimated to the central office in Washington, followed in its march by the telegraph, and timely warning of its approach is sent to the coasts which it will visit. The United States of America is thus favourably circumstanced for carrying out effectually the system of storm-warnings.

23. On the contrary, Great Britain, France, and the rest of Western Europe, are unfavourably situated to allow of timely warning being given of coming storms. If no warning is sent till the storm has made its appearance, it is too late for the western seaports. In Europe, however, stormy weather is accompanied by a diminution in the atmospheric pressure, the centre of which, after traversing more or less of the Atlantic, arrives on the coast of Europe. The existence of this diminished pressure is made known by the barometer when the maximum depression is still at a considerable distance out in the ocean; and collateral information pointing to an advancing storm is to be obtained from the direction of the wind and the cirrus cloud. Here, then, we have the materials for foretelling the approach of storms on the west coast of Europe. For though we have not the same advantages towards arriving at the degree of certainty of the American predictions, and so of telegraphing to ports on the west coast that a storm is *actually seen advancing on them,* yet from the premonitions afforded by the barometer, the wind, and the cirrus cloud, we can warn them to prepare for a storm likely to visit them. The giving effect to this idea constitutes the splendid contribution to practical meteorology made by Admiral Fitzroy in February 1861, by the system of storm-warnings which has since been adopted by almost every country in Europe – a service

which has made his name a household word, and entitled him to be considered as a public benefactor.

24. Considering the practical application of the knowledge of storms towards the saving of life and property, it is a duty incumbent on European meteorologists (being the most important practical problem they have to solve) to examine, analyse, and carefully study in detail the storms which have traversed Europe during the last few years, – when, from the growing popularity of the science, meteorological stations have largely increased – with the view of ascertaining the course storms usually follow, and the causes by which that course is determined, so as to deduce, from meteorological phenomena observed, not only the certain approach of a storm, but also the particular course it will take in its passage over Europe. In carrying out this extensive investigation, the 'Bulletin International' of Le Verrier, published daily, if supplemented by additional observations from places in the British Islands, and in Northern, Eastern, and South-Eastern Europe, will furnish the required materials. This admirable publication, which must be regarded as the latest important step taken in the progress of meteorology, shows graphically the atmospheric pressure, and the direction and force of the winds, each morning over Europe, together with tables of temperature, rainfall, cloud, and sea disturbance.

25. The next important step to be taken in the development of the capabilities of the science to promote the interests of humanity, will, we trust, be the giving practical effect to a suggestion made by a writer in the 'Athenaeum' of the 28th September 1867, to the effect that Mr Reuter should add to his telegraphic messages a line, which need not consist of more than six letters and as many figures, announcing the state of the weather from all points of his world-wide correspondence. Thus, B. 30.15, T. 51, N.E., with R., S., F., would indicate that the barometer was 30.15 inches, the temperature 51°, the wind from the N.E., and R., S., and F., rain, snow, or fair, as the case may be. The benefit of such a practice would, it is justly added, be incalculable, for ere long all who are interested in the weather would be better prepared to take advantage of the warnings of storms on all occasions, particularly during the seasons of the year when gales most frequently occur. Much might also be contributed to the meteorology of the globe by missionaries sent out to the islands of the Pacific, and other parts of the world remote from civilised society. If possessed of the principal meteorological instruments, and of a moderate knowledge of the science, the missionary could not fail to benefit his fellow-men, and legitimately increase his influence over them, by often warning them of approaching storms and other changes of the weather. This can be done more easily and with great certainty in tropical and subtropical countries, where a fall of the barometer slightly greater than usual, and a change of the wind from its ordinary direction, indicate the presence of a storm at no great distance. By closely watching the wind and the barometer, the course the storm is taking can be readily known.

26. In the schools of the United States of America, Meteorological Observations, and the keeping of Meteorological Registers, form a part of the common

education of the people. Also in the higher schools of France, and some other European countries, systematic instruction is communicated on this subject. But in this country few even of the liberally educated classes are able to read from a vernier; they are ignorant of the use of the movable cistern of a barometer; they have not the elementary knowledge to give an intelligible interpretation to the fluctuations of the barometer as indicative of coming changes of the weather; and when required to send their barometers to a distance for repair, so ignorant are they of their construction, that they forward them by rail as ordinary parcels, thus almost to a certainty securing their destruction. This state of things is the necessary consequence of the general neglect which meteorology receives in our educational system. There are, however, a few noteworthy exceptions. Meteorology has been taught for upwards of thirty years in the Dollar Institution, which has long been distinguished for the lead it has taken in incorporating science into its curriculum of study. This example has recently been followed by the Roman Catholic College at Stonyhurst, the Grammar School of Aberdeen, the High School of Inverness, Lerwick Educational Institute, Elgin Institution, Larchfield Academy, and other schools in the country. But the objects of meteorology can never hold that place in the public mind to which they are entitled, till the science becomes, as in America, a recognised branch of education. As contributing towards this desired result, the name of Mr Thomas H. Core, now of the International College, London, deserves special mention, in respect of his having, while mathematical master in the Church of Scotland's Normal School, Edinburgh, systematically trained the students to regular habits of meteorological observation. When the law of storms begins to be generally understood, and, as a consequence, the value of observations of the barometer, of the direction of the wind, and of the appearances of the clouds, comes to be appreciated as heralding changes of the weather, we cannot doubt that this very practical age will take steps to provide for the instruction of the people in the elementary facts of meteorology, and in the use of the different meteorological instruments.

[. . .]

STORM-WARNINGS.

659. It is in tropical and sub-tropical countries that an isolated observer may, with the greatest degree of certainty, predict the approach of gales and hurricanes. In these regions barometric pressure and the other meteorological elements are so constant from day to day, that any deviation, even a little, from the average of the hour and season in respect of pressure, the direction and strength of the wind, and the direction and amount of cloud, supposes the existence of a storm as the disturbing agent at no great distance. This important practical problem has been worked out with great success by Mr Meldrum at Mauritius. A few words will explain the method of proceeding.

660. At Port Louis, Mauritius, in July, the mean pressure at 32° and sea-level is 30.274 inches at 9 A.M., from which it falls to 30.186 inches at 3 P.M., it then

rises to 30.236 inches at 10 P.M., and again falls to 30.192 at 4 A.M. The mean direction of the wind for the same month is from S.E. to E., generally E.S.E.; and its force about 0.41 lb. to the square foot, subject to an increase or decrease at different times of the day. Suppose now that the barometer was observed to fall after 9 A.M. with a greater rapidity than was due to the usual daily range; or that in the afternoon it did not rise to the second maximum, or that it fell instead of rising; or suppose, in short, any deviation from the observed daily march, then it is certain that there is *somewhere* an atmospherical disturbance, near enough to Mauritius to influence the pressure.

661. The direction in which the disturbance is from Mauritius is readily known from the wind. Thus, during the gale which occurred from the 11th to the 17th January 1860, the wind veered from S.E. to S., S.W., and W. This was a great revolving storm, the centre of which passed on the east side of the island, its nearest distance at any time being 129 miles. Here it will be observed that the storm with its low pressure being in the east, and travelling to the south-westward, the wind at Mauritius veered from E.S.E., its usual direction, into the west; or, to express the relation from the opposite point of view, the veering of the wind pointed to the east and south as the seat of the atmospheric disturbance. In all cases the wind is found to veer round so as to blow approximately at right angles (60° to 80°) to the bearing of the storm. The distance of the storm may be closely approximated to by observing the rate and amount of fall of the barometer, taken in connection with the observations of the wind and the clouds. The rate and direction of the progressive motion of the storm are known chiefly from the veerings of the wind.

662. Mr Meldrum worked at the solution of this problem for several years, and took a very effective course to test the justness of his conclusions. He laid down the path of the storm from the Mauritius observations alone; and on afterwards receiving, from captains of vessels who had encountered the storm, a note of the latitudes and longitudes, stated, to their surprise, when and where they had the storm and the direction and veerings of the wind during its continuance. After sufficient experience had been acquired, a note was sent to the daily newspapers when it was concluded from the Mauritius observations that a storm was abroad, stating its position and probable course from day to day. These notifications, inferences (or predictions, as they may be called when the storm afterwards reached Mauritius), were carefully compared with the logs of ships which afterwards touched at the island. No case of failure has occurred since these notices began to be sent to the daily press.

663. This gratifying result is of great value in various ways. It shows what may be done at an isolated station in the ocean; in other words, what may be done in ships at sea. Since these deductions regarding storms are made at Mauritius, from the observed deviations from the average height of the barometer and the direction and force of the wind, the reduction of the ocean statistics of meteorology, and the charting of the results for the use of sailors, becomes a question of pressing importance. Maury's Sailing Charts have already been referred to for the signal

service they have rendered to navigation. The Isobarometric Charts given at the end of this book will, it is hoped, prove a valuable auxiliary to navigation; and when the whole thirteen have been prepared, their value will be much increased.

664. In discussing Meteorological Ocean Statistics referring chiefly to tropical and sub-tropical regions, the most important points to be determined are the following: 1. The daily range of the barometer for the different months, as applicable to different parts of the ocean; or, better still, the law of its variation according to season, latitude, and proximity to large masses of land. 2. The mean monthly barometric pressure of the different seas chiefly traversed by ships, given for every 0.050 inch, or, if possible, every 0.025 inch of mean pressure. 3. The deviation from the mean pressure of the month which may be regarded as an amount of disturbance sufficient to be considered as indicative of a storm at no great distance. 4. The mean direction and force of the wind for the different months. This is no doubt a vast problem which it will take many years to work out. Mr Meldrum's paper on the Meteorology of Port Louis in the Island of Mauritius, printed *in extenso,* with its elaborately prepared 42 Tables, in the Report of the British Association for 1867, is one of the most valuable contributions towards its solution yet made. Much may in the mean time be done by ship-captains to enable them to steer clear of storms, if they would carefully note the occurrence of deviations from the average daily barometric range, and barometric fluctuations generally, taken in connection with the direction, force, and veering of the wind.

665. In regions such as Great Britain, an isolated observer cannot, with a like certainty, draw conclusions from his own observations regarding the approach of storms and other weather-changes. But he may almost, if not altogether, attain to the same degree of certainty, if he be assisted by a sufficient staff of observers, well distributed over Western Europe; and be able, when he considers it necessary, to communicate with them through the telegraph. The possibility of this is evident from what has been already stated in the chapter on Storms, and its practicability is seen from what was done by Admiral Fitzroy, and his assistant, Mr Babington, previous to December 1866.

666. In his letter of 15th June 1865 to the Board of Trade, General Sabine states that he had examined the warnings given on our coasts during the two years ending 31st March 1865, and found that in the first year 50 per cent, and in the second year 73 per cent, were right; and he adds, "That it seems not unreasonable to attribute to increased experience the marked improvement of these results upon those of the preceding year, and to anticipate still further improvement." The above percentages of warnings which turned out correct, were formed from the warnings sent to all the seaports. But, since a number of these places were not very favourably situated for receiving timely warning, being too near the west coast, if the warnings sent to places more favourably situated were examined, fewer failures would appear. The Channel may be considered as thus favourably situated. Warnings of storms were sent by Admiral Fitzroy to the north and west coasts of France. Captain de Rostaing, while at the head of the French Marine Meteorological Department, compared the warnings sent from the Meteorological Office in

London with the gales *(vents forts,* or winds represented by 8 on Beaufort's scale) which occurred in France during the two winters 1864–65 and 1865–66. The results were published in the *Revue Marine et Coloniale.* From this investigation it appears that, out of 100 warnings sent to the north coast of France during the first of these winters, 71 were realised, and during the second winter 76; and out of 100 storms which occurred, 89 were signalled during the first winter, and 94 during the second winter. This result is remarkable, as showing that of the storms which occurred in the north of France during these two winters, warning of the approach of 11 out of every 12 was sent.

667. A word or two may be here said regarding the causes of a considerable number of the failures. The first and chief source of failure was the want of a day-and-night watch maintained on the west coast of Ireland. This, which an examination of British storms shows at once to be the first requisite of any system of storm-warnings, was pointed out by R. Russell some years since, but with no practical result. Suppose only two observations a-day made in the west of Ireland, one at 9 A.M. and the other at 2 P.M., and that shortly after the latter hour the barometer began rapidly to fall and a storm to arise – then, since the average rate of the progressive motion of our British storms is nearly 20 miles an hour, rising at times to 30, and on rare occasions to 45 miles an hour, it would happen that by the time the telegrams were received next morning, and the warnings issued, the storm had already passed many of the seaports, especially in the west; was passing some places; whilst to only a very few places in the east was it yet advancing. Thus the storm outstripped the warning at all except a very few places, to which only, therefore, it was of use as a warning.

668. The occasionally late arrival or non-arrival of the telegrams was also a source of failure. For thus the cautionary signals were sometimes ordered to be hoisted at places to which no warnings might have been sent, if the telegrams had arrived up to time. The direction and force of the wind were determined by estimation only, by the telegraph clerks intrusted with sending the telegrams; and these are known to have been in some cases defective. The same liability to error will to some extent continue until each station be provided with a vane to show the direction of the wind, and a simple pressure anemometer to indicate the force on the scale 0 to 12, together with the maximum force. There were too few stations on the Continent to the east and south-east of Great Britain, so that an important element in estimating the probable force of easterly and southerly gales, especially on the east coast, was wanting. But the barometric observations, which are the basis of every system of storm-warnings, were remarkably accurate.

669. If, then, allowance be made for the failures which arose from these causes, it can scarcely be doubted that we have the explanation of nearly every one of the 10 storms which were not signalled out of the 93 which occurred on the north coast of France during the winter of 1864–65, and of the 6 not signalled out of the 99 which occurred during the following winter. This leads to an important result. Since experience has shown it to be practicable for the north of France – and what

is practicable for the north of France applies equally to the greater part of Great Britain – it follows that, practically, the approach of every storm which visits Great Britain may be signalled some time before it reaches the different ports.

670. There is thus no difficulty in carrying out a system of storm-warnings in this direction. The practical difficulty lies not in securing that a warning of each storm will be received before it breaks out, but in securing that no warning be sent except to those places at which the wind will rise to the force of a gale. If this happened uniformly over the whole region included at any time within the low barometric pressure marking out the atmospheric disturbance, of which the wind-storm is one of the accompaniments; or if a certain wind-force invariably attended on a certain pressure in each storm – then the problem of storm-warnings might be regarded as completely solved. But it has been shown that in an atmospheric disturbance the force of the wind at a particular place is proportioned neither to the amount of fall of the barometer, nor absolutely to its rate of fall at any time, but to the relative differences of pressure as observed simultaneously at two other places between which it lies. To determine these relations for any place for the next 30 hours, so that from the result warnings might be framed for that place, is, it needs scarcely be said, a problem which admits only of an approximate solution; any exact solution and exact rules framed from such solution to guide in issuing storm-warnings being impossible. But in this as in many other cases where science is applied to the practical affairs of life, no such exactness is required. It is only necessary that the number of successful warnings be sufficient to warrant the practice of issuing them.

671. It is evident that the first requisites in a system of storm-warnings are these: – 1. A day-and-night watch at Valentia in the west of Ireland; 2. The telegraphing to the office in London notice of storms in the north-west of France and Spain, the west of Ireland, and the west and north of Scotland, immediately on their occurrence; 3. Power to the person intrusted with the issuing of the warnings to use the telegraphic wires at his discretion, so that he may ascertain how the storm advances. Till this be done, a considerable percentage of failures must occur, whatever be the system of warnings practised. The only difficulty in carrying it out is the expense. Any sum hitherto granted by the British Government for this purpose is altogether inadequate for efficiently carrying it out. The sooner it is understood the better, that if early and more certain warning or notification of the approach of storms be wanted, it must be paid for; and that if not paid for, it must follow that storms will occasionally burst upon us unprepared.

672. The importance of stations in Scotland at Oban and Golspie, or the extreme north-western and northern points to which the telegraph wires extend, is very great, since these places are best situated for giving the earliest possible intelligence of the approach of heavy gales from N.W., N., and N.E. The storm of the 1st and 2d December, when the Ivanhoe of Leith and other 60 vessels were lost, and 201 vessels damaged, was of this description. It came on suddenly. The rate and manner in which it advanced over Great Britain is seen from the following

figures, showing the hour at which the wind had risen to at least a gale at the different places: –

Hour.	Gale had begun at
5 A.M.	North Unst (4.15 a.m.); the Hebrides and nearly all Shetland.
7 „	Skye, Wester Ross, Sutherland, and Orkney.
9 „	Ardnamurchan and Tarbet Ness.
11 „	Kinnaird Head and Islay.
1 P.M.	Isle of May; Pladda, south of Arran and Cork.
3 „	Hawick; Douglas, Isle of Man; and Dublin.
5 „	Liverpool.
7 „	Yarmouth and Oxford.
11 „	Christiania in Norway.

Hence, then, this storm might have been telegraphed from the north-west of Scotland in time for ample warning of its approach to be sent to every principal seaport in Great Britain.

673. If there had been telegraphic communication with Stornoway, where the storm began as early as 5 A.M., its approach could have been notified four or five hours earlier than from any place at present in connection with the telegraph. Hence the great importance of Stornoway in relation to storm-warnings. The advantages which would result to the shipping interests from weather-telegrams from this place would soon pay the laying of a cable to the Lewis. The great shipping interests of the Clyde and the Mersey are chiefly interested in the carrying out of this suggestion, because of their greater proximity to the west, and the small extent of the arms of the sea, which peculiarly expose vessels to be wrecked. Dr Buys Ballot has for some years strongly advocated the Azores as an outpost which could not fail to prove of the utmost importance in warning of the approach of storms. The observations which the Scottish Meteorological Society has received from their observers in Faro and Iceland conclusively show the great importance of these islands in connection with storm-warnings. Indeed, with daily telegrams from the Azores and Iceland, two and often three days' intimation of almost every storm which visits Great Britain could be had. It is to be hoped that in laying down any new Atlantic cables, the advantages to navigation and commerce of either of these islands being in the route of the telegraph will not be lost sight of.

674. As the laws which regulate the general movements of the atmosphere become better known, the knowledge thus acquired will no doubt be turned to excellent account in its application to storm-warnings. Of these laws the most important are those which are concerned in the production of storms, and in their translation from place to place over the earth's surface till they die out. It is almost universally held as a truism among meteorologists that storms are produced by the mutual action of the polar and the equatorial currents, the chief difference of opinion being as to how the one acts on the other. Have we evidence, in the storms which pass over Western Europe, of the coexistence side by side of these great

currents? Or is it the fact that in these storms the moist equatorial current is invariably succeeded and swept eastward by a polar current? On this point I would offer a few suggestions.

675. It will be necessary first to state what is here meant by polar and equatorial currents. A polar current is a movement of the atmosphere, which the distribution of the barometric pressure and the winds at the time connect with the polar regions as the source out of which it is flowing. An equatorial current, on the other hand, is shown by the pressure and the winds to have its origin in southern latitudes – to be, in short, a vast atmospheric stream issuing from tropical or subtropical regions, and thence spreading itself continuously towards the north. In the case of a polar current, the barometer is highest in the north, from which it falls on proceeding southward as far as the current extends; but with an equatorial current pressures are highest in the south, from which they diminish on advancing northward, till the northern limit of the current is reached. These two currents, as regards moisture and temperature, are directly opposed to each other; but they are not opposed to each other as regards the absolute pressure which accompanies them at any place – the opposition being only in the manner of the distribution of the pressure. Thus it not unfrequently happens in Great Britain that the barometer is low during the prevalence of a polar current, in which case Great Britain is near the southern limits of the current; and high during the prevalence of the equatorial current; indeed, some of the highest pressures which occur advance on Great Britain *viâ* Spain and France.

[. . .]

678. It rarely, if ever, happens in British storms that the pressure rises equally all round from the region of lowest pressure as a centre, but the pressure is generally observed to be much higher on one side than on any of the other aides of the storm. This is a point of great importance, since it directs attention to that side of the storm where the greatest force of the wind will be felt, and shows at the same time the probable degree to which the temperature will fall in the rear of the storm. It also sometimes happens that high, and even rapidly rising, barometers are observed immediately to the E. and S.E. of the storm, as it passes over Great Britain; and in such cases the storm is most violent when the wind is from S.S.E. to S.S.W. The great hurricane which swept over Scotland on the 24th January 1868 was of this description.

679. If what has been suggested in par. 481 as the law regulating the movements of the lower and upper currents of the atmosphere be correct, the following will be the explanation of these facts regarding storms: The increase of pressure over any region is brought about by means of the upper currents. When a storm is passing over Great Britain at the time atmospheric pressure remains persistently low in Iceland, as happened in the beginning of December 1863, it follows that the ascending current which rises from the storm cannot, when it reaches a great height, flow towards Iceland, but will flow over the region where the pressure at that height is least, or generally where the mean temperature and humidity of the whole atmospheric strata is least. Now, observation shows that this usually

takes place in the rear of the storm; in other words, the upper current generally flows from the central region of the storm backward towards the S.W. and W., thus increasing the pressure there. But when, as in the beginning of December 1867, high pressures from Iceland follow the central depression, the temperature and humidity there are much lower than elsewhere, and consequently the upper current will flow over towards the N.W. Lastly, on the 24th January 1868 the temperature was lower in Scotland than in Farö, while the storm was yet advancing on Scotland; and thus part, at least, of the upper current flowed over to the eastward as the storm advanced; in consequence of which the heaviest part of the storm occurred when the wind was from S.S.E. to S.S.W., or in the direction from which the wind in storms is usually light. The temperature in Scotland was from 10° to 17° lower on the 23d, when the storm was advancing, than it was on the morning of the 25th, when its centre had passed, and barometers were everywhere rising, and the wind had veered to the N.W.

680. Whatever may be thought of this explanation of the facts of storms, there can be no difference of opinion regarding the facts themselves. The practical conclusion to be drawn from them is this: It is most desirable that the region from which daily telegrams of the weather are received be extended farther into the continent of Europe, especially toward the S.E., E., and N.E., in order that the system of storm-warnings may be made more definite and precise.

681. The following are the regions of the globe which are chiefly characterised as stormy: – The region of the typhoons in China, which lies between the high pressure of the North Pacific and the low pressure of Asia: their general course is from E.N.E. to W.S.W. The typhoons of Hindostan occur between the high pressure in the Indian Ocean and the low pressure of Asia: their general course is from S.S.E. to N.N.W. The storms in the north-west part of the Indian Ocean have their origin in the low pressure towards which the trades blow, which lies between regions of high pressure to the north and south; their general course is towards W.S.W., after which they frequently turn round to S.E. The storms in the south of the Indian Ocean occur between the high pressures there and the lower pressures farther south; their course is, roughly speaking, eastward, the exact track of the greatest barometric disturbance being yet unknown. The West India hurricanes occur between the high pressure in the Atlantic north of the equator, and the lower pressure in the Gulf of Mexico, the United States, and farther north in the Atlantic; their course is nearly along the isobarometric line, 30.1 inches (Plate I.), which it will be seen approximates to a parabola. In North-Western America storms advance from the south-west, following a course between the high and low pressures there. In the United States in winter, they principally occur about the isobarometric line 30.15 inches (Plate IL), having a high pressure over the Gulf of Mexico to the south. In Europe they generally pass between the isobarometric lines 29.9 and 29.7 inches (Plate II.), being thus between the high pressure to the south in the Atlantic and the low pressure to the north. In all these cases the storm-tracks are contiguous to a high and comparatively steady barometer over a large evaporating surface on the one hand, and a much lower pressure on the

other hand. When the whole thirteen Isobarometric Charts have been prepared, the interesting connection here suggested may be better discussed.

682. It is seen from Plates I. and II. that the highest atmospheric pressures are chiefly distributed over the northern hemisphere in January, and the lowest in July. If the pressure due to the elasticity of the vapour of the atmosphere was deducted from that of the whole barometric pressure, the pressure of the dry air which remains would present this relation in a much stronger light. The causes by which the dry air should have a minimum pressure in the hottest months of the year have been already referred to in par. 96.

683. Isobarometric Charts of dry air for the globe, taken in connection with charts representing the whole pressure, may be regarded as the last step to be taken towards attaining a position whence the great movements of the atmosphere over the globe may be observed, and the causes of these movements satisfactorily explained. But we are very far from this result, owing to the scantiness of observations, and reduced means, of the elastic force of atmospheric vapour. Considering the great importance of the inquiry, it is hoped that meteorologists in different parts of the world will lend their assistance, by reducing and publishing their observations of the vapour of the atmosphere.

684. Whilst the rule is, that the minimum pressure of dry air occurs in the hottest months, there are important exceptions. The following table gives the mean pressures of the dry air at five places in the British Islands from Sandwick in Orkney to Greenwich, calculated for the ten years from 1857 to 1866; they are not reduced to sea-level: –

	Sandwick.	Culloden.	Aberdeen.	Glasgow.	Greenwich.
	Inches.	Inches.	Inches.	Inches.	Inches.
January,	29.327	29.339	29.402	29.337	29.584
July,	29.403	29.364	29.400	29.293	29.415

Hence in the north of Scotland the pressure of dry air is greater in summer than in winter; at Aberdeen nearly equal; but at Glasgow the pressure of the dry air is greater in winter than in summer, and at Greenwich the difference in favour of winter is still greater. It does not admit of a doubt, that in Iceland the pressure of dry air in summer exceeds that in winter in a still greater degree than in Orkney. Deducting the mean elastic force of vapour, as observed in Iceland during the three winter and summer months of 1866 and 1867, from the whole barometric pressure of thirteen years, it is seen that the probable pressure of the dry air in winter is 29.215 inches, whereas in summer it is 29.402 inches. At Sitka, in the north-west of North America, the pressure of the dry air on a mean of ten years is 29.393 inches in January, and 29.515 inches in July; whereas to the south, at New Westminster, in British Columbia, the pressure of the dry air for two years

was 29.885 inches in January and 29.596 inches in July – thus showing the same peculiarity to prevail in the North Pacific that prevails in the North Atlantic.

685. This low winter pressure is explained by the ascent, and thence flowing away, through the upper currents, of the moist warm air which is poured into these regions from the respective oceans to the south of them; and the outflow is maintained at the high rate necessary to reduce the winter pressure to so marked an extent by the extraordinarily high pressures in Asia, Europe, and North America, by which they are surrounded. As the pressure falls in summer over these continents, the force, arising from the difference of the pressures, which originates and maintains the outflow, is diminished; and consequently less air being drawn off from the North Pacific and North Atlantic in summer, the pressure is more nearly maintained at a normal height. This is doubtless the explanation of the higher pressure of the dry air in these regions in summer – viz., a falling-off of the outflow owing to a diminished pressure all around, and not, as has been supposed, to an overflow into these regions by the upper currents from regions contiguous.

686. It is evident that very great differences between the winter and summer pressures of dry air will take place in dry climates, where the annual diminution of pressure which occurs arises solely from the high summer temperature. It is to the centre of Asia we must look for an example of this; and accordingly the pressure of the dry air at Barnaul, in Siberia, on a mean of ten years, is 29.760 inches in January, and 28.666 inches in July, – thus showing the enormous difference of nearly eleven tenths of an inch.

687. This question has important bearings on storms, particularly on the storms of Great Britain, which is situated within the southern limits of the region of most anomalous barometric pressure, which has its centre in Iceland. As the summer temperature falls in September and October the pressure is increased over the continents of Asia, Europe, and North America, and consequently air-currents are poured into the northern part of the Atlantic, in larger volume and with accelerated velocity; and since at the same time the temperature is everywhere falling, the whole atmosphere becomes saturated with vapour. This accumulated vapour is discharged in those furious storms of wind and rain which strew our shores with wrecks, and carry misery and desolation into the hearths of our seafaring population. Considering the important interests at stake, no amount of labour and no expense can be regarded as too great which might lead to a better understanding of these great atmospheric disturbances.

3.3

INTERNATIONAL STANDARDISATION

Editorial Headnote

3.3 International Standardisation

A new type of synoptic meteorology began to emerge in Europe and America in the 1870s as an official form of international meteorological institutionalisation, based on the idea that the combination of isobaric maps and statistical science could serve as a basis for legitimising and consolidating national weather services. Tracing the origins of this institutionalisation can be interesting, since in the 1850s and 1860s various international statistical congresses served as precursors to the first international meteorological meetings. The first two international meteorological meetings, held in Leipzig and Vienna in 1872 and 1873 respectively, illustrate this organisational influence of statistics on meteorology. The new synoptic meteorology was projected as an official, state-sponsored science, with the aim of standardising definitions and methods, and coordinating and systematising observations.[1] In particular, the concept of barometric gradient became an indicator of approaching storms. The four texts selected in section 3.3 describe the projection of this new type of meteorology.

A disciple of the physicist Heinrich Wilhelm Brandes and director of the Preussische Meteorologische Institut from 1849 to 1879, Heinrich Wilhelm Dove (1803–1879) was one of those responsible for the institutionalisation of international meteorology. Best known today for his 'law of gyration of the wind' (*Drehungsgesetz*) and his theory of the global wind system, Dove was a prolific writer, publishing more than 330 papers on physics and meteorology and editing the 8-volume *Repertorium der Physik*.[2] He was attracted to the international meteorological organisation because he saw a need among meteorologists for standardisation of procedures and uniformity of notation. Without organisations to enforce these requirements, meteorologists would not be able to develop general laws about weather changes and atmospheric phenomena. These needs led him, in 1867, to attempt to set up a 'common organisation for terrestrial meteorological observations', inviting representatives of the meteorological services of Austria, France, Italy and Spain, while the Swiss Naturalists' Society was meeting in Geneva.[3]

Although the 1867 initiative did not succeed, Dove was able to analyse in detail the confusion of scales and systems of measurement that existed in different countries and, at the International Meteorological Conference held in Leipzig in 1872, which he did not attend, he sent a paper that was presented by the event's director, Carl Bruhns. The paper is dated 24 July 1872 and was written in Berlin. Dove shows how measurements were often made according to local or national rules rather than uniform and global standards. He also shows how the thermometric and barometric scales illustrate the confusion of the time: while the British and

Americans used the Fahrenheit and English scales, Europeans preferred the centigrade or Réaumur scale and the metric scale. In this respect, he is not in favour of the introduction of the centigrade scale until the former abandoned the Fahrenheit scale. Finally, he advocates the adoption of a common system for storm signals and English denominations for wind directions.

Scientific journals were an important vehicle for disseminating the ideas of the new institutionalised synoptic meteorology. The article on the 'Meteorological Conference at Leipzig during August 1872', published in *Nature* on 28 August 1873, by Robert Henry Scott (1833–1916), was one of the most widely read. Director of the Meteorological Office (from 1867), Scott always showed an unusual capacity for administration rather than original research, as evidenced by the fact that he was Secretary of the International Meteorological Committee from 1874 to 1900.[4] The article is a translation of the official report of the *Proceedings* of this conference, Protocols and Appendices, published in Volume VII of the journal *Zeitschrift für Meteorologie*. The conference, which was intended to promote coordination and standardisation, was only consultative.[5] In June 1872, the initiators of the event – Carl Jelinek (Austria), Carl Bruhns (Saxony) and Heinrich Wild (Russia) – invited meteorologists to this preliminary meeting to prepare the programme for the Congress to be held the following year in Vienna. The invitation was accompanied by a set of twenty-six questions for consideration by the more than fifty meteorologists who attended the event, plus the fourteen who could not be present (including Dove).[6] Their views on the issues raised are detailed in the conference proceedings. The subjects treated are grouped under the headings of instruments, their position, and the methods of discussing, publishing and using observations, which in Scott's article are in the following order: barometers, maximum and minimum thermometers, radiation instruments, hygrometers, wind, rain and evapometers.

An illustrative concept of this new type of synoptic meteorology is the *barometric gradient*. Its introducer, Thomas Stevenson (1818–1887), was a Scottish engineer who specialised in lighthouse construction.[7] He is best known today for his studies on improving the lighting of lighthouses, and for designing a thermometer screen in 1864, which is still in use today. He also published meteorological studies and was one of the founders of the Scottish Meteorological Society in 1855. The barometric gradient was probably the first numerical indicator showing the degree of approach of a storm. Proposed in a study of storm intensity published in the *Journal of the Scottish Meteorological Society* in 1869, the paper provides a numerical basis that would serve as a reference for weather forecasting. This concept is central to the synoptic analyses of Buchan, who shows how wind speed is proportional to the closeness of isobars, as evidenced by his mention of the Stevenson gradient in the second edition of his *Handy Book of Meteorology* (1868).[8] Believing in the importance of expressing storm intensity numerically (which he later extended to temperature and humidity), Stevenson conceptualises the relationship between the barometric difference at two stations located on a line perpendicular to the isobars and the distance between these two stations. At the end of the article, he suggests that by using telegraphic communication between observatories and observing differences in barometric readings, the meteorologist could subsequently calculate hourly velocities and barometric gradients in future storms. Perhaps as a result of this premonition, though without explicitly mentioning Stevenson's work, the British meteorologist Ernest Gold discussed the relationship between the two physical variables several decades later in his *Barometric Gradient and Wind Force* (1908), a report 'On the calculation of wind velocity from pressure distribution and on the variation of meteorological elements with altitude', which was submitted to the Meteorological Office.[9]

Standardisation, instrumental uniformity and, in particular, isobaric mapping became an important part of the new institutionalised synoptic meteorology at the Vienna Meteorological Congress in 1873, when they were transformed from consultative issues decided by professional and amateur meteorologists (as in Leipzig in 1872) into 'scientific policy issues' discussed by the official delegates of the national meteorological institutes present at the event.[10] These delegates tended to support the control policies of their states; thus, while they supported the creation of a permanent committee to oversee international cooperation, they rejected the proposal to create an independent international institute. The *Report Weather Telegraphy and Storm Warnings* (1874), presented to the Vienna Congress by a committee appointed at the Leipzig Conference, reflects the extent of these changes.[11] This committee, consisting of Christoph H.D. Buys Ballot (Holland), Scott (Great Britain) and Georg von Neumayer (Germany), had prepared a questionnaire on the feasibility and usefulness of weather and storm warnings in Europe. Six questions were sent to the world's leading meteorologists, who submitted their answers for discussion in Vienna. The Vienna agenda contained twenty-eight items, twenty-six of which had been discussed in Leipzig.[12] After analysing the meteorologists' answers, the committee on weather-telegraphy and storm-signalling concludes that this service is a necessity for all countries. It is then accepted that barometric gradients should always be indicated and that a reduction of the barometric reading to mean sea-level is acceptable for stations below 300 metres. The reason adduced by the committee is that these measurements are important for 'the drawing of isobars for the purposes of storm warnings' (p. 13). This core affirmation is then supplemented by examining the responses on the role of the other meteorological elements (temperature, humidity, clouds, etc.). The majority of respondents recommend studying these elements 'for the investigation of the causes of storms' (p. 13), not for warning or forecasting. With a consensus on the best way to produce storm warnings, isobar plotting is strongly endorsed by delegates from national meteorological institutes, while the other elements are relegated to the research domain.

Notes

1 Aitor Anduaga, *Politics, Statistics, and Weather Forecasting, 1840–1910. Taming the Weather* (London and New York: Routledge, 2020), pp. 230–39.

2 Gisela Kutzbach, *The Thermal Theory of Cyclones: A History of Meteorological Thought in the Nineteenth Century* (Boston, MA: American Meteorological Society, 1979), pp. 230–31; Kurt Møller Pedersen, 'Dove, Heinrich Wilhelm', in Charles Coulston Gillispie (ed.), *Dictionary of Scientific Biography* (New York: Charles Scribner's Sons, 1971), vol. 4, pp. 174–75.

3 Heinrich Wilhelm Dove, 'Ueber den Meteorologen-Congress zu Wien', *Zeitschrift der Österreichischen Gesellschaft für Meteorologie in Wien*, 7 (1872), pp. 297–98; Evgeny Konstantinovich Fedorov and V. Böhme, 'The Leipzig Meteorological Conference, 1872', *World Meteorological Organization, Bulletin*, 21:3 (1972), pp. 136–41, on p. 137.

4 Jim Burton, 'Scott, Robert Henry (1833–1916)', in H.C.G. Matthew and Brian Harrison, *Oxford Dictionary of National Biography* (Oxford; Oxford University Press, 2004), vol. 49, pp. 465–66; Alastair D. Wood, 'Scott, Robert Henry (1833–1916)', in Bernard Lightman (ed.), *The Dictionary of Nineteenth-Century British Scientists* (Bristol: Thoemes Continuum, 2004), vol. 4, pp. 1790–91.

5 Wilfried Schröder and Karl Heinrich Wiederkehr, 'Georg von Neumayer (1826–1909) und die internationale Entwicklung der Geophysik', *Gesnerus, Swiss Journal of the History of Medicine and Sciences*, 49 (1992), pp. 45–62, on pp. 49–51.

6 Carl Bruhns, Carl Jelinek and Heinrich Wild, 'Einladung zu einer im August d. J. in Leipzig abzuhaltenden Meteorologen – Versammlung', *Zeitschrift der für Meteorologie*, 7:12 (1872), pp. 193–96, on pp. 194–95.

7 William J. McPeak, 'Stevenson, Thomas (1818–87)', in Bernard Lightman (ed.), *The Dictionary of Nineteenth-Century British Scientists* (Bristol: Thoemes Continuum, 2004), vol. 4, pp. 1908–1909; Roland Paxton, 'Stevenson, Thomas (1818–1887)', in H.C.G. Matthew and Brian Harrison, *Oxford Dictionary of National Biography* (Oxford; Oxford University Press, 2004), vol. 52, pp. 606–08.

8 Alexander Buchan, *A Handy Book of Meteorology* (Edinburgh and London: William Blackwood and Sons, 1867, 2nd ed. 1868), pp. 531–32.

9 Ernest Gold, *Barometric Gradient and Wind Force: Report to the Director of the Meteorological Office on the Calculation of Wind Velocity from Pressure Distribution and on the Variation of Meteorological Elements with Altitude* (London: H.M.S.O., 1908).

10 *Report of the Proceedings of the Meteorological Congress at Vienna. Protocols and Appendices* (London: G.E. Eyre and W. Spottiswoode for H.M. Stationery Office, 1874); 'The Meteorological Congress at Vienna', *Nature*, (7 May 1874), pp. 17–8.

11 Katharine Anderson, *Predicting the Weather: Victorians and the Science of Meteorology* (Chicago: The University of Chicago Press, 2005), pp. 245–46; J. Malcolm Walker, *History of the Meteorological Office* (Cambridge: Cambridge University Press, 2012), pp. 109–12.

12 Hendrik Gerrit Cannegieter, 'The History of the International Meteorological Organization, 1872–1951', *Annalen der Meteorologie*, 1 (1963), pp. 7–280, on pp. 10–11.

26

GEH. REGIERUNGSRATH H. W. DOVE TO DIRECTOR C. BRUHNS, LEIPZIG

Heinrich Wilhelm Dove

Source: Heinrich Wilhelm Dove, 'Geh. Regierungsrath H. W. Dove to Director C. Bruhns, Leipzig', in *Report of the Proceedings of the Meteorological Conference at Leipzig. Protocols and Appendices* (London: Georges E. Eyre and William Spottiswoode, 1873), pp. 45–46.

APPENDIX.

APPENDIX A.

COMMUNICATIONS ADDRESSED TO THE METEOROLOGICAL CONFERENCE IN LEIPZIG.

I.

Geh. Regierungsrath H. W. Dove to Director C. Bruhus, Leipzig.

Berlin, July 24th, 1872.

I BEG you to excuse my attendance in Leipzig as I must get some mountain air in August. As for the questions to be laid before the Conference with the hope of arriving at some general agreement, I should wish that unanimity were attained on the following heads: –

1. *Division of Time.* – For Non-Periodic Variations the hours 6, 2, 10, seem to me the best arrangement, with Five-day means for the yearly period, beginning with January 1st. For the years and months I think it best to hold to the old division, as a unanimous agreement of all observers to a change is hardly probable; of course Old Style is inadmissible, we must use New Style. Uniform hours for the daily means for all countries are scarcely to be hoped for, as the hours convenient in one country are not so elsewhere. At all events it seems to me necessary to state what correction has been applied to reduce the mean of the hours of observation to a true mean. It seems to

DOI: 10.4324/9781003650294-41

me very desirable to give, in addition to the means for the individual stations, the reduction to a normal station, as I have always done in the case of my stations. If no such normal station exists, observations at similar hours are the only ones we can employ in isothermal inquiries. Inasmuch as the hours which suit one instrument are not necessarily suitable for other instruments, it appears to me advisable to consider the Thermometer as the instrument which rules the choice of hours, owing to the magnitude of the diurnal variation of temperature.

By time I mean local time; this certainly interferes with the investigation of really simultaneous phenomena. To refer everything to the time of one place, as was done by Gauss in his inquiries on Terrestrial Magnetism, seems to me impracticable for Meteorological investigations, and rather disadvantageous than otherwise.

2. I have already proposed, and still think it desirable, to adopt the English denominations for the direction of the wind, E. for east and W. for west, for then there is no possibility of error for Germans and Frenchmen. The plan employed in the Swiss system seems to me thoroughly unpractical, where we have, in the same book, different terms for the stations where French or German are spoken respectively. The employment of 8 or 16 points for the wind may be left to the observer if the vane is not graduated so as to give degrees. If there be degrees, however, it must be settled what degrees correspond to the several points, *e.g.* N.N.E. from to The degrees should be counted as on a clock face, and also counted in the same way in the Northern and Southern Hemispheres, even though the direction of gyration of the wind is different.

   I think Robinson's principle the best for anemometrical inquiries. The vane must possess an absolutely free exposure. The determination of the distribution of Pressure, Temperature, and Vapour in the windrose is desirable, together with the law of variations of the instruments, which has only been ascertained for a few stations, by comparing the observation before the shift of the wind to a certain point with that after the shift.

3. *The Scale.* – The introduction of the Centigrade scale into meteorological inquiries does not appear advisable to me until the English and Americans give up their Fahrenheit's scale. The division to tenths is universally referred to the unit, whatever the decree is. In all the older investigations Fahrenheit and Reaumur's scales are so decidedly predominant that I have always preferred Reaumur's scale for my inquiries. The attempt of Forbes at Chiswick to introduce the Centigrade scale instead of Fahrenheit's failed completely. Fahrenheit's scale offers the advantage of giving only positive quantities almost all over the Temperate Zone. Negative degrees could certainly be avoided, if a new scale were introduced, with 100 added to it. This addition could be made to the older observations. The sole inconvenience in the Fahrenheit scale is found in inquiries on the distribution of plants, as we then only use readings above the freezing point.

The introduction of the Metrical scale for Barometers meets with a difficulty in the fact that maritime nations use the English scale almost universally. In this the unit is the inch, not, as in the Russian system, the half inch.

The remarks on the Barometer apply also to Rain, which I am convinced should be given, not in the cubic quantity of rain which falls, referred to a definite unit of area, but in depth. It is absolutely necessary to reduce, by melting it, the amount collected in a solid form to the liquid state. It seems to me, however, desirable to make a distinction between snow and rain in giving the quantity of rain.

4. For *storm signals* the adoption of a common system appears to me absolutely necessary, so that the sailor may expect the same language in all parts.

   Whether only daily means or the observations of individual hours should be published, must be so entirely decided by the amount of means at disposal that no general arrangements can be made on the subject. The sole requirement which must be complied with is to calculate observations of the several hours to daily (five-day) and monthly means, as it is not fair to have to carry out these calculations when a journal comes to be discussed.

5. *Moisture* is to be given relatively and absolutely, *i.e.,* as per-centages of Saturation and by Tension of Vapour.

27

METEOROLOGICAL CONFERENCE AT LEIPZIG DURING AUGUST 1872[1]

Robert H. Scott

Source: Robert H. Scott, 'Meteorological Conference at Leipzig during August 1872', *Nature*, 28 August 1873, 341–343.

OF the Congresses which have recently been held, none were more urgently called for than an International Congress of Meteorologists. Doubtless even under the diverse systems of observation which have been in use at national observatories and among meteorologists of different countries, large and valuable contributions have been made to Climatology and other departments of Meteorology. We need only refer to the various charts which have been published, showing the geographical distribution of atmospheric and oceanic temperature, of atmospheric pressure, of humidity, of prevailing winds, and of rainfall, and to the enormous amount of materials now being amassed, illustrative of the nature and course of storms, to show the important results which have been obtained. It must, however, be confessed that, as respects nearly the whole of this information, it can be regarded as valuable only in the sense of its being sufficiently approximate so as to meet the requirements of some of the more pressing practical questions of the science, and not because it is precise.

It is when we attempt inquiries into such questions as the diurnal and annual march of the different meteorological elements, and the relations of these elements *inter se*, and of weather on a large scale, that the general unsatisfactoriness of the systems by which observations are made in different countries comes to be forcibly felt, owing to their want of precision and uniformity. The want of uniformity is most conspicuous as respects temperature, humidity, and wind – or just those fundamental facts which must be scientifically observed and discussed before we can hope to solve the problem of weather changes.

In order to bring about a greater uniformity of procedure in different countries, it was proposed to hold a Meteorological Congress at Vienna in 1874. In June last, Bruhns of Leipsig, Wild of St. Petersburg, and Jelinek of Vienna, issued aft invitation to meteorologists to attend a preliminary conference to be held at Leipsig in August, for the purpose of preparing the programme for the Vienna Congress, to instigate preparatory experiments on some of the more important questions, and

 DOI: 10.4324/9781003650294-42

thereby render it possible for the Congress to arrive at immediate conclusions on many points. The Conference was thus only consultative. Accompanying the invitation were a series of twenty-six questions, which it was proposed to submit to the consideration of the Conference.

Upwards of fifty persons attended the meetings of the Conference, which lasted three days. The opinions of the different speakers on the points raised by the 26 questions are detailed in the pamphlet before us, which contains also the written opinions of 14 meteorologists who were unable to be present, including the well-known names of Dove, Rubenson, Mohn, Mühry, and Wolf, as well as the results of the deliberations of the French meteorologists at Bordeaux in September. The subjects treated of may be conveniently classed under the heads of instruments, their position, the methods of discussing, publishing, and utilising the observations.

Barometers. – To those who have attempted to discuss weather, it is well known that nothing exact or satisfactory need be looked for in the result, unless observations from numerous barometers well distributed be available. It is thus desirable that barometers be procurable at a moderate price for stations of the second order. Are Board of Trade barometers – barometers fitted with a float – or aneroids, suited for such stations; or is there any other cheap form of barometer that would serve the purpose? After a lengthened discussion it was referred to Dr. Hann of Vienna to prepare a report for the Vienna Congress. The most diverse opinions were expressed regarding the aneroid, arising probably from the experience of the different writers and speakers – some aneroids going well for years with no permanent alterations occurring in their indications; some going well so long as a small range of pressure is recorded, but undergoing alterations after every great barometrical depression; some constantly altering in one direction, others in either direction, &c. Since, however, it can be safely affirmed of no aneroid, how good soever it may have proved itself to have been, that it will continue to indicate correctly for even a brief time to come, the Conference came to the sound conclusion that the aneroid should not be used instead of the mercurial barometer, but only as an interpolation instrument, to fill up blank when the mercurial barometer is out of order, or when it cannot be observed on board ships in rough weather.

Maximum and Minimum Thermometers. – Rutherford's minimum spirit thermometer was regarded as satisfactory. On it being pointed out by several members that this thermometer is liable to go out of order by the spirit evaporating and condensing in the upper end of the tube, Ebermeyer, of Aschaffenburg, stated that this objection could be removed if the tube were at its entrance into the bulb inserted nearly up to the inner side of the bulb. We commend this suggestion to opticians; for if Ebermeyer's experience be confirmed, a source of serious and not infrequent error will be removed. On the other hand the performance of no *maximum thermometer* was considered to be so satisfactory that a uniform construction could be generally recommended; and the opinion was expressed that it was very desirable that a trustworthy maximum thermometer was devised, not liable for instance to have the mercury disturbed during high winds like Negretti and Zambra's, or the index portion go out of order as Rutherford's or Phillip's.

Instruments for Radiation. – Mr. Symons, who has paid much attention to this question, has been requested to give a report to next Congress on the modes of observation adopted in England for radiation. But it must be confessed that the methods of observation in this important inquiry are still in a very primitive state. Mr. Salt well pointed out that at present the results obtained with different instruments were not comparable with each other, and one hardly knew with the instruments now in use what was really observed.

Hygrometers. – Since the dry and wet bulb hygrometer is not trustworthy at low temperatures and in cases of extreme dryness, and the hair hygrometer fails also at the dew points, and since there is no hygrometer yet devised, at least for regular observations at stations, which gives approximately exact results as to moisture in all cases, it was recommended to make further experiments and collect the experience of meteorologists on the subject. From the favourable opinions expressed by Wild and others of the action of the hair hygrometer, further experiments with this instrument are very desirable, so that it might be made available for more accurate observations on the hygrometry of the air at temperatures below the freezing-point than the dry-and-wet hygrometer admits of. Another desideratum is an extensive series of experiments with Regnault's hygrometer in conjunction with the dry-and-wet bulb hygrometer in dry hot climates such as N.W. India, for the purpose of ascertaining how far the readings of the dry-and-wet bulbs can be used as data from which the dew-point may become known; and determining the requisite data for the correction and completion of the present hygrometric tables, particularly at points below freezing, and at high temperatures combined with great dryness.

Wind. – Curiously enough, the question of proper instruments for measuring the velocity and force of the winds does not seem to have been under discussion, even though it is one of the most important and pressing questions of the science. Anemometers, both for velocity and pressure, are indispensable to properly equipped observatories. Now it cannot yet be said that the anemometers for velocity give quite correct indications that they are comparable, *inter se*, and that we have a practicable means of ascertaining their errors from time to time.

Equally remarkable was the omission in the discussions, to consider what are the required conditions which anemometrical stations ought to fulfil, so that the instrument shall indicate the true movement of the air over the region where it is placed; or, if this cannot be accomplished, what observations should be instituted in order to ascertain how far the direction of the wind is deflected by the physical configuration of the surface, and its force diminished (or in rare cases accelerated) as compared with the general movement of the air over the place.

Pressure anemometers at a moderate cost are a great desideratum. Little satisfactory is known of the relation of pressure to velocity.

Rain. – The Committee proposed that a report of all the experience regarding the position, size, height above ground, and time of reading the rain gauge which has been yet gained should be prepared for the Vienna Congress. For the preparation of such a report the great storehouse of facts at hand are those collected by Mr. Symons in the successive parts of his "British Rainfall" and "Meteorological Magazine," which the members of the Congress would do well to consult.

Evapometer. – The present state of the evapometer is one of the least satisfactory of all the meteorological instruments. Considering the importance of the drying property of the air in relation to meteorology generally, but especially as one of the most important constituents of climate, it is to be hoped that some method will be devised by which results, at least roughly comparable to begin with, may be obtained.

The difficult, but vital question of the position of the thermometer does not seem to have been faced by the conference. It is earnestly hoped that the Vienna Congress will not shirk this question, but will seriously discuss it and arrive at some decision, or suggest some steps to be taken, that will ultimately lead to the degree of uniformity which is so imperatively called for. Till this be secured, the expensive systems of horary or continuous registration of temperature carried on at the great observatories of this and other countries, cannot supply the data for the determination of temperature "constants," seeing that they are incomparable with each other, as well as with the observations made at those numerous stations of the secondary order to which we must look for the working out of the great national question of local climates in their bearing on the health, productions, and commerce of the country. The question would be of comparatively easy solution were it possible, in the interests of cosmical inquiries, to ignore the past. But it is essential in the case of the older observatories to adhere to the same system of observing hitherto in use; until at least four or five years' observations have been made simultaneously with a second set of instruments placed in uniformity with those of other observatories.

The question of the practicability and utility of Weather and Storm Signals in Europe was considered, and it was remitted to Messrs. Buys Ballot, Scott, and Neumeyer, to collect the opinions of meteorologists on this important question, and draw up a report for the Vienna Congress. As it is understood that the committee have collected a good deal of information, some valuable results may be expected.

In the "Sequel to the Suggestions," Dr. Buys Ballot has suggested for the consideration of the Congress, the establishment, by societies, of stations in regions which are at present a blank. The Smithsonian Institution, the Dutch Meteorological Institute, and, in our country, the Scottish Meteorological Society have, with the means at their disposal, done a good deal in this direction, with results which have aided much in the furtherance of the science. But to fill up the enormous blanks which still disgrace British America, South America, most of Africa, and the Pacific, some concerted action on the part of meteorologists is indispensable. In connection with this proposed development, reference may be made to the scheme in contemplation by the Chinese Government, in carrying out which Mr. Campbell has been sent to this country to request advice from scientific authorities as to the general organisation of the stations, and to procure the necessary instruments, registers, &c. Towards the carrying out of this plan, the Congress will doubtless give Mr. Campbell very hearty support.

Source Endnote

1 Report of the Proceedings of the Meteorological Conference at Leipsig. Protocols and Appendices. Translated from the Official Report, Appendix to Vol. vii. of the "Zeitschrift für Meteorologic." Published by the authority of the Meteorological Committee. London, 1873.

28

ON ASCERTAINING THE INTENSITY OF STORMS BY THE CALCULATION OF BAROMETRIC GRADIENTS

Thomas Stevenson

Source: Thomas Stevenson, 'On Ascertaining the Intensity of Storms by the Calculation of Barometric Gradients', *Journal of the Scottish Meteorological Society*, 2, 1869, 132–136.

(Extract from paper read at the General Meeting of the Scottish Meteorological Society, June 1867.)

A STORM is a violent aerial disturbance due to horizontal inequality of density in our atmosphere, and that inequality of density, or want of homogeneity, is produced by an inequality of temperature. The terrestrial atmosphere is, of course, never homogeneous in the vertical plane, but varies in density with the height to which we ascend; every 100 feet of elevation corresponding to nearly $^{1}/_{10}$ inch of depression in the barometer. But if the statical equilibrium of the atmosphere be disturbed by heating a portion of one of the spherical layers of air near the earth's surface, such heated air will ascend, and the colder and therefore denser portions which surround it will rush in to supply its place. This is the primary cause of all atmospheric currents, which are therefore due wholly to those *differences* of density, and the unequal distribution of aqueous vapour, which are occasioned by an inequality of temperature. So long, then, as the normal density due to gravity is preserved, no disturbance can possibly arise.

Most of our wheel barometers, and many of the dials of aneroids, are marked with the word 'storm' opposite 28 inches; 'much rain,' opposite 28·50; 'rain,' opposite 29; 'change,' opposite 29·50; 'fair,' opposite 30; and 'set fair,' opposite 30·50. Such a system of nomenclature is wholly inadmissible, for if the instrument be taken to the summit of a mountain, the zero or starting-point of the scale will be altered, and the nomenclature will be in like manner affected. But besides this cause of error, such predictions of weather are absurd, and should therefore never be marked on the dial, because, although the barometer stood, all round the globe, as low as 28 inches at the sea level, that fact would not afford ground for expecting either the 'storm' which is announced on the dial, or any breath of

DOI: 10.4324/9781003650294-43

wind, however gentle. But, on the other hand, if two barometers placed at the same altitude, but separated at some distance from each other, are found to differ *simultaneously* in their readings, atmospheric disturbance must, by the laws of equilibrium of fluids, inevitably ensue.

It follows, then, from these principles, *that it is impossible, by means of a single barometer, to foretell with certainty the occurrence of wind*, although low readings and sudden changes in the height of the mercury no doubt frequently precede gales, and generally indicate an unsettled state of the atmosphere. To predict a storm on sound physical principles, we must be able to prove that there is an inequality of pressure at places which are at equal altitudes above the sea level. Before we can attain the knowledge of such a physical relation, *two* barometers placed at some considerable distance from each other, and equally elevated, must be observed simultaneously.

What has been stated is, however, not only curious, and perhaps interesting, but in a maritime country like this may be turned to good practical account, for westerly storms could with certainty be predicted if an observatory were established on the west coast of Ireland. This suggestion was made many years ago by Mr. Russell, Pilmuir, who has long studied the subject of hurricanes. Suppose, for example, that such an observatory had existed on 7th January 1839, when the well-known tempest swept over Scotland from the west, – the readings of the barometer on the west coast could have been easily telegraphed to ports upon the east. I have examined the Scotch Lighthouse returns for that year, and have found, after making corrections for height and temperature, that at 9 A.M. on the 7th, the barometer at Bara Head, in the Hebrides, stood at 28·409, while at Lismore, near Oban, it read 28·148, and at the Bell Rock, situate in the German Ocean, it was as low as 27·402. Had such a telegraphic system been in existence, the startling fact would have been elicited, that the barometer stood at the same moment $^{7}/_{10}$ of an inch higher on the west than on the east coast, and nothing could have foretold more significantly the approach of the storm. Even such information as this, is however insufficient, if we desire not only to predict the approach of wind, but to judge whether, when it reaches us, it will prove a gentle breeze or a tempestuous gale. For this purpose we must take into account the distances between the points of observation. For example, if the stations where the instruments were placed were many hundreds of miles apart, even a considerable difference of reading would not produce any very violent disturbance.

In further corroboration of these views, reference may be made to storm charts, on which are laid down the velocities of the wind at different places, and the lines of equal barometric pressure. Thus it appears from a chart of the storm of 6th February 1867, that comparatively *light* winds prevailed where the barometer stood *lowest*, while the *strongest* winds prevailed where the barometer was nearly at its *highest*, and where, according to the popular notion, there ought to have been the smallest disturbance. It is very likely that the scepticism that still prevails in many quarters as to the inutility of meteorological inquiries, and as to the accuracy of the results afforded by the instruments which are employed, is largely due to the existence of such-like apparently anomalous phenomena. A closer inspection of the chart, however, at once explains the difficulty. Over that portion of the earth's surface where the barometers were lowest, the isobarometric lines were farther apart than where the readings were higher, and thus the longer distances between these lines account for the smaller velocities of the wind.

The general principles which have now been advanced are not unknown to meteorologists. Sir John Herschell, for example, says, 'It is only when sensible differences subsist between places near each other that violent phenomena arise;' and Mr. Buchan, in his very excellent book, says, that 'when the isobarometric lines, or lines of equal pressure, approach near each other, or crowd together, the difference of pressure, or the atmospheric disturbance, was greatest, and least when they were most apart.'

It has occurred to me that we may get a very simple expression (embracing the differences in reading, and the distances betwixt the points of observation), which will give *numerical* values of the relative intensities of storms. Such ratios are represented by the inclination to the horizontal, of a line drawn from the top of the mercury in the barometer at one station to the top of the mercury at the other, and are obtained numerically by the simple arithmetical division of the distance, stated in nautical miles, by the difference of pressure stated in inches of mercury. This slope, which I propose to call the *barometric gradient*, is, however, not strictly a straight but a curve line. Moreover, it furnishes us only with the *ratios* of intensity of different aerial currents, but does not give the absolute intensity of any one gale, expressed in velocity of current by the number of miles per hour – or in impulse against a surface exposed to its action, by the pressure in pounds per square foot. If a series of simultaneous observations on the force or velocity of the wind at different places were coupled with a simultaneous series of barometric readings, we should then be able to construct a formula by which to predict, approximately at least, the velocity or impulse due to any gradient.[1]

Let us now proceed to calculate some of those gradients during storms, of which we have authentic records. Taking our first example from that of 6th February 1867, we find in the space where the wind was not very strong, that between the isobarometric line of 28·3, which was the line of maximum depression, and the line 28·5, there was a mean gradient of 1 inch of mercury in 1818 miles. But between the lines 29·1 and 28·9, which were much higher readings, yet, where the wind was the strongest, there was a mean gradient of 1 inch in 250 miles, thus clearly showing that the velocity depended *not on the lowness of the reading, but on the steepness of the gradient.* Turning now to the tempest of 7th January 1839, we find a mean gradient between Bara and Lismore of 1 inch in 280 miles, and of 1 inch in 264 miles between Lismore and the Bell Rock – slopes steep enough to account for the damage done to buildings and other property. No doubt, had intermediate points of comparison been in existence, we should have found still steeper gradients. Tropical storms, from their greater violence, might be expected to occur in connection with very steep gradients, and such appears to be the case. Thus, on calculating that of the Bahama tempest of

September 1866, we get 1 inch in 143 miles, while, during the hurricane on the Malabar coast, recorded by Colonel Sykes, the observations indicate a gradient as steep as 1 inch in only 50 miles!

These data show that *in our northern latitudes barometric mean gradients of* 1 *inch of mercury to* 170 *miles have occurred,*[2] *and that such gradients produce atmospheric disturbances sufficient to endanger many of our structures.*

EDINBURGH, *June* 1867.

P.S. – Since the above was written, the violent tempest of last month swept over Scotland, when much damage was done; more especially in Edinburgh, where it raged most furiously, and where sixteen chimney stalks were blown down, occasioning a loss of four lives. A large part of the gables of two buildings were also levelled with the ground, and a hearse and several cabs were blown over in the streets. Mr. Buchan has calculated the gradients at different times of the day, and is still engaged in getting new facts and observations, which will be afterwards published. Meantime the following table has been drawn up, which, so far as it goes, most strikingly corroborates the views expressed in the foregoing paper. Mr. Buchan has still, however, to verify the readings by personally inspecting the instruments at Thirlestane Castle; but the results given below may be accepted as very near the truth.

OBSERVATIONS MADE DURING THE TEMPEST OF 24TH JANUARY 1868.

Hour.	Barometer at Thirlestane Castle, near Lauder.	Barometer at Edinburgh.	Barometric Gradients.	Records of the times when damage occurred to buildings.
			Miles.	1.20 P.M., chimney stalk in Duke Street fell, killing four persons; at 2 P.M. gable of Morningside U.P. Church was blown down; 2 P.M. gable at Melville Road; 2 P.M. stalk at N. Frederick Street; 2 P.M. do. Thistle Street; 2.15 P.M. do. Hope Street; 2.15 P.M. do. S. Frederick Street and Queen Street; and about 3.30, chimney at Holyrood Palace was blown down.[3]
9 A.M.	29·915	29·909	1 inch in 3333	
12 NOON.	29·713	29·552	" 124	
2 P.M.	29·583	29·327	" 78	
3 P.M.	29·483	29·244	" 84	
4 P.M.	29·403	29·216	" 107	
8 P.M.	29·300	29·198	" 196	
9 P.M.	29·190	29·151	" 513	

The table shows the extraordinary maximum gradient of one inch of mercury in 78 miles, while it is supposed the hurricane which visited the Malabar Coast was one inch in 50 miles. Many persons familiar with tropical tempests who were in Edinburgh on 24th January last, declared that that storm was equal to most of those which occur in the tropics. Mr. Hay, architect, who was examined in court in a law-plea arising out of the fall of Morningside Church, stated in his evidence: 'The gale of 24th January would have been called a hurricane in Bermuda, though I have seen cyclones more severe.' Perhaps as notable a fact as any is the existence of high barometric readings – about 29·5, which corresponds to 'change' on the dials of wheel barometers, – and which prevailed during the worst of the gale. Even the lowest reading, 29·198, corresponds only to what is termed 'rain' on the dials of wheel barometers; while the name of 'storm' is in these instruments placed opposite 28·00, or nearly 1 inch lower than was recorded in this hurricane. Surely nothing could prove better the impropriety of such a system of nomenclature, or the erroneousness of the notion that a low barometer is any measure of the force of a storm; for the readings were high during the whole day, and the strength of the wind had passed long before the lowest reading occurred. The average of the times when buildings were damaged was 2 hours 10 minutes, which corresponds with the time when the steepest gradient existed.

I have only, in conclusion, to make three suggestions with reference to the subject of storms. 1*st.* Might it not be possible to have special electric wires connecting an observatory on the west coast of Ireland and one on the north of Scotland with that at Kew, so as to show *constantly*, by an automatic appliance, the *differences* of the barometric readings? Those differences of reading might perhaps be announced to the observer by the sound of a bell or some other similar alarm, so that warnings of approaching tempests might be timeously given. 2*nd.* It is most desirable that in all future storms, such observations should be made as will enable hourly velocities and barometric gradients to be afterwards calculated. 3*d.* The safest telegraphic indications of storms would be such as the following: '*Bar.* 28·72; *Bar. Gradient S.S. W., one in* 150 *miles.*'

EDINBURGH, *February* 19, 1868.

Source Endnotes

1 Such a formula will probably, as in the case of a current of water, assume the simple form of $x = a \sqrt{s}$, where x represents the velocity of the wind, s the barometric gradient, and a the constant, ascertained by long-continued observations of the velocities of winds due to different gradients. The formula will, however, be only approximate, as the result cannot be accurate unless when the directions of the wind were exactly at right angles to the isobarometric lines. This, however, cannot be the case, owing to the diurnal rotation of the earth on its axis, the progressive horizontal movement of the centre of the storm, the deflections produced by the geographical formation of the land, and the necessarily spiral course of the currents, which in this respect resemble

what takes place in the filaments in a whirlpool of water, or in the movement of the air in a whirlwind. – *Feb.* 1868.

2 A gradient has occurred since this, between Lauder and Edinburgh, of 1 *inch in* 78 *miles*. *Vide* postscript to this paper.

3 The above column contains only a selection from the damages that occurred. So far as I can learn, all the masonry that was thrown down fell between 12.15 and 4.30 P.M.

29

REPORT WEATHER TELEGRAPHY AND STORM WARNINGS, PRESENTED TO THE METEOROLOGICAL CONGRESS AT VIENNA, BY A COMMITTEE APPOINTED AT THE LEIPZIG CONFERENCE

Source: *Report Weather Telegraphy and Storm Warnings, Presented to the Meteorological Congress at Vienna, by a Committee Appointed at the Leipzig Conference* (London: Her Majesty's Stationery Office, 1874), pp. 7–18.

INTRODUCTORY REMARKS.

THE Programme originally submitted for the consideration of the Meteorological Conference at Leipzig in 1872 contained the following question as its 25th paragraph: –

> "Does the interchange of Weather Telegrams appear so useful that it should receive a fuller development and a firmer organization?"

This question, although not expressly mentioning the subject of Storm Warnings, gave rise to a lengthened discussion at the meeting of the Sub-Committee on Maritime Meteorology, which resulted in the following resolution; No. 5 in their Report to the Conference: –

> "The Sub-Committee would recommend that the Conference should, to be in unison with most authorities of science, express itself to the effect that the question of the practicability and utility of Weather and Storm Signals in Europe must still be considered as an open one from many points of view. They would recognise the importance of more thoroughly sifting the material which is in existence, and for this purpose have named a Sub-Committee which should consist of Messrs. Buys Ballot, Scott, and Neumayer, with power to add to their number. All authorities

DOI: 10.4324/9781003650294-44

> in this branch of science, among whom M. Mohn was especially named, should be requested to give their opinion, and the results of this investigation should be drawn up in a report which should be published at least two months before the meeting of the Congress in Vienna."

This resolution was adopted by the Conference at its meeting of the 16th August, 1872, and in consequence the above-named Committee was appointed, consisting of Professor Buys Ballot, Dr. Neumayer, and Mr. Scott.

Several circumstances concurred to prevent the Committee from taking any actual steps towards the solution of the problem entrusted to them until the commencement of the year 1873. When this period arrived the question was raised whether or not it was advisable to make use of the right of co-optation. This was answered in the negative, chiefly on the ground that all which was requisite was to procure as many competent opinions as was possible, and that such could most effectually be obtained by the issue of a circular, in which the principal points at issue should be submitted to the consideration of such scientific men as, from their position or from the character of their scientific labours, were known to take an interest in the subject.

The next step was the preparation of the questions to be contained in the circular. In respect of these a discussion arose within the Committee concerning the scope of the inquiry which was to be instituted, viz., whether the subject should be treated from a more theoretical point of view, not, however, neglecting its practical bearing, or whether the practical side of the subject of Weather Telegraphy and Storm Warnings should alone be taken into account. The latter view was adopted and the following six questions were embodied in a circular issued to elicit the opinions of competent meteorologists. The questions were issued in English, German, and French.

I.

What are your opinions in respect of the action of the system of Storm Signals hitherto in use, either from your own experience, or from a consideration of the Bulletins published in the United States, in England, and in France?

II.

Is it your opinion that in addition to communications of the direction and force of the wind which is actually blowing, the Barometrical "Gradients" should be given, in order to give warning of approaching wind?

To what should these gradients be referred? viz.: –

a. To the differences between the actual readings at the different stations.
b. To 30 inches, at sea level.
c. To the mean normal heights of the barometer at the stations taken into consideration.

III.

In what way should we take into consideration Temperature, Vapour, &c., &c.?

IV.

If we assume that each Director will institute the proper arrangements in his own district, what do you recommend as the minimum that each Director should communicate to the Director of another district, and should receive from him in return?

V.

Are you of opinion that the present condition of our weather knowledge justifies us in issuing distinct prophecies or forecasts of weather in contradistinction to the simple telegraphic communication of facts, or should we confine ourselves to indications of the condition of the atmosphere in adjacent countries, from which the recipients of the Report must draw their own conclusions? In the latter case should we communicate the general conditions by signals visible from a distance, posting up, at the place itself, the detailed information received from other localities?

VI.

In what way can it be arranged that intelligence of the conditions of weather can be conveyed to ships at sea, by means of Semaphores or by signals from Lighthouses?

The fact that the negotiations had to be conducted entirely by letter caused an unavoidable delay in the issue of the circular, and it was not until the beginning of May that this could be effected. At that period the circulars were issued to men of note in the special branch of science in all countries, the following being the division of labour adopted within the Committee:

M. Buys Ballot to communicate with France, Spain, Portugal, Holland, Belgium, Italy, and the United States of America.

Herr Neumayer to communicate with Germany, Austria, Switzerland, and Norway.

Mr. Scott to communicate with Russia, Sweden, Denmark, Hamburg, the United Kingdom and its Colonies, &c.

The number of replies received was comparatively small, except in the case of England, as evidently many meteorologists did not feel themselves justified in entering on a discussion on a subject of which they did not possess the practical experience which is requisite for the satisfactory handling of the questions proposed in the circular.

Some of the replies did not arrive until a late period, as was a natural consequence of the great distances over which the correspondence had to be conducted, as *e.g.* to Mauritius, &c.

In order to draw up a resume of all the answers received, and embody the same in a general report to the Congress about to meet at Vienna, it would have been indispensable that the three gentlemen forming the Committee should have met and arrived at definite conclusions concerning this report. This was, however, impracticable, and on this account the stipulation of the last sentence of the Leipzig resolution, to the effect that the report should be published at least two months before the meeting of the Congress at Vienna, was necessarily left unfulfilled, and it was not until the actual time of the assembling of the Congress that the report was drawn up by the Committee and finally adopted.

Although the report of the Committee (p. 11) as received and adopted by the Congress contains the sum and substance of the evidence collected in the manner just described, it has been deemed expedient, for the development of meteorological science, that the various opinions should be printed in full or in abstract, and circulated as widely as circumstances admit, in order that all those who may wish to judge for themselves shall have the opportunity of doing so. Furthermore, it was one of the first acts of the Permanent Committee, appointed by the Congress at its meeting of the 16th of September, to express an opinion in favour of the publication of the original documents, which were held to afford valuable materials for forming an exact idea of the present state of Weather Telegraphy and Storm Warnings.

In conclusion, little remains to be said in regard of the general arrangement, and the principle on which in some cases action was taken when a *selection* or an *abstract* had to be made. As to the former it is strictly alphabetical; as to the latter, it suffices to say that in the small number of cases of that nature the test has been whether or not, in any special paper, the remarks were pertinent to the questions of the circular. Wherever the answers deviated too widely from the mark a curtailment was deemed justifiable.

(Signed)
Dr. G. NEUMAYER,
Secretary.

Berlin, March 1874.

REPORT OF SUB-COMMITTEE V. ON WEATHER TELEGRAPHY AND STORM WARNINGS.

The Sub-Committee has adopted as a basis of its discussion the six questions which have been proposed by MM. Buys Ballot, Neumayer, and Scott, who had been entrusted by the Leipzig Conference with the duty of discussing the question of the practicability and utility of weather telegrams and of storm prognosis, and it has arrived at the following results: –

Question 1: –

What are your opinions in respect of the action of the systems of Storm Signals hitherto in use, either from your own experience, or from a consideration of the Bulletins published in the United States, in England, and in France?

It is sufficiently evident, from the various answers of the scientific men who have been asked for their opinion, as well as from the other official and non-official reports of the individual Directors of Meteorological Institutes, that there is a general conviction of the importance of Weather Telegraphy for the purposes of practical life, and that the existence of such a system is considered a necessity. This is shown from the results obtained hitherto from the systems already in existence.

The Sub-Committee therefore expresses its opinion in the following words: –

It seems desirable that the system of Weather Telegraphy should be developed as generally as possible, and on as uniform a plan as possible. In all countries in which up to the present time such systems have not been organized, steps leading to such an organization should be taken as soon as possible.

The Sub-Committee is further of opinion that it would be desirable to institute this organization on the following principles: –

1. The systems of observation which are already organized or to be organized in the various countries should be in as close relation to each other as possible, and this is of importance, not only with reference to the general arrangement, but also especially with reference to the telegraphic communications.
2. The system of each country must have a Central Office, which, if possible, should be combined with the central office for Maritime Meteorology; but, in any case, like the latter, must be subordinate to the Central Institute in each country. The observing stations belonging to the system should be placed under the central office for Weather Telegraphy and Storm Warnings, and their officers must be always ready to send to this Central Office weather reports, or to answer its inquiries for information.
3. The observers at the stations belonging to the system must be well instructed, for the purposes of Weather Telegraphy.
4. The choice of a locality as a station for Telegraphic Weather Reports must be made with consideration to all local circumstances and requirements, and with especial reference to the Wind; furthermore,

with reference to the series of observations which have already been recorded and reduced for the station, and with reference to the general system. Regard must also be had for the interests of the adjacent systems. But the final decision must always rest in the hands of the respective directors.

5. In order to make the system of warnings as perfect as possible, according to the present state of the Science, the Sub-Committee must indicate that it is desirable that the observations and reports based thereon should be made as complete and continuous (as regards the former) as is possible, *i.e.,* neither Sundays nor holidays should make a difference to them, and there should not be a complete interruption during the night.
6. Representations should be made to the respective Governments to facilitate Weather Telegraphy as much as possible. This is of importance, not only with reference to the expense, but also with reference to time, in respect of the telegrams within each system, and from one central office to another. If Telegraphic Weather Reports are delayed from the consideration of other despatches, the operation of the system will, in the opinion of the Sub-Committee, be seriously interfered with.

It is desirable, in order to carry the arrangements which are required according to the above views, that the respective Governments in granting a concession to a private telegraph company should take care that most thorough regard should be paid to the interests of Weather Telegraphy.

Question 2: –

Is it your opinion that in addition to communications of the Direction and Force of the Wind which is actually blowing, the Barometrical "Gradients" should be given, in order to give warning of approaching wind?

To what should these Gradients be referred? Viz.:

A – To the differences between the actual readings at the different stations.
B – To 30 ins., at sea level.
C – To the mean normal heights of the barometer at the stations taken into consideration.

With reference to this question the Sub-Committee is of opinion that besides communications of the Direction and Force of the Wind, the Barometric Gradients at the time of observation should also be added, in order to give warning of the wind which is coming.

Although the concordance of opinion within the Sub-Committee, with respect to the first part of the question, is great, this cannot be asserted with reference to the second part, inasmuch as the opinions of the scientific men who have been consulted are partly for the mode of reduction given under B, and partly for that given under C.

However, the various opinions within the Sub-Committee have been brought to agreement on this point, that for stations which are not above 300 metres above the sea, the reduction of the barometer readings to the Mean Sea-Level will be admissible. This agreement was attained with special reference to the drawing of isobars for the purposes of Storm Warnings, and in recognition of the theoretical difficulty of carrying out this method of reduction for elevations greater than those named.

Question 3: –

> In what way should we take into consideration Temperature, Vapour, &c., &c.?

With regard to Temperature and Moisture, as well as with regard to Cloud, Appearance of the Sky, State of the Sea and Tides, and Precipitation, it was recommended by most of the gentlemen consulted that some attention should be paid to these subjects, but in what way attention should be paid has not been indicated by any of them; therefore the Sub-Committee is of opinion that this question should be considered as not answered, and it therefore recommends its examination most specially to the notice of scientific men who occupy themselves with the study of storms and their causes, and draws attention to the investigations in this direction by Dove, Mohn, Clement Ley, Peslin, Buchan, and Reye.

Question 4: –

> If we assume that each Director will institute the proper arrangements in his own district, what do you recommend as a minimum that each Director should communicate to the Director of another district, and should receive from him in return?

The answers to this question must be considered as a *résumé* to those given to Questions 2 and 3, and they have frequent reference to communications within one and the same system, but not to the mutual communications between the directors of the individual systems. The Sub-Committee is of opinion that a distinction should be drawn between the theoretical and practical treatment of the question. For theoretical investigations all the elements are requisite and they should all be taken into consideration, but this can take place at any subsequent period. For practical purposes, with regard to which each director must consider the special requirements in his own country, they must be confined to a minimum.

It is sufficient that each director of a district should receive from his neighbour, as a minimum of what is to be communicated, the following elements: – Wind, according to Direction and Force, Pressure (including the Pressure for the evening of the previous day); Air-Temperature, Humidity, amount of Rain, Cloud, Weather, Sea, and when atmospheric disturbances are approaching, the magnitude and direction of the most important and most recent Gradient.

The Sub-Committee is further of opinion that these data should all be sent reduced, in order that a more rapid treatment of the materials should be rendered possible.

It must be left to the individual directors to determine among themselves the arrangements of a more special nature. This has special reference to the magnitude of the gradient which should be considered as an indication of atmospheric disturbance.

Whenever the International Institution, which is considered as very desirable by the Congress, is called into existence, all the evening observations should be sent to it, in order that a general view of the conditions of the weather of the whole of Europe can be given early in the morning to the individual systems.

Question 5: –

> Are you of opinion that the present condition of our weather knowledge justifies us in issuing distinct prophecies or forecasts of weather in contradistinction to the simple Telegraphic communication of facts, or should we confine ourselves to indications of the condition of the atmosphere in adjacent countries, from which the recipients of the Report must draw their own conclusions? In the latter case should we communicate the general conditions by Signals visible from a distance, posting up, at the place itself, the detailed information received from other localities?

Question 6: –

> In what way can it be arranged that intelligence of the conditions of weather can be conveyed to ships at sea, by means of Semaphores or by Signals from Lighthouses?

From an analysis of the various opinions on the practicability of storm-warnings which have been received, it appears that the great majority have declared *for* Storm-Warnings.

If we give to the opinions of individual societies, as well as to those of prominent authorities on the subject of Storm-Warnings, a weight = 10, to the other practised investigators of this subject a weight = 5, while less experienced men receive a weight of 1 or 2, the result is as follows: –

> For Storm-Warnings of a more or less detailed and decided character, but certainly *above* the simple communication of facts, we find 72 votes.

Against Storm-Warnings, and for restriction to a communication of facts, only 24 votes.

The majority of the opinions which have been received tends to the view that these warnings should be given in connexion with communications of facts, only a few hold these communications as quite unnecessary, while other and important voices are only for communication of facts without warnings; but it must be kept in mind that almost all the opinions expressed refer to the north and north-west of Europe, and are only in the first place to be understood as applicable to these districts; in fact the limits should be drawn even more narrowly, it being only the coasts of the districts described which are meant.

On these grounds the Sub-Committee expresses its opinion that Telegraphic Weather Reports have become a necessity, as has been proved by the answers to the first question, and furthermore that the results of the systems which have been or are to be organized for this purpose, can and must be utilized by the communication of simple facts of weather, with such developments, in the sense of storm-warnings, as may be justifiable by the further development of our knowledge of atmospheric disturbances. The results which have been obtained empirically, and by the statistics of weather knowledge, can be utilized for the benefit of navigation and agriculture, &c.

In the present condition of the development of weather prognosis, the matter of most importance is to define as accurately as possible the character of that which comes under the head of communication of facts. Among all the opinions which are submitted there is not a single one which does *not* reject for weather prognosis the title of *prophecies* or *predictions*. The most decided protest is lodged by everyone against the idea of infallibility being implied in this use of such a term. All that is said is, that "an opinion should be given on the weather to be expected, – or probability of the approach of a storm should be given – or possibility of prediction of wind, direction, and force, &c."

In the opinion of the Sub-Committee, a considerable difficulty in the practice of storm warnings consists in the careful avoidance of too special indication of probable wind and weather. The objection which is made by several of the gentlemen whose opinion was asked; that the reports would be discredited, owing to non-fulfilment of the "prophecies," loses its force if only such atmospheric disturbances are communicated, as will probably be accompanied by serious results. Accordingly, warnings by signals should not be issued for winds of forces 5 to 7 of Beaufort's scale, but above those numbers. Such warnings would meet with fulfilment with incomparably greater certainty. On the other hand, according to the experience which has yet been made, only few dangerous phenomena would escape notice. A general rule for storm gradients cannot be given, as many of them will be of a local nature; the director must therefore gradually form his opinion on this question, according to his experience, so that in the opinion of the Sub-Committee it appears desirable that actual storm warnings should not be attempted in newly organised systems immediately after the establishment of the system, but that the director should first gain for himself the experience necessary for the purpose.

If we were in a position to give from time to time, by means of signals, the accurate definition of the communications on conditions of weather, and on the indications of coming weather, the greatest portion of the difficulty would be removed; still we must endeavour to meet this difficulty by notices posted up on flagstaffs, containing explanations of the weather, and the opinion of the director thereon, which must somehow or other be more widely communicated.

This, however, can only have reference to communications in harbours, while for ships passing at a distance, and at sea, means must yet be found to give them an explanation of the signals.

In consideration of all these circumstances, and on the supposition that the conditions indicated in the answers to Question 1, will be strictly fulfilled, the Sub-Committee expresses its views on this very difficult question in the following words: –

1. Communications on the state of the weather should be given by telegraph by the central office to the public daily, and, wherever possible, twice a day, in the first place with reference to the objects of practical life, and especially of navigation.

These communications should be distributed as widely and as rapidly as possible by publication in the newspapers, or by synoptic charts, &c., with opinions on the conditions of the weather. The extent of these communications must be regulated by the means at the disposal of each system, and therefore nothing determinate can be settled on the question.

In case of disturbances in the atmosphere which are expected or have already commenced, telegrams should go to the signal stations, harbour-masters, &c., and rapid distribution must be afforded to them.

2. With reference to these latter telegrams – in cases of atmospheric disturbances – the following distinction may be drawn, according to the opinion of the Sub-Committee.

   a. If they are intended for ships lying in larger harbours, in order to afford their captains the opportunity of forming an opinion of a disturbance which is expected, and perhaps has already commenced to be developed, it is possible by a suitable apparatus, perhaps by the *aeroclinoscope* in a somewhat altered form, to bring to the knowledge of the public the conditions of pressure, motion of the wind, &c. Moreover, the receipt of the information must also be announced at the signal stations, and at the open spaces of the harbour, in forms prepared for it, and must be posted up.

   While the public, thus interested, obtains an idea of the progress of the disturbance not only at the locality itself, captains, from the movements of their own instruments on board, and from comparison with the telegrams, can form for themselves an opinion on the state of affairs.

b. If such telegrams are intended for harbours for small shipping, fishing, &c., in which case meteorological instruments cannot be supposed to be in the hands of the receivers, any more than a complete comprehension of the rules according to which the disturbances take their course, the apparatus mentioned above can be of little use.

Explanatory notices from the central office, posted up, are therefore above all things necessary, while that office has also to take care that at any very little port of this kind, a barometer and a thermometer are erected at easily accessible places.

3. In consideration of the circumstance that we cannot pre-suppose in the receivers of the communications the degree of meteorological knowledge which is requisite for drawing correct conclusions from the conditions of wind and weather, and from the movement of the meteorological instruments, and, moreover, that these receivers necessarily cannot come into possession of the requisite information until many hours later than the central office, – and inasmuch as the rapid progress of the development of a disturbance hardly leaves them any time for independent prognosis, – and in further consideration that our knowledge, being gained in an empirical way, puts us in a position of being able to foresee for certain districts the course of the atmospheric disturbance, the Sub-Committee is of opinion that the Director of the Central Office should give his opinion on this probable course of atmospheric disturbances which are expected, or have already commenced, not in the form of a prophecy, but in the form of *probabilities.* This opinion must be brought to the knowledge of the public by the Director in question, according to his own circumstances, by the printing and posting up of notices.

It is only when from the conditions of pressure, serious storms of 7–8, according to Beaufort's scale and upwards, are to be expected, according to the opinion of the Sub-Committee, that the views of the directors of the central offices on the direction, course, and force of the storm should be announced by signal apparatus on prominent points of the coast and in harbours and roadsteads.

The Sub-Committee is further of opinion that complicated apparatus cannot be recommended for this object, and proposes, therefore, the use of the drum and cone by day, and the corresponding signal lanterns by night, and recognizes that an expansion of this system of signals, as by the addition of a truncated cone, will be desirable; but under any circumstances the warning signals must be International.

4. In order to insure for these Storm Warnings, which are a benefit for shipping, as great effectiveness as possible, and to avoid misunderstanding, the Sub-Committee considers that the issue of explanations for the comprehension of the signals is very desirable. These explanations should contain, in a series

of propositions, explanations of the signals and their meanings for various cases serving as illustrative examples, which propositions can be applied to the several signals in some manner which has yet to be determined.

The Sub-committee is of opinion that while some such course of action would afford protection against misapprehension and discredit of the warnings issued, the public will gradually be educated by it to a clear comprehension of the reality and meaning of weather prognosis.

6.* Such an explanation cannot, according to the nature of the case, be the same for all coasts and waters, and must, therefore, be prepared in consideration of that, but should certainly be international, and added to every Code of Signals.

7.* Finally the Sub-Committee requests the Congress to express its opinion that on the part of the respective Governments, action should be taken to perfect as far as possible the storm-signal system, especially the apparatus and the explanations for the comprehension of the signals, and that special investigations and inquiries which are calculated to promote the signal system should be most warmly supported.[1]

Reporter, M. NEUMAYER.

According to a resolution of the Congress at the eleventh meeting on the 16th September 1873, the following proposal of M. Hoffmeyer is to be introduced into the foregoing report of the Sub-Committee.

"As it is, in my opinion, of the greatest importance to be able to give the daily meteorological communications as early in the morning as possible, I permit myself to draw the attention of the Congress to the question whether it would not be desirable that these communications should be principally based on the observations of the previous evening. If a well-constituted central institute could receive telegraphic despatches every evening from a sufficiently large number of European stations, in order to be able to deduce from them a resumé of the general meteorological conditions, these resumés could be communicated to the respective national systems during the night, or else very early in the morning, in order to be utilized in communication with the despatches of the individual stations received in the morning.

"The special national systems would, therefore, not only obtain a better and more complete grasp of the circumstances, and of the changes of the great barometric maxima and minima, but by such a centralization of evening observations, it would be possible to content ourselves in each individual system with morning telegrams from a more limited area than at present, when almost every meteorological institute is compelled gradually to extend its meteorological network on all sides, far over the frontiers of its own country, in order to obtain the requisite knowledge of the weather conditions. The telegrams from foreign and distant stations arrive so irregularly and so late, that the daily weather summaries, as well as the storm warnings, cannot be published before noon, or even later, while they could be ready at least two hours earlier if they depended only on the telegrams of a smaller area."

Source Endnote

1 So in original; No. 5 is omitted.

3.4

EULERIAN EPITOME

Editorial Headnote

3.4 Eulerian Epitome

The attribution of weather types to certain isobar forms became an increasingly common practice among meteorologists in the 1880s and 1890s, and a practice that could (and did) lead forecasters to believe that past synoptic situations could be repeated in the near future. Those who fully embraced these methods and techniques relied almost exclusively on surface observations and omitted any reference to any physical structure or theoretical principle that would underlie the physical solutions to be formulated from the 1900s onwards. As borne out in this section, the tone of institutionalised synoptic meteorology was clearly Eulerian, and this approach played a hegemonic role in weather forecasting in the late nineteenth century. Ralph Abercromby's isobaric geometry epitomised the use of the Eulerian approach to weather forecasting and laid the foundations of the art for subsequent generations of meteorologists.[1]

Ralph Abercromby (1842–1897) was a key figure in British official meteorology in the last quarter of the nineteenth century.[2] A descendant of an aristocratic lineage of soldiers, who had to abandon his military career because of ill health, he was awarded a prize by the Scottish Meteorological Society (of which he became a fellow in 1869), for his first essay 'A Scientific Explanation of the Popular Weather Prognostics of Scotland' (1864).[3] In his thirties, he fought against those who attacked and even denigrated meteorology as not being a science, arguing forcefully that knowledge of forecasts could not be expressed in mathematical formulae or simple maxims. As attested by his short note 'Is Meteorology a Science?' published in *Nature* on 12 April 1877, Abercromby was an outspoken and often non-theoretical critic of the opinions of eminent scientists such as the astronomer George B. Airy and the physicist William Thomson, debunking their claims that meteorology was not a science and extolling the virtues and strengths of synoptic meteorology. Central to this defence of synoptic meteorology is the desire to show that meteorology is not an exact science – 'mean values or harmonic series can never much advance meteorology' (p. 510) – and that, instead, it is as much a science as geology (p. 510). This view does not prevent Abercromby from recognising the limitations of meteorology in its present state, especially how to 'determine their future courses and changes', given the position of cyclones and anticyclones at any instant (p. 510).

As a member of the Meteorological Council (the governing body of the Meteorological Office), Abercromby was an uncompromising empiricist who combined the old weather lore with statistical evaluation of synoptic situations. From the early 1880s, he championed the method of isobaric geometry of weather changes and patterns in a series of articles and works, as evinced by his study 'Weather

Prognostics and Weather Types', published in *Nature* on 2 August 1883. In this article, Abercromby sets out his vision of synoptic meteorology, underpinned by the idea of the dominant role of the baric field in weather change – an exposition which he defines as the best-known popular prognostics by means of the most recent discoveries. He is explicit in stating that the weather 'is entirely dependent upon the form and closeness of the isobars' and that the wind speed in a storm also depends on the proximity of the isobars (p. 331). Deeply rooted in his synoptic-geometric vision is the notion of "weather typing", a concept that involves classifying isobars by their shape and then grouping certain phenomena of wind and weather round these forms. Abercromby has no objection whatsoever to admit that 'the great advantage of this method is that it avoids any theoretical considerations'.[4]

A self-confessed and self-taught amateur, Abercromby was eclectic in his attitude of combining aspects of Buchan's method of synoptic analysis with the ideas of the old weather wisdom.[5] The fullest expression of his isobaric geometry can be found in his book *Principles of Forecasting by Means of Weather Charts* (1885), which was the first book authorised by the Meteorological Council that had been written by non-Meteorological Office staff. Abercromby's title attempts to place his work alongside two very influential books on forecasting: Robert FitzRoy's *Barometer Manual* (London, 1858, first ed.) and Reverend Clement Ley's *Aids to the Study and Forecast of Weather* (London, 1880). Originally written as the second part of the revision of the FitzRoy *Manual* (the first part having been previously commissioned from Ley), and in view of the fact that the two drafts seemed completely incompatible in approach and methodology, the Council decided to publish the two parts as separate books.[6] Both are considered to be the two classic nineteenth-century English texts on meteorology. The two also provided two new competing cyclone models of the time, one (Ley's) which showed cloud distribution and high cloud motion in cyclones with round isobars (a forerunner of the flow cyclone models of the 1930s, but which gained little acceptance at the time), and the other (Abercromby's) which was based on isobaric geometry and whose diagrams did not seek to represent any physical structure, but only to relate the old weather wisdom to areas of low pressure (see Vol. III, 3.4 'Cyclone model').

Abercromby's *Principles of Forecasting* is an imaginative approach to how a varied configuration of isobaric shapes and figures can be used to prognosticate the weather. This comes across clearly in the extract included here, which links seven types of isobaric shapes to certain weather patterns that supposedly occur in conjunction with them. The pictorial style of the extract is typical of Abercromby's enthusiastic and amateurish attitude to the science of meteorology. Its configurations are multiform: *straight* isobars, circular or oval isobars encircling *cyclones* and *anticyclones*, converging isobars of high-pressure *wedges* and *V-shaped depressions* (low-pressure troughs), *cols* (low-pressure saddles), and *secondary* cyclones. Consequently, one became a good predictor insofar as one was able to foretell the appearance and movement of these configurations.

Abercromby's *Principles of Forecasting* became a highly influential work in meteorological circles, not only in Britain but also in Central Europe and the USA. The Met Office adopted the cyclone model proposed by Abercromby in his book instead of Ley's flow cyclone model. Historian James Burton claims that Abercromby's work exerted 'a massive influence on the development of weather forecasting within Britain'. Indeed, 'extensive verbatim extracts from it were still being quoted as late as 1940'.[7] Burton was probably referring to the prominent British meteorologist Napier Shaw, who in all editions of his *Forecasting Weather* (1911, 1923 and 1940) included 'over sixteen pages, literally and within quotation marks, from Abercrombie's *Principles* to form the substance' of the chapter on the relationship of temperature and weather to barometric pressure.[8] His influence in central Europe was also considerable: the first four editions of Julius von Hann's classic *Lehrbuch der Meteorologie* (first ed. 1901, fourth ed. 1926) include a cyclone model similar to Abercromby's.[9] In the USA, this influence reached the official spheres of meteorology. In the well-known *Weather Forecasting in the United States* (Government Printing Office, 1916), written by Alfred J. Henry and other members of the Weather Bureau, Abercromby's classification of isobaric shapes is regarded as 'standard' and a detailed explanation of their configurations is given.[10]

The Eulerian approach to weather prediction faded as a dominant force in the early decades of the twentieth century. However, the success of Abercromby's *Principles of Forecasting* allows us to trace its widespread influence back to 1900 and beyond, showing how these principles underpinned synoptic meteorology based on isobaric geometry in the last decades of the century. In reality, these principles were the almost inevitable result of a deeply ingrained mindset in the Met Office and other official circles that advocated the dominant role of the baric field in weather changes. As these three texts attest, Abercromby's ideas laid crucial groundwork for synoptic meteorology for forecasting purposes that prioritised isobaric geometry and weather typing over the physical structures and air trajectories in the atmosphere.

Notes

1 Aitor Anduaga, *Politics, Statistics, and Weather Forecasting, 1840–1910. Taming the Weather* (London and New York: Routledge, 2020), pp. 248–97.
2 Robert Henry Scott, 'Hon. Ralph Abercromby', *Nature*, 57 (1897), p. 55.
3 Mick Wood, 'Meteorologist's Profile – Ralph Abercromby', *Weather*, 56 (2001), pp. 102–06, on p. 102.
4 Frank Henry Ludlam, *The Cyclone Problem: A History of Models of the Cyclonic Storm* (London: Imperial College of Science and Technology, 1967), p. 28.
5 J. Malcolm Walker, *History of the Meteorological Office* (Cambridge: Cambridge University Press, 2012), p. 124.
6 James Burton, 'The History of the British Meteorological Office to 1905', PhD diss., (The Open University, 1988), pp. 146–47.
7 Ibid, p. 146.

8 Tor Bergeron, 'Synoptic Meteorology: An Historical Review', *Pure and Applied Geophysics*, 119 (1981), pp. 443–73, on p. 453.

9 Julius von Hann, *Lehrbuch der Meteorologie* (Leipzig: C.H. Tauchnitz, 1901), pp. 509, 545–47.

10 Alfred Judson Henry et al., *Weather Forecasting in the United States* (Washington, DC: Government Printing Office, 1916), pp. 74–5.

30

IS METEOROLOGY A SCIENCE?

Ralph Abercromby

Source: Ralph Abercromby, 'Is Meteorology a Science?', *Nature*, 15, 1877, p. 510.

THE recent article in NATURE on the Treasury Blue Book relating to meteorology brings into unpleasant prominence the opinion of the eminent astronomer, Sir G. Airy, that meteorology is not a science; and the evidence of the eminent physicist, Sir William Thomson, to the same effect; while a celebrated philosopher in the columns of the *Fortnightly Review* has not long since described meteorology as "a formless registry of facts."

But surely these eminent authorities have hardly realised the great change which has come over the whole aspect of meteorology since the introduction of synoptic charts?

Synoptic meteorology shows that the world is, broadly speaking, covered with shifting cyclones and anti-cyclones, which have each, subject to local, diurnal, and other variations, a characteristic weather, and physical appearance, and one great problem of meteorology is to explain the observed weather over any area, at any instant, by defining the position of these cyclones and anti-cyclones. It is in fact analogous to that branch of geology which explains the scenery and contours of any country by the position of areas of upheaval, crumbling, and subsequent denudation.

But there is another problem for the meteorologist to solve, viz., Given the position of the cyclones and anti-cyclones at any instant, to determine their future course and changes; and this can even now be done in certain cases. As if the geologist were asked what the future course of the present state of the earth's surface will be, where fresh upheavals or crumblings will occur, and what the corresponding changes in scenery will be? In this case the position of the meteorologist is far in advance of the geologist.

But still another reproach is cast upon meteorology – that the knowledge requisite to issue forecasts cannot be expressed in mathematical formulæ or in simple maxims. Here, too, the analogy of geology may show that neither formulæ nor maxims are necessary to make a science. Just as a number of skilled geologists, from long experience, agree as to the structure of a complicated piece of country, so will a number of meteorologists agree as to the probable course of any series of cyclones or anti-cyclones.

DOI: 10.4324/9781003650294-46

The limits of a letter do not permit me to show why mean values, or harmonic series can never much advance meteorology as a science, if any better argument were needed than their failure after a trial of forty years.

But I think we are justified in saying that, since the introduction of synoptic methods, meteorology is as much a science as geology; that both are pure observational sciences, and that their methods have much in common, while in some points meteorology is even the further advanced of the two.

21, Chapel Street, S.W.
RALPH ABERCROMBY

31

WEATHER PROGNOSTICS AND WEATHER TYPES[1]

Ralph Abercromby

Source: Ralph Abercromby, 'Weather Prognostics and Weather Types'. *Nature*, 28, 1883, pp. 330–34.

THE object of the first paper was to explain the best known popular prognostics by means of the most recent discoveries in meteorological science.

A great advance has been made in meteorology during the last twenty years owing to the introduction of daily synoptic charts of the distribution of atmospheric pressure, temperature, wind, rain, &c. From these it is evident that there is a distinct relation existing between the distribution of pressure and the direction and force of the wind, and temperature and weather generally. A glance at a number of the charts shows that there is nearly always present either an area of low pressure called a cyclone, usually having a circular form, and as a rule moving in an easterly or north-easterly direction; or an area of high pressure, called an anticyclone, also nearly circular in form but almost stationary in position. The wind in all cases also blows nearly parallel with the isobars, having the region of lowest pressure on the left hand. This has given rise to the following simple law propounded by Dr. Buys Ballot for the northern hemisphere, viz. "Stand with your back to the wind, and the barometer will be lower on your left hand than on your right." In cyclones the wind circulates round the isobars in the opposite way to which the hands of a watch move, but exhibits usually a little indraft; while in anticyclones the wind circulates round them in the same way as the hands of a watch, but exhibits usually a little outward motion. The velocity of the wind in all cases depends mainly upon the closeness of the isobars; for the closer the isobars the greater is the difference in pressure, and consequently the stronger the wind.

Since therefore nearly all our weather is of the cyclonic or anti-cyclonic type, and is entirely dependent upon the form and closeness of the isobars, it is by the aid of isobaric charts that the authors have attempted to explain a number of popular prognostics, and to associate them with certain kinds of weather. The method of research actually adopted has been for many years past to take notes of any good observation of any prognostic and put them by in a portfolio with the nearest synoptic chart available; or preferably with the nearest both before and after. When a sufficient number had been collected they were analysed, and

DOI: 10.4324/9781003650294-47

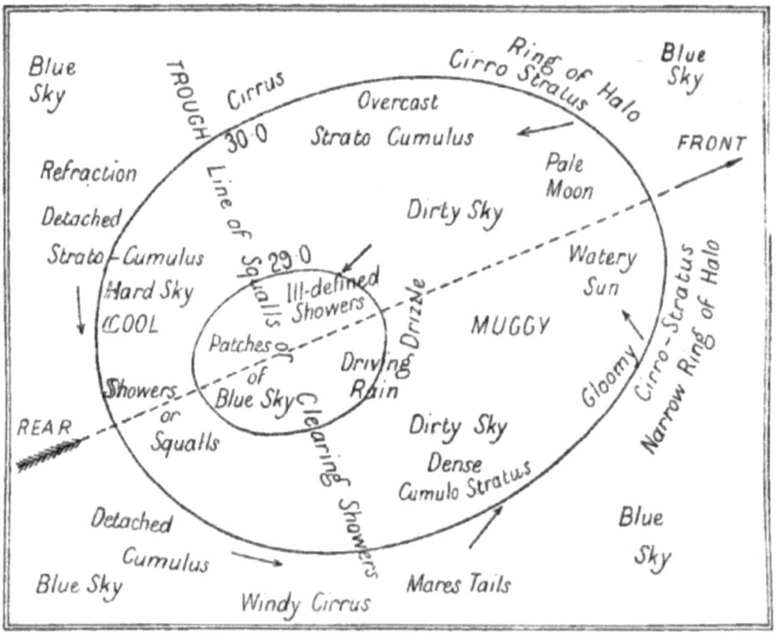

Fig. 1. Cyclone Prognostics.

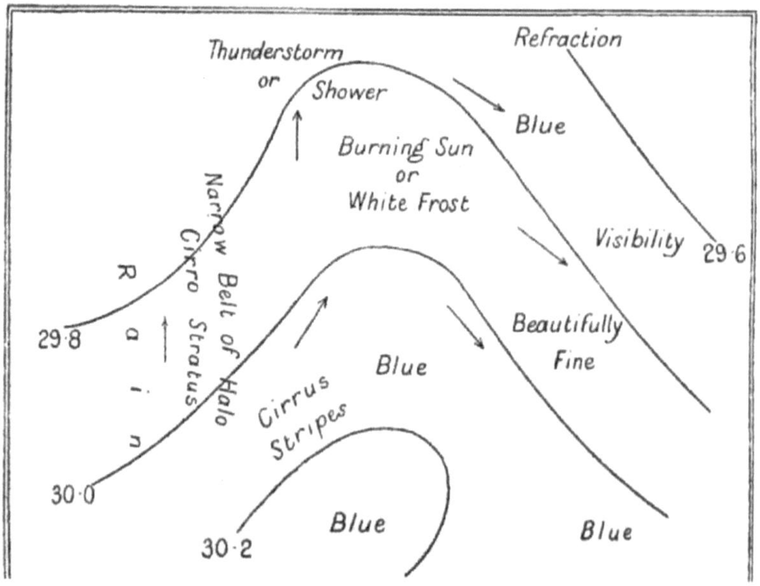

Fig. 2. Wedge-Shaped Isobar Prognostics.

the remarkable result has been arrived at that the greater number of prognostics are simply descriptive of the weather and appearance of the sky in the different portions of the various shapes of isobars seen on synoptic charts; and that they indicate foul or fair weather just as they precede the shifting areas of rain or blue sky which are mapped out by the isobaric lines.

These charts not only show the success of the prognostics, but also explain wherein they sometimes fail, by tracing the changes of each particular condition of weather. Hitherto the only prognostics which have been accounted for have been those due to excessive damp, but by means of isobaric charts many others can be readily explained. It must not be supposed that the modern methods diminish the value of prognostics, for even in forecasting weather from synoptic charts they are of great value, and will always be exceedingly useful to solitary observers who have only a single barometer to depend upon besides these prognostics, as for instance on board ship.

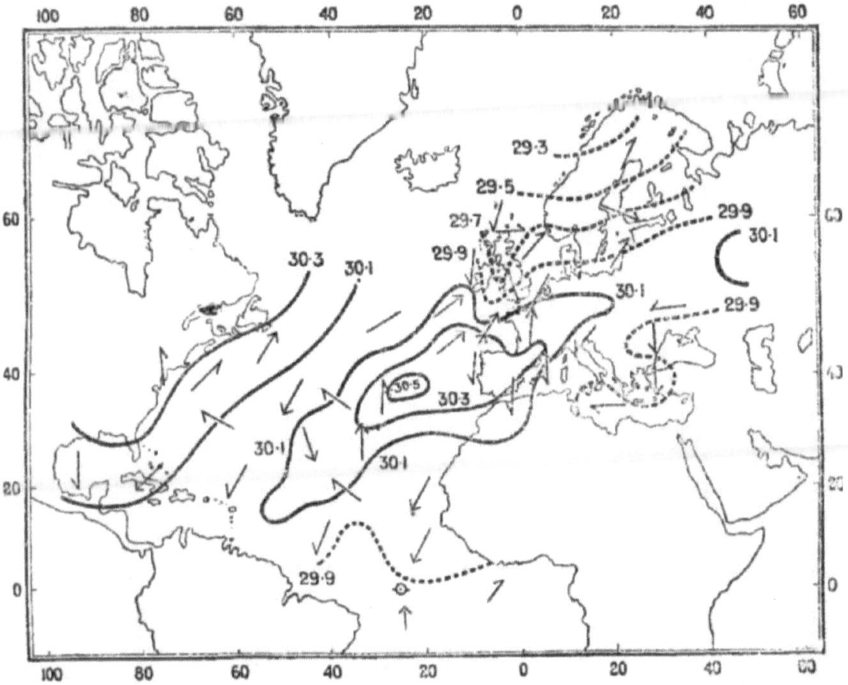

Fig. 3.

Though this way of treating prognostics is a great advance on the older methods, still there remains what may be called a higher line of explanation. There is no doubt that the different shapes of isobars are the product of different phases of atmospheric circulation, just like the eddies and backwaters of a river, and that the appearance of the weather is the product of the complex vertical and lateral movements thus set up. For instance, there is no doubt that the principal cause of rain in a cyclone is the condensation of the ascentional current of air round its centre, while Ley and others have shown that many of the well known forms of clouds are due to the action of upper currents moving in a different direction to those on the surface, and with a different velocity. Similarly the bright dry weather of an anticyclone is certainly due to the descending current found round its centre, and so on for every shape of isobars. Any reference to these movements was, however, intentionally omitted by the authors, as these movements are still to a certain extent only partially understood, and it was their desire to rest the explanations which they gave exclusively on observation without reference to any theoretical considerations.

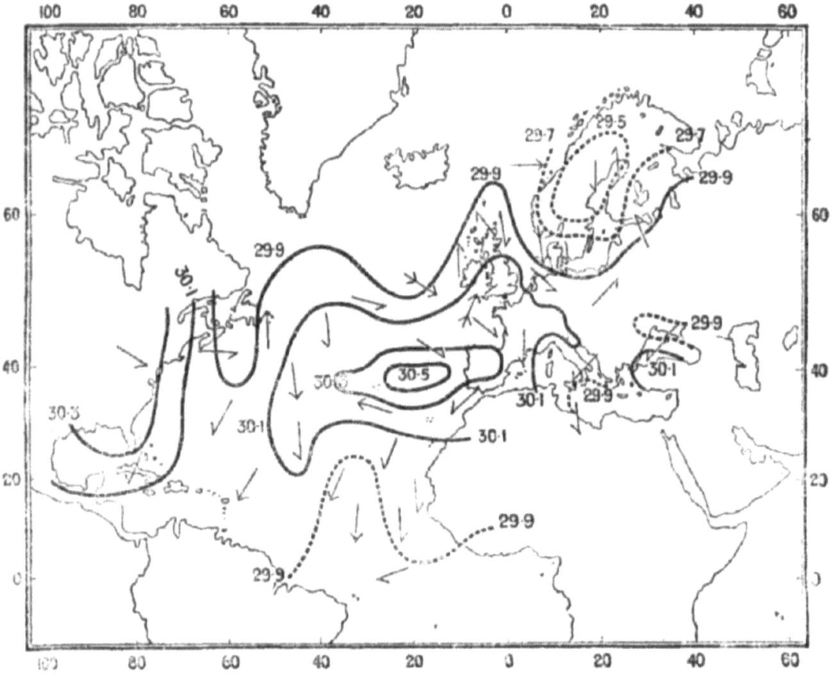

Fig. 4.

In a cyclone the broad features of the weather are a patch of rain near the centre, surrounded by a ring of cloud. But if we write down on a diagram, as in Fig. 1, the details of weather and kind of cloud in the different portions of the cyclone, we find that many of the best-known prognostics owe their value to the fact that they are characteristic of the front of a cyclone, and that after they have been observed, the rainy portion must pass over the observer before the sky becomes clear again. Sometimes a cyclone, after crossing a portion of the British Isles, dies out, and then the prognostics will fail in some districts.

The prognostics of settled fine weather are shown to be characteristic of anticyclones, which are nearly stationary for several days, and even weeks, together.

Though the bulk of British weather is made up of cyclones and anticyclones, there are two other distributions of pressure, marked out by wedge-shaped isobars and straight isobars respectively, which are associated with many well-known sayings. The chief interest in these prognostics consists in the contrast which they present to cyclone prognostics, as in many cases they are associated with fine and dry weather as opposed to the damp of an approaching cyclone.

In the front of wedge-shaped isobars (which are frequently found between a retreating and advancing cyclone) the weather is beautifully fine, of the sort of which we should say that it was "too fine to last"; or, if it listed a whole day, we should call it a "pet day."

During the day the sun is hot, at night white frost forms. Great visibility, with a blue sky, and unusual refraction, are often observed.

On the west side of the wedge-shaped area, as the new cyclone comes on, the blue sky gradually assumes a dirty appearance, accompanied by a halo, and gathers into cloud, and later on rain begins to fall; while in the southern portion the rain is often preceded by cirrus stripes, either lying with the wind, or sometimes at right angles to it.

> "Cirrus at right angles to the wind is a sign of rain."

These are all shown in the diagram (Fig. 2).

Some very interesting rain prognostics are also associated with straight isobars. While those in a cyclone are preceded by an almost ominous calm, and a dirty, murky sky, these are associated with a hard sky and blustery wind, of which it would be ordinarily remarked that "the wind keeps down the rain," or that, "when the wind falls, it will rain." While also the prognostics which precede cyclone rain hold good for the reason that they are seen in front of the rainy portion, those associated with straight isobars hold good because, though there is little rain actually with them, the area which they cover to-day will probably be covered by a cyclone tomorrow – the conditions being favourable for the passage or formation of cyclones.

Altogether, about 100 prognostics are associated with these four shapes of isobars.

The use and position of prognostics relative to forecasting from synoptic charts was stated thus: –

> Theoretically, when the isobars are well-defined, we ought to be able to write down the prognostics which might be visible, but practically we cannot do so. Besides, there are sometimes cases of isobars which have no well-defined shape, but with which thunderstorms or heavy showers often occur. These, as is well known, hardly affect the barometer, but are abundantly forewarned by the commonest prognostics, and as the rainfall is usually heavy in them, the failure of the forecast which omits to notice them is very conspicuous.

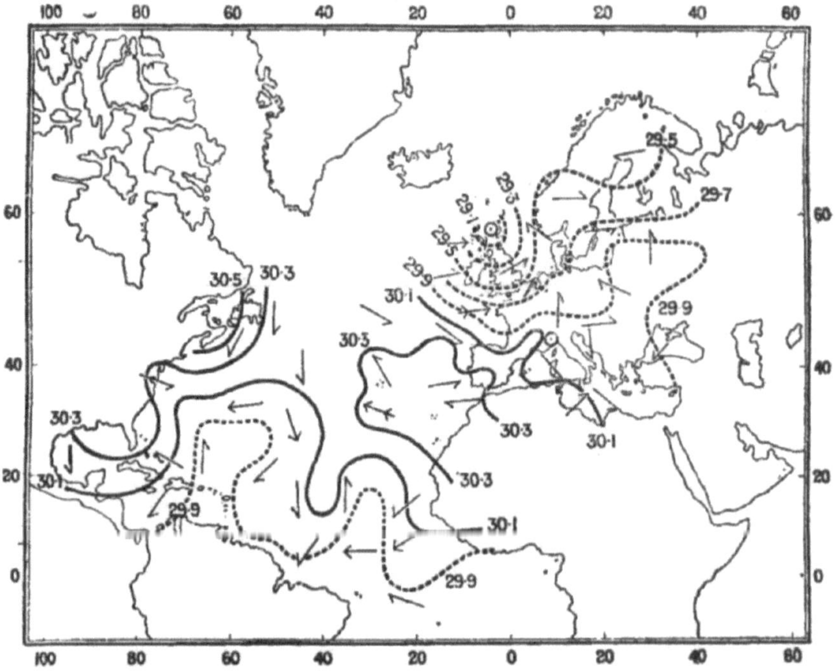

Fig. 5.

The scope of the paper precluded entering into the complicated question of the non-cyclonic rainfalls in this country. It was only stated that the prognostics which precede them are rather those associated with broken weather, such as bright sunrise or heavy clouds banking up without the barometer falling, than the muggy, dirty weather of a cyclone front. The warning they give is also much shorter, rarely more than three or four hours, if so long.

The other paper is an attempt to classify certain types of British weather.

It is familiar to many observers that the weather in this country frequently occurs in spells of several weeks' duration, during which there is a remarkable persistence of the general type of weather overriding both a considerable fluctuation from day to day, and a considerable local variation from place to place.

For instance, the wind will often back to some point of south with a high temperature, a dull sky and rain, and then veer to some point of west with a cooler air and brighter sky; and after a day or so of fine weather it will back again to the south with bad weather, perhaps this time rising to the intensity of a gale, and subsequently veer towards the west with finer weather, and so on for weeks together.

The changes only vary in intensity and detail, not in general character, while the feel of the weather and the look of the sky remain through all of them what are customarily associated with westerly winds.

Similarly the wind will often blow persistently from some point of east, fluctuating between south-east for fouler weather and north-east for finer weather, and back again with many variations for several weeks, during which the predominant features of the weather are always characteristic of east winds. The frequent recurrence of particular types of weather at particular seasons of the year is also a matter of common observation; the north-east winds of March, the cold north winds of the middle of June, and the wet west winds of September are well-known instances.

If we examine a large number of synoptic charts we find that relatively to Europe the general position of the great areas of high pressure frequently remained constant for a lengthened period. Further examination shows that the constancy of these positions coincides with persistent types of weather similar to those above mentioned, the fluctuation of type being due to the passage of cyclones, while the local variation depends on the position of the cyclone centres and on the innumerable local conditions which modify any general type.

Over the North Atlantic and Europe the distribution of atmospheric pressure presents certain constant features, namely –

1. An equatorial belt of nearly uniform low pressure.
2. A tropical belt of high pressure rising at intervals into great irregular elevations or anticyclones.
3. A temperate and Arctic region of generally low pressure, but in which occasionally areas of high pressure appear for a considerable period.

The equatorial belt constantly covers the Sahara and the Amazon valley, and always narrows over the Atlantic at about 30° west longitude, where it often does not reach higher than 10° north latitude. The shape and depth of this area are tolerably constant.

The tropical belt comprises a region of high pressure rising at variable intervals into great anticyclones. Their position is generally variable, with the exception of one, which is always found over the central Atlantic. This anticyclone forms a very important factor of the weather of western Europe, and will be constantly referred to as "the Atlantic anticyclone." Its extension south and west is tolerably constant, while towards north and east it is variable, sometimes rising as far as 60° north and stretching over Great Britain and continental Europe.

The temperate and Arctic region extends from the tropical high pressure belt to the pole. The pressure, though ordinarily low, is perpetually fluctuating by reason of the incessant passage of cyclones; yet occasionally persistent areas of high pressure appear in certain portions of it.

With reference to western Europe there are at least four persistent types of weather –

1. The southerly, in which an anticyclone lies to the east or south-east of Great Britain, while cyclones coming in from the Atlantic either beat up against it or pass towards north-east.

2. The westerly, in which a tropical belt of anticyclones is found to the south of Great Britain, and the cyclones which are formed in the central Atlantic pass towards east or north-east.
3. The northerly, in which the Atlantic anticyclone stretches far to the west and north-west of Great Britain, roughly covering the ocean. In this case cyclones spring up on the north or east side, and either work round the anticyclone to the south-east, or leave it and travel rapidly towards the east.
4. The easterly, in which an apparently non-tropical anticyclone (or one disconnected with the tropical high-pressure belt) appears in the north-east of Europe, rarely extending beyond the coast-line, while the Atlantic anticyclone is occasionally totally absent from the Bay of Biscay. The cyclones, then, either come in from the Atlantic and pass south-east between the two anticyclones, or else, their progress being impeded, they are arrested or deflected by the north-east anticyclone. Sometimes they are formed to the south of the north-east anticyclone, and advance slowly towards the east, or in very rare instances towards the west.

The details of the southerly and westerly types are given in the paper. Here we can only reproduce the three diagrams of the westerly type, Figs. 3, 4, and 5, in which the general characteristics of the type, just mentioned, are readily seen.

The value of the recognition of type groups is shown in the following ways: –

1. They explain many phenomena of weather, and many popular prognostics.

   For instance, besides showing the nature of spells of good, bad, dry weather, &c., they explain by reason of their persistence such prognostics as why "grouse coming down into farmyards are a sign of snow." Also why the prognostics, "When a river like the Tweed rises without any rain having fallen," or "Irregular tides are signs of rain," have a significance for the future; for though both are caused by past bad weather at a distance, yet the persistent type will almost certainly sooner or later bring more bad weather over the place of observation.

   Then the recurrence of hot and cold periods, many of them well known, are shown to be clue to the recurrence of a similar type of pressure distribution about the same season of the year. Particulars of seventeen such are given, and the manner in which the knowledge of them can be utilised in forecasting is stated thus: that though the forecaster is not justified in stating that any period will occur absolutely, still when about the time of its usual recurrence the synoptic charts show signs of the expected type, then the forecasts for a few days ahead can be issued with greater confidence. For instance, suppose that about November 6 – a cold period – the charts begin to show traces of the northerly type, then, but not before, there would be good grounds for saying that a period of cold weather, which usually occurs at this season, has already set in, and may be expected to last for five or six days, the forecaster being thus enabled to issue a much longer forecast than can as a rule be safely attempted.

2. Type groups are of the utmost value in forecasting, for when the existence of the type is fairly recognised then the general features of the weather are at once given, as well as the probable motion of the cyclones which are formed during the continuance of the type. Unfortunately in many cases no certain indications can be given of an approaching change of type.
3. Statistical results can be corrected by their means, for they give a true test of identity of recurrent weather, which no single item, such as heat, cold, rain, &c., can do.
4. They enable geological questions to be treated, such as the influence of changing distribution of land and sea on climate, in a more satisfactory manner than any other method.

The general principles of prognostics and types hold all over the world, but the details in these papers apply to Great Britain only.

RALPH ABERCROMBY

Source Endnote

1 Abstract of two papers read before the Meteorological Society: "On Weather Prognostics," by Hon. Ralph Abercomby and W. Marriott; "On certain Types of British Weather," by Hon. R. Abercromby. (*Quarterly Journal of the Meteorological Society,* vol. ix. No. 45.)

32

PRINCIPLES OF FORECASTING BY MEANS OF WEATHER CHARTS

Ralph Abercromby

Source: Ralph Abercromby, *Principles of Forecasting by Means of Weather Charts* (London: Printed for H.M. Stationery Off., 1885), pp. 1–3, 69–81, 99–100.

CHAPTER I.

SYNOPTIC CHARTS.

THE earliest stage of Meteorology in every country has always been the collection of popular weather prognostics, referring to the appearance of the sky or the movements of animals; and everywhere these observations lead to a considerable success in forecasting weather. For the last 150 years the indications of the barometer, as observed by each solitary observer, have been added in all civilized countries to the older weather lore, and with increased power of forecasting. But within the last 25 years a new method has been introduced of combining on a chart the observations made at the same instant at a large number of places, especially with regard to the relative height of the barometer. These maps are called synoptic charts, and round them a totally new science has grown up, which has developed a vastly increased power of forecasting. The problem of forecasting weather by means of prognostics and the observer's own barometer will always be of value at sea and in rural places, but will only be casually alluded to in this work.

Our object is to place before the reader, in an elementary form, a brief account of the modern method of forecasting weather by means of synoptic charts; and though the general principles laid down hold all over the world, the details will refer to Great Britain only. It may also be remarked that the whole system of synoptic forecasting, as this may be called, depends entirely on the observed association of different sets of phenomena, and is totally independent of any theory of atmospheric circulation.

Synoptic charte are so called from a Greek word meaning "*a general view,*" that is, over a large tract of country, and those prepared at the Meteorological Office

 DOI: 10.4324/9781003650294-48

are constructed in the following manner. Every morning at 8 o'clock punctually, telegraphic reports are sent up to London from about 30 stations in the United Kingdom, giving the height of the barometer and thermometer, with the direction and force of the wind, together with the actual state of the weather. Practically the weather changes are so rapid that these reports must he sent by telegraph to be of any use.

Here is a portion of such a report: –

Stations.	Barometer at 32° F. and Sea Level.	Temp. in Shade.	Wind.		Weather.
			Direction.	Force (0 to 12).	
Pembroke	29·62	46	W.N.W.	5	bc
Scilly	29·71	48	N.W.	6	c
Prawle Point	29·67	46	W.N.W.	4	bc
Hurst Castle	29·56	44	N.W.	4	b
Dover	29·43	45	W.	5	b
London	29·48	44	W.N.W.	2	bc
Oxford	29·56	42	N.W.	4	b
Cambridge	29·39	44	S.W.	2	b
Yarmouth	29·28	42	W.S.W.	5	cq
The Helder	29·25	45	W.S.W.	5	r

A blank map is then taken, and the first thing done is to mark at each station the height of the barometer. Lines are then drawn through all those places where the pressure is equal. If two stations have exactly the pressure required, they are simply joined, but if neither has the pressure wanted, we draw a line between them so as to divide the distance proportionally. Thus from the data just given, if we wanted to draw the line of 29·5 inches, we should have to draw it proportionally between London, where the pressure is 29·48 inches, and Oxford, where it is 29·56 inches, and it would pass much nearer London than before. Lines so drawn are called "isobaric lines" or shortly "isobars," from two Greek words signifying "equal weight," because they pass through places where the barometric pressure is equal. They are by far the most important element in forecasting, and the greater portion of this work will be occupied in showing the relation of wind and weather to the different shapes which these lines assume.

After the isobars, the direction of the wind is marked by arrows, which are always supposed to fly with the wind and have a number of "flèches" proportional to the force of the wind.

The force of the wind is estimated by the Meteorological Office on what is universally known as "Beaufort's scale." On this system all possible forces or velocities are estimated on an arbitrary scale of 0–12, or from calm to hurricane force. The intermediate numbers refer to the number of knots that a man-of-war can make per hour, or the extent of canvas which the wind permits her to carry. The scale is as follows; and we have added in the last column the estimated equivalent velocity in miles per hour of the arbitrary numbers in the first column, being that adopted in the Meteorological Office.

BEAUFORT'S SCALE.

				Velocity of wind in miles per hour.
0	Calm.			3
1	Light air	Or, just sufficient to gire steerage way.		8
2	Light breeze	Or, that in which a well-conditioned	1 to 2 knots.	13
3	Gentle breeze	man-of-war, with all sail set, and clean	3 to 4 knots.	18
4	Moderate breeze.	full, would go in smooth water from	5 to 6 knots.	23
5	Fresh breeze		Royals, &c.	28
6	Strong breeze		Single-reefed topsails, and topgallant sails.	34
7	Moderate gale	Or, that to which she could just carry in chase, full and by	Double-reefed topsails, jib, Ac.	40
8	Fresh gale		Triple reefed topsails, &c.	48
9	Strong gale		Close-reefed topsails, and courses.	56
10	Whole gale	Or, that with which she could scarcely bear close-reefed main-top-sail and reefed foresail.		65
11	Storm	Or, that which would reduce her to storm-stay-sails.		75
12	Hurricane	Or, that which no canvas could with-stand.		90

Then the weather is marked. In this country a system of abbreviations is used known as Beaufort's notation, which is –

b	blue sky.	f	fog.
c	detached clouds.	g	dark, gloomy.
d	drizzling rain.	h	hail.

l	lightning.	S	snow.
m	misty (hazy).	T	thunder.
o	overcast.	U	ugly, threatening.
P	passing showers.	V	visibility, unusual transparency.
Q	squally.		
R	rain.	W	dew.

Lastly, the readings of the thermometer are marked, and lines are drawn through those places where the temperature is equal, just as was done for pressure; these lines are called "isotherms," that is, lines of equal heat. They play, at present, a very small part in forecasting. The chart is then complete.

We shall next proceed to point out the relation which the examination of a very large number of similar maps has shown to subsist between the wind and weather and isobars.

[. . .]

CHAPTER V.

WEATHER FORECASTING.

So far we have only explained the nature of weather and weather sequence, and only incidentally alluded to forecasting proper, so that we must now go more minutely into that portion of the subject. We have already mentioned that the greatest assistance the forecaster has, is the recognition of certain types of weather; but we shall now give some of the general principles of cyclone motion and weather changes on which he has to fall back when he cannot recognise any known type. We shall then explain the use of various aids to forecasting, and conclude with some detailed examples of successful and unsuccessful forecasts.

Cyclone Paths.

In the first place, as the motion of cyclone centres is always towards some point of east, except in certain cases of the Easterly type which we have already explained, in an uncertain case we may always assume some Eastward drift.

Another very noticeable point about cyclones, is the tendency of their centres to follow a coast line rather than to strike inland. Also when they do cross the land, they seem to take by preference the lines of valleys or at all events of lowest ground, which seem to be lines of least resistance to cyclone motion. In this country cyclone centres have a tendency to pass up channel, or round the north-west coasts of Ireland and Scotland; and when they cross the country they frequently select the line of the Caledonian Canal, or the line of the Forth and Clyde Canal.

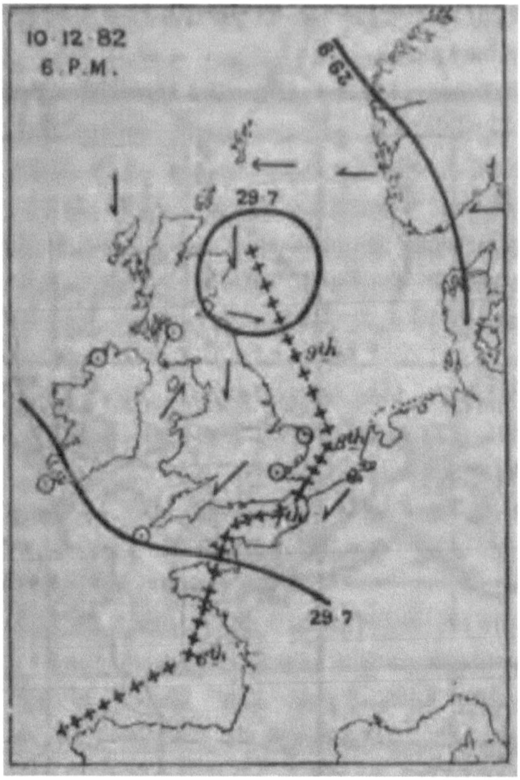

Fig. 42.

In Fig. 42, for December 10th, 1882, at 6 p.m., we give an example of a cyclone centre working up channel, and then along the east coast of England, so that the centre seems to hug the shore, but to avoid passing over the land. Though these cases do often occur, still the knowledge of the fact is of but little use in forecasting, for the centres often take other paths, and it is only in very rare cases that the knowledge is of any practical use.

Storms Crossing the Atlantic.

Public opinion has been much fascinated by the discovery that occasionally cyclonic storms have been traced from America across the Atlantic to the coasts of Europe, and exaggerated hopes have been entertained of thus issuing Storm Warnings to Great Britain several days in advance. We propose, therefore, to devote a few paragraphs to the consideration of this subject.

First, as to the number which actually do cross, their tracks, and their time of passage. In the year 1865, as far as the French charts show, that is to say a little north of 50° lat., only four were traced as follows: –

Date.	Distance traced.	Days in Transit.
1865, April 19–May 1	Traced from Labrador to Ireland	12
„ May 26–31	„ Gulf St. Lawrence to Finland	5
„ Oct. 2–14	„ Cape Cod „ „	10
„ „ 13–23	„ Newfoundland „ „	10

The tracks of all of these are marked in Fig. 43, which shows at a glance that their paths were so different, and their rates of progress so irregular, that any attempt to forecast British weather from the mere knowledge of the fact that a cyclone has left the United States and gone into the Atlantic Ocean, must end in disastrous failure.

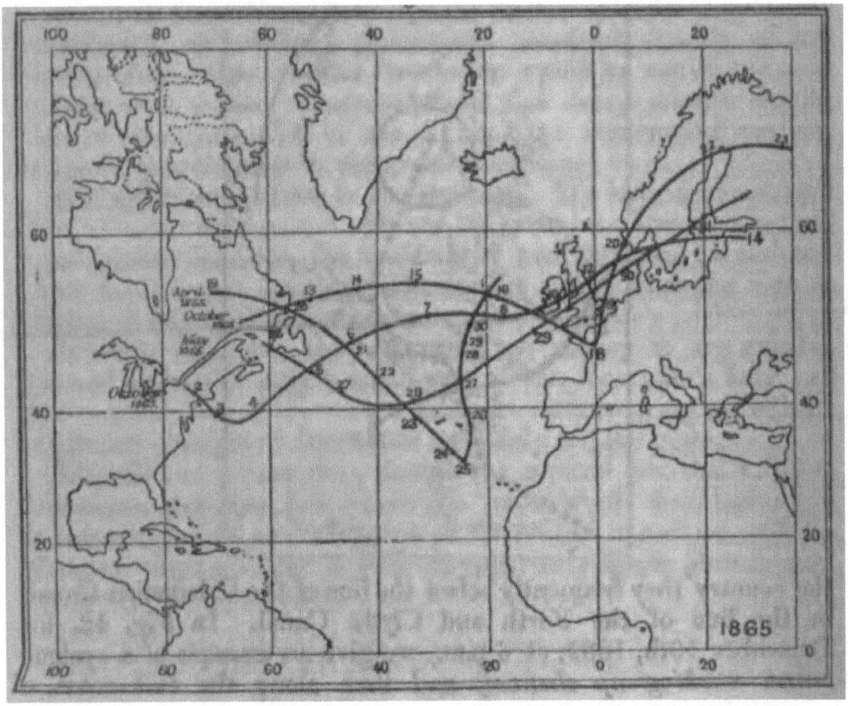

Fig. 43.

In the year 1874, the following six were traced by Professor Loom is, the greater number probably by means of Hoffmeyer's charts, giving observations up to lat. 70° N.: –

Date.	Distance traced.	Days in Transit.
1874, March 1–5	Traced from Hudson's Bay to North Cape	4
„ April 14–17	„ „ „ to Norway	3
„ „ 16–23	„ Gulf of St. Lawrence „ „	7
„ May 23–30	„ „ „ „ „	7
„ Aug. 1–4	„ „ „ „ North Cape	3
„ „ 12–17	„ Hudson's Hay „ Norway	5

In this case the passages are a little less irregular than in the year just described, but still they have no definite feature.

Some recent years, for which more complete materials exist, have given larger numbers, and one or two cyclones have even been traced from the Pacific coast to Russia. But in many cases there is much uncertainty as to how far the cyclone which reaches us is really the same one which left the United States. When a cyclone which has left Canada combines with another which has come up from Bermuda, how are we to identify the resulting depression with either?

Turning, now, to what we may learn from the illustrations already given as to the more ordinary course of cyclones which affect the United States, in Figs. 28 to 31 (pp. 50–51), we see a cyclone leaving Nova Scotia and going partially across the Atlantic; while in Figs. 32 and 33 (p. 54) we see the commonest case of all, in which cyclones, which have crossed some portion of the American continent, die out when they have reached the Atlantic seaboard.

Besides all these, we have seen cyclones originating in mid-Atlantic, which have become so large that they completely filled that ocean, both Europe and the United States being simultaneously, instead of successively, under the influence of the same cyclone.

We must also bear in mind that even when a centre crosses, we have no cognizance of either the size or intensity of the cyclone. A cyclone which has hardly brought rain in Canada may develop into a dangerous storm in its course across the Atlantic; while a storm which has ravaged Newfoundland may give rise to a moderate breeze in Western Europe.

From all these considerations the conclusion is inevitable, that though there is an undoubted connexion between the weather on both sides of the Atlantic, the nature of the changes are so complicated and variable, that no practical use can be made of the knowledge for forecasting purposes without observations over the intervening ocean. The Meteorological Office, in conjunction with the Bureau Central Météorologique, Paris, has since the beginning of the year 1885 made arrangements with the Chief Signal Office, Washington, for the receipt of telegraphic intelligence of storms which had been encountered by ships in the Atlantic. The service is now in activity.

Unequal Change of the Barometer.

So far, then, as general knowledge of cyclone motion is concerned, we see that pretty nearly all it enables us to do is to say that, except during the Easterly type, the motion of a depression will be towards the east; but we have already shown that there are a great many weather changes which are not associated with well-defined cyclones, and that even when they are, the changes are on so large a scale that we cannot trace them on a small area. This brings us to what is practically the forecaster's greatest stand-by when he has no definite isobaric shape to trace, viz., to watch over his limited area of observation where the barometer is falling, and where it is stationary or rising. By combining these changes, with the existing shape of the isobars, he can usually determine whether the wind will increase in force; and also whether the weather is likely to improve or otherwise, for, we have shown that in a general way, increasing gradients are associated with weather getting worse. He is also enabled by this means to form an opinion whether an anticyclone is likely to break up, or as to the likelihood of any other very great change. We shall defer our illustrations of this subject till we have explained the system on which the United Kingdom is divided into districts for forecasting purposes.

Periodicities.

We have already alluded to the question of periodicities when speaking about recurrent types of weather, but it will now be convenient to discuss the subject a little further, with, special reference to the case of the supposed connexion between sun-spots and weather, concerning the value of which most exaggerated ideas have been circulated. The best method of explaining the true nature of all meteorological periodicities will be to take the most obvious one – the daily range of temperature – and then from a consideration of its nature, we shall be enabled to understand the value in weather forecasting both of annual periodicities and of those of longer period.

In Fig. 44, where we give thermograms at Stonyhurst for the four days May 16th–19th, 1874, we have a very fair specimen of the variation of the thermometer on a fine day in England. A chart for May 17th is given in Fig. 15 (p. 27) as illustrating anticyclone weather. We see that the temperature is usually highest about 1 p.m., and lowest about 4 a.m. But how should we fare if we attempted to say that on every day it will be hottest at 1 p.m. and coldest at 4 a.m.? Look at the thermograms in Fig. 45 for December 7th–10th of the same year. On the first day the *minimum* is at 2 p.m. and the *maximum* about noon, the whole curve being the result of a dull, cloudy, day with cold showers. On the next day the thermometer rises till about 6 p.m. and then falls so continuously for the next 48 hours, that during the 9th December there is hardly any hour which is not colder than the

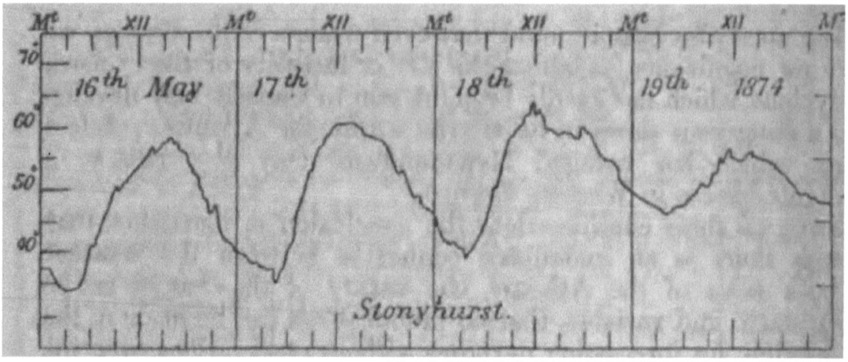

Fig. 44

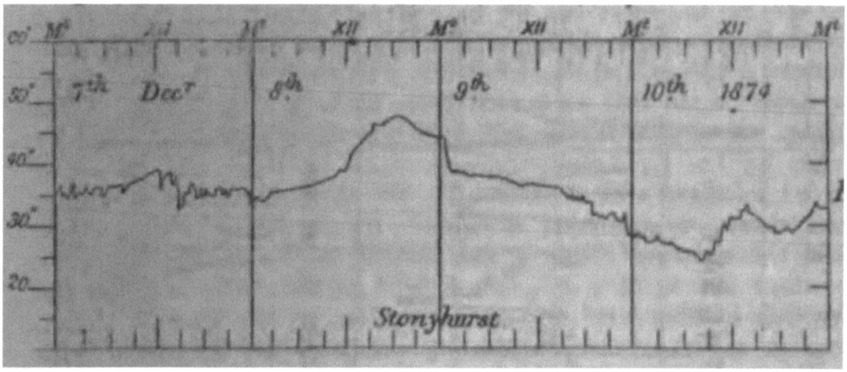

Fig. 45.

preceding one. To explain this apparent anomaly, we must describe the charts for December 8th and 9th, 1874, at 8 a.m. In the first a cyclone was coming on, which accounts for the rise of temperature seen in the thermogram for that day; while in the next chart it was seen that Stonyhurst was under the influence of the rear of the same cyclone, which accounts for the great cold, that was sufficient to override the diurnal variation. On the last day of our thermogram we see that after the mid-day maximum the thermometer rose again to an equal height, under the influence of another oncoming cyclone.

Considering, then, all these illustrations, it is evident that the observed temperature is the result of a number of different causes, and that because on any particular day we do not find the ordinary maximum, it does not prove that there is no such thing as a solar diurnal variation, but only that the power of the sun has been overridden by stronger influences. For instance, in this country during the summer months, the diurnal variation of temperature is always greater than what we may call the general changes from cyclonic or other causes. At that season we always find the nights colder than the days, however irregular the curves may be. In winter, on the contrary, as we have just seen, the general changes are so much stronger than the diurnal that the nights are sometimes hotter than the days. The same argument holds good for every other diurnal variation.

With respect to annual variations, we have already shown, when speaking about weather types, that we cannot assert absolutely that any periodicity will recur every year at the same time, but that when it does return, and we recognize its nature, then, and not till then, can we utilise it to give greater certainty to our forecasts.

Sun-Spots and Weather.

Now, with regard to cycles of longer period, the only one which has attracted much attention, is a supposed relation between sun-spots and weather, especially with rainfall The nature of the cycle is said to be that the maximum of solar spots has a period of about 11·1 years pretty certainly, and that there is an ill-defined maximum of rain with the same period, which falls a year or two behind the solar maximum.

To enable our readers to judge of the amount of connexion, in Kg. 46 we gîte on the lower line a diagram of the relative number of sun-spots, from Wolff's and other observations, as given by Professor Balfour Stewart, and on the upper line we have added the observed rainfall at Rothesay, County Bute, Scotland, for 80 years. A glance at this curve will show that the connexion between the two is very uncertain, and that it is extremely doubtful whether there is really any periodicity in the rainfall curve.

When we come, moreover, to examine the details of rainfall over Great Britain in any maximum year, similar uncertainties are discovered. Mr, Buchan has plotted the rainfall for the northern portion of our islands, for the year 1872, a year of maximum sun-spots. In his map he finds that while near Aberdeen the rainfall was 75 per cent. in excess of the average, the amount near Cape Wrath, about 100 miles distant, was below the average.

But even supposing that the existence of this periodicity was ever clearly established, what use could we make of it in forecasting? It is very obvious that we could not predict an increase of rain even in a general way, and that even if we could, we should be unable to say where it would fall, speaking only of yearly totals.

Still less, when we come to predict the weather for any particular day, can we avail ourselves of even a well-defined periodicity. It is obvious, from all the illustrations we have given already, that when we have to determine the probable course or intensity of any cyclone, such knowledge would be of no assistance to us.

Other stations have given signs of a more likely connexion between sun-spots and rainfall than Rothesay, and it is probable that if curves of the number of rainy

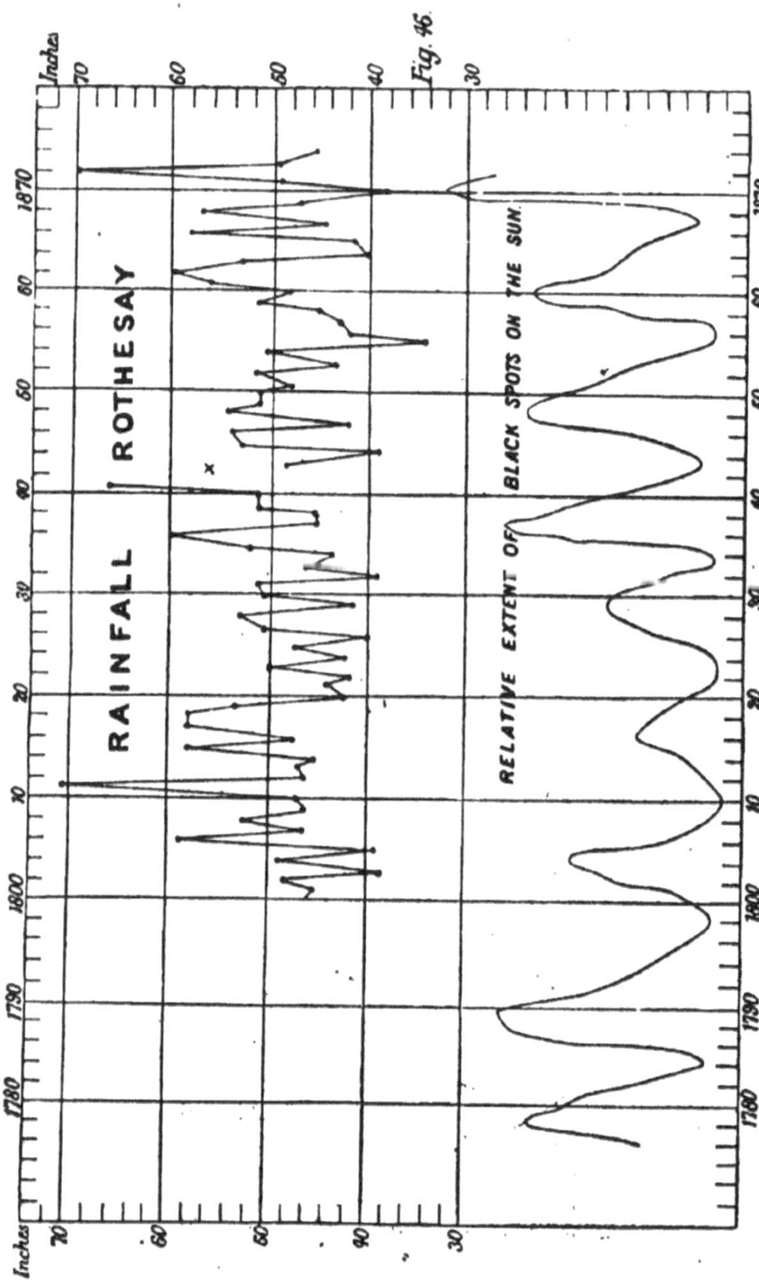
Fig. 46.
Inches
RAINFALL ROTHESAY
RELATIVE EXTENT OF BLACK SPOTS ON THE SUN.

Fig. 46.

days in a year, instead of the quantity of rain were employed, any relation between the two might be more clearly shown. But even if such a connexion should be ever fully established, it is evident that the knowledge would be of little service in forecasting.

Statistics.

A similar line of argument will explain why we are unable to use average values or any statistical results in weather forecasting. We have already explained why such a curve, as that for mean temperature, does not enable us to forecast the fluctuations of the thermometer for any single day, and when we apply average values to the motions of isobaric lines, the impossibility of using statistical results is even more obvious. For instance, we have seen that the motion of a cyclone may vary from 70 miles an hour eastwards to 10 miles an hour westwards. In some cases, by taking each depression on its own merits, we can form a very good opinion of the probable course, but the mean velocity of British cyclones, 20 miles an hour east, is only the arithmetical sum of all that have been observed, and represents no physical law, nor any number which can be used in forecasting.[1]

Prognostics.

We shall now discuss the relation of popular weather prognostics to modern synoptic methods, as it is a point of great importance. We will show both the relation of the old and new meteorology, and explain how well-known sayings can now be utilised for synoptic forecasting. We have already seen, in Fig. 18 (p. 30), that prognostics are merely the detailed description of the appearance of the sky in different parts of a cyclone, and further on we have shown that every shape of isobars has a set of prognostics belonging to itself. This is the fundamental principle of all synoptic meteorology. Theoretically, then, when the isobars are well defined, we ought to be able to write down the prognostics which might be visible, but practically we cannot do so completely; and also theoretically, all that any prognostic does, is to enable a solitary observer to identify his position in any phase of atmopheric circulation. The mere observation of the prognostic can give little or no clue to the probable path of the cyclone; that depends on the position of the surrounding areas of high pressure, and can be judged by synoptic charts alone.

On the other hand, there are sometimes cases of isobars which have no well-defined shape, but with which thunderstorms or heavy showers often occur. These, as is well known, only affect the barometer slightly, but are abundantly forewarned by the commonest prognostics, and as the rainfall is usually heavy in them, the failure of the forecast which omits to notice them is very conspicuous. The scope of this work precludes us from entering into the complicated question of the non-cyclonic rainfalls in this country. It will be enough to state that the prognostics which precede them are those associated with broken weather, such as bright sun-rises or heavy clouds banking up without a fall of the barometer, rather than the muggy, dirty weather of a cyclone front. The warning they give is also much shorter, rarely more than three or four hours, if so long.

We shall now illustrate all these points. First, for a prognostic and synoptic forecast both succeeding. Turning to the cyclone shown in Fig. 11 (p. 20), the author observed a very well-defined halo of 46° diameter at Brighton about 6 p.m. the previous evening. The prognostic held good in this instance, because a halo is characteristic of the front of a cyclone, and here the cyclone pursued the ordinary course towards some point of east. The Meteorological Office also warned all coasts successfully of a South-westerly gale, because the charts indicated an approaching cyclone of the Westerly type and of great intensity.

But then for a halo failing when a synoptic forecaster might have succeeded. On February 5th, 1883, at Folkestone, the writer saw a halo off and on, from 9.30 a.m. to 4 p.m., and this was due to the front of the cyclone which is seen lying off our north-west coasts in the 8 a.m. chart The whole of that day and night, as well as the succeeding day, was very fine, so that the prognostic might seem to have failed. Before the days of synoptic charts, this is all that we could have said about the matter, but the chart for 6 p.m. the same day showed that, owing to the increasing pressure over the North Sea, the cyclone we saw at 8 a.m. had not been able to move in its usual eastward course. In this case, though the synoptic forecaster can explain the failure of the prognostic, he would not have been able to state at 8 a.m. that rain would not have fallen in that part of England, for the isobars then presented the difficult case of a "col."

Next as an example of prognostics indicating rain, when it would not have been suspected by synoptic charts, in Fig. 47 we give a chart for February 11th 1883, at 6 p.m. In it the edge of an anticyclone lies over the south-east of England, and the weather there was generally fine. At Folkestone, however, after a wildish day, but while the harometer was rising, a brilliant halo of 46° diameter was observed at 4 p.m.; a pale sunset followed. A little after 6 p.m. the clouds showed evident signs of rain, which fell heavily from 7 to 10.30 p.m., during the crest of the barometric trace which began to fall just afterwards. By 11.50 p.m. the stars were shining, but next day was warm, squally, and threatening. In the chart for the morning of February 12, Fig 48, the low pressures lying over our western coasts on the previous evening have gathered into a cyclone over the north-west of Ireland, with almost straight isobars over the east of England. Combining all these observations, it seems that the rain which fell at Folkestone was either associated with the crest of the ill-defined wedge seen in Fig. 47, a phenomenon which we have already noticed when describing wedges generally; or else with a secondary so small as not to be detected on a synoptic chart, which was based on observations at stations at least 70 miles apart. The forecast issued by the Meteorological Office at 6 p.m. that evening, the 11th February, for the district in which Folkestone lies, was: – 5, S. (London and Channel.) – Southerly to south-westerly or westerly winds, strong to a gale; fair at first, then very rainy, and squally.

This was only partially successful; but it is manifest that if the observer at Dover, 7 miles from Folkestone, had sent up with his ordinary 6 p.m. telegraphic

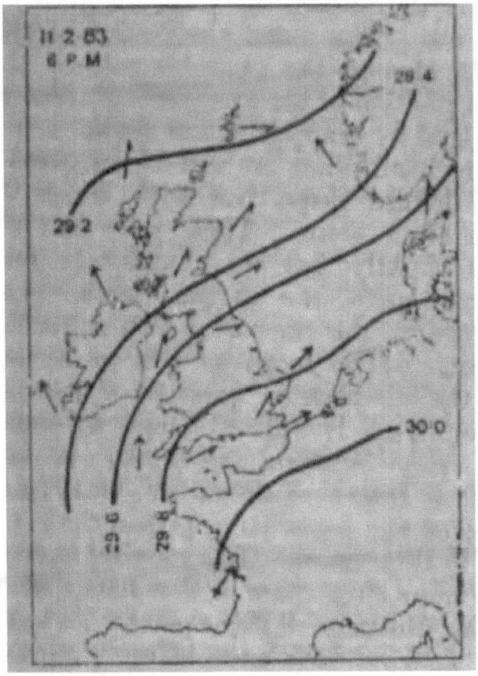
11 2 83
6 P.M.
29·4
29·2
30·0
29·6
29·8

Fig. 47.

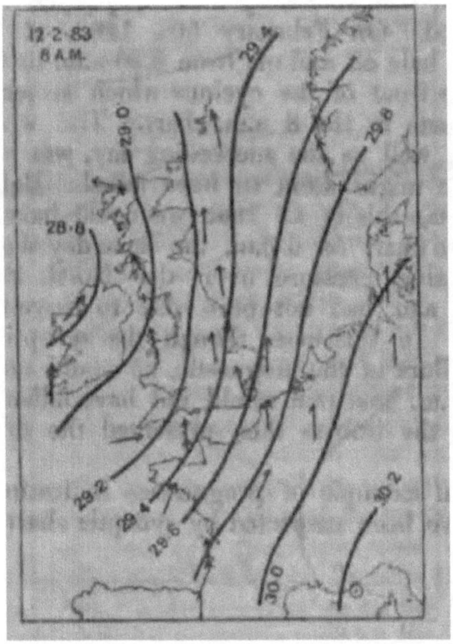
12·2·83
8 A.M.
29·4
29·0
29·6
28·8
29·2
29·4
29·6
30·2
30·0

Fig. 48.

report a notice that heavy rain clouds were banking up to windward, the accuracy of the forecast would have been much increased.

Lastly, as an example of the failure of a prognostic when synoptic charts succeeded. In London at 4 p.m. on the 22nd of February 1883, a solar halo was visible, though the surrounding clouds were not threatening. By 6 p.m. the sky was nearly blue, and the night and all next day were very fine. Here an observer, who had only prognostics to guide him, might have expected bad weather, but a synoptic forecaster could not have made such a mistake. A glance at Fig. 49 for 6 p.m. that evening will show that the halo was due to the presence of a very small secondary, hardly indicated on the chart except by the deflection of the wind to the north-west at Portsmouth. The forecaster, with a full knowledge of the prognostic, would have seen at once that with the barometer rising in the west and north, there could be no danger of bad weather in any form, and the chart shown in Fig. 50 for 8 a.m.

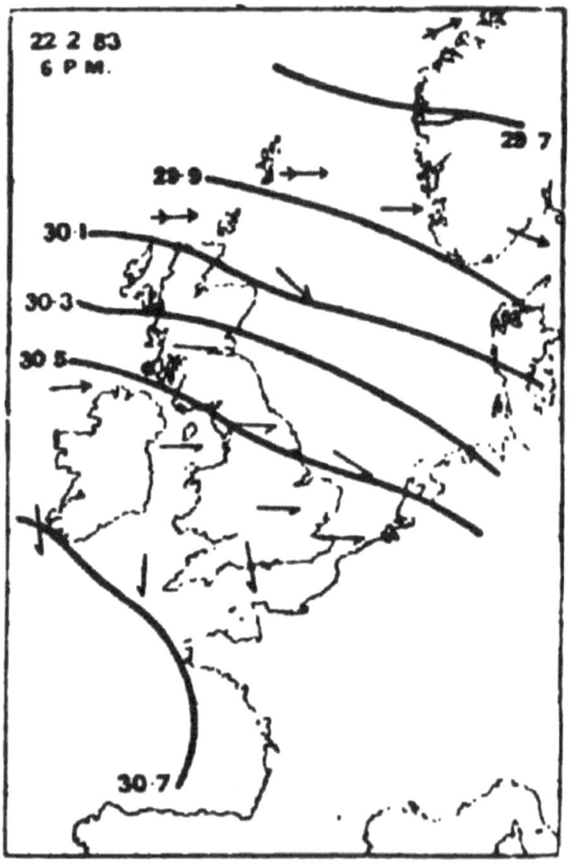

Fig. 49.

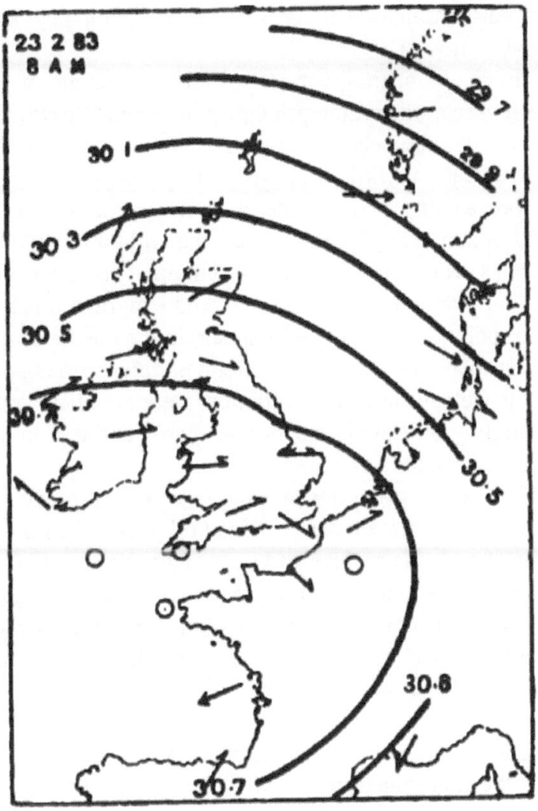

Fig. 50.

the next morning, shows that an anticyclone had stretched over England from the west. The forecast issued by the Meteorological Office was: "District 5, England, S. (London and Channel), North-westerly and perhaps Northerly airs and light breezes; dry and colder, with local fogs," and was thus abundantly justified.

Observations on Cirrus Clouds.

Allied to prognostics, and indeed only a particular case of them, are observations on the highest cloud called "*cirrus.*"

The forms of certain kinds of *cirrus,* such as "mare's tails," "wisp," "mackerel sky," and many others, are household words; but a great development of the subject has been made during the last few years, principally by the Rev. W. C. Ley. By studying the movements of the upper currents over cyclones and anticyclones, shown in Figs. 6 and 16, which have practically been worked out by himself and Professor Hildebrandsson, of Upsala, he has discovered that in many instances, *cirrus* gives very early intimation not only of an approaching cyclone, but in some cases of its probable motion.

It is beyond the scope of this work to go into the details of the subject, but we may make a few remarks on the principles involved. *Cirrus,* like every other prognostic, bears a definite relation to some shape of isobars. When these are well defined, *cirrus* tells us little that we do not know before; in fact in most cases we could forecast that *cirrus* would make its appearance in certain districts. For instance, if at 8 a.m. a wedge just covers the United Kingdom, the sky over the Eastern counties would be cloudless. A synoptic forecaster would see that a cyclone was certainly approaching, and he would telegraph down to an observer in that district not only that *cirrus* would appear in the course of the day, but that the cloud would probably take the form of long parallel stripes converging in the South-east. But, as we have already mentioned, there are many conditions of rain which are very slightly indicated on synoptic charts, and it is in those cases that *cirrus* gives the most valuable indications.

The power of sometimes pointing out the direction of a cyclone's motion is derived from the fact, which we have explained fully when dealing with cyclones, that the direction of both the surface and upper currents depends on the position of the front of the cyclone, that is, on the direction in which it is going. It is, however, doubtful whether the upper currents are of more use than the lower ones, and we have abundantly shown that the direction of the latter are of quite secondary importance. In neither have we any clue as to whether the cyclone will stand still, die out, or move onwards.

The greatest use of *cirrus* observations undoubtedly is in assisting the forecaster to foretell the rain which occurs with the small secondaries before mentioned, and to give more precision to forecasts which would otherwise be rather vague, as for generally unsettled weather, &c.

The importance of these observations is, however, so great, that the Meteorological Office have been endeavouring for the last three or four years to organize a system of *cirrus* observations. A very perfect scheme of observing and telegraphing has been worked out by Mr. Ley, and is given in Appendix I. Unfortunately, the art of making these observations is so difficult and incommunicable, that the Office has not been able to find a sufficient number of observers to fully utilize the method; besides, the observations cannot be taken at a fixed hour, when wanted, but only when the *cirrus* is visible, and there is great difficulty in finding men with sufficient leisure to be always on the look out. It is also very important to notice that no prognostic or *cirrus* observation gives much clue to the intensity of the cyclone in which it is observed, in fact they are more valuable for rain than for wind.

[. . .]

FUTURE PROSPECTS.

Having already detailed the present condition of weather forecasting, as far as regards the percentages of successes, we shall now devote a few paragraphs to the consideration of what prospect there is of increasing the number of successful predictions. To do so we shall briefly sketch several lines of research which promise useful results, but which have not yet been sufficiently investigated to enable us to form a definite opinion of their value.

We have already shown how often we are dependent on a knowledge of the difference of rate at which the barometer is rising or falling in different districts. At present we take the differences between the present readings and those at the preceding report; that is, at an interval of 12 or 14 hours. There is no doubt that during such a long interval many minor changes may escape detection, and that if we could have the barometric difference for a shorter period, we should be in a better position to consider its import.

Moreover, we have seen that the barometer may rise or fall from a variety of different causes, and when we look at any barogram we find that the curves of rise or fall differ very much in small details, and it seems more than probable that some day we may be able to derive far more information from barograms than we do at present.

Again, though we have shown that Great Britain is at a great disadvantage as regards some other countries in respect to tracing cyclones before they arrive at our coasts, still there seems to be a great field of research open in the study of the relation of the pressure to leeward of an advancing cyclone to its future path. The most casual inspection of the charts which have already been given will show that when a cyclone is once formed its path is far more controlled by the pressure which lies in front of it, than by any other consideration. The study of these relations, therefore, specially concerns British meteorologists, and we are firmly convinced that much progress is possible in that direction.

The importance of the four great types of weather which have been described has been previously pointed out, but it is extremely probable that further research may prove the existence of sub-types which will remove many difficulties which are at present apparent, by showing real differences to exist where now we only see uncertainty.

We have already alluded to the advantage to be obtained by the receipt of reports at shorter intervals than 12 hours, and though the difficulties and expense, both of sending up the observations and distributing the information derived from them, are at present great obstacles there is no doubt that when they can be overcome, great improvement will result from more frequent observations.

We have also pointed out the assistance which may sometimes be derived from the use of popular prognostics in conjunction with synoptic charts, and also from the observation, not only of the appearance, but also of the motion of cirrus clouds in upper currents. So far, however, the principal use of observations on cirrus is to increase our knowledge of the mechanism of a cyclone, that is, to show the nature of the complex circulation which makes up that kind of depression. But there is another view which is supported by some evidence. In thunderstorms we often find a very complete circulation of wind under an upper current which remains constant throughout the storm, and in almost every case the storm moves in the direction of this upper current. Now, it has been suggested, that in the case of ordinary cyclones, the circulation is confined to the lower strata only, and that there is an upper current which gives the drift of the cyclone as a whole. Even if this is the case, however, it is quite certain that such an elevated current can rarely be observed, and of course the term "upper current" is a very vague one, and really means clouds at almost any height above the surface.

Allied to cirrus observations are records derived from high-level stations. Like cirrus, their greatest value is in giving the mechanism of atmospheric circulation, and in some respects they are not so valuable as cirrus. It is obvious that when a surface wind meets an obstacle, such as a mountain, a portion of the current must be deflected upwards, and that the wind and moisture which we measure on the top of the hill by no means represent the condition of the atmosphere as it would be over an undisturbed level surface.

Finally, the Office has collected data over the whole North Atlantic for the period from August 1, 1882, to August 31, 1883, and charts have been prepared showing the state of the barometer, the wind, the temperature of the air and sea, and the weather over the whole of that ocean in their relation to the data published in weather charts for Europe and America. These charts contain on an average more than 400 observations per day over the sea surface. We may hope that these charts when published will help to elucidate the relations between wind and pressure, when there is no land to interfere with air motion, and to throw light on the causes which affect the tracks of storms across the Atlantic.

It is evident, therefore, that even within sight, there are many lines of research which will probably improve our power of weather forecasting, and as the history of all science points to the fact that research always leads to greater results than were expected, we are fully justified in the belief that within ten or twenty years the proportion of successful forecasts will be considerably increased.

Source Endnote

1 For further details of the application of statistics to synoptic meteorology, see Abercromby, "On the Application of Harmonic Analysis to the Reduction of Meteorological Observations, and on the general Methods of Meteorology," Q. J. Meteor. Soc., Vol. IV., p. 141.

Part 4

STATE AND THE PRIMACY OF PUBLIC SERVICE

4.1

GALTON REPORT

Editorial Headnote

4.1 Galton Report

Robert FitzRoy's methods and results were a matter of intense controversy during the 1860s in Britain. Members of the Royal Society of London criticised both the scientific principles on which his forecasting method rested and the fact that he focused on daily forecasts rather than storm warnings and research into the laws of weather – the research programme originally established by the Royal Society. In January 1867, following FitzRoy's death in 1865, the Royal Society, at the instigation of the Board of Trade, formed a Meteorological Committee – consisting of Francis Galton (chairman), Thomas Henry Farrer (nominated by the Board of Trade) and Frederick John Evans (nominated by the Admiralty) – to make recommendations on the future Meteorological Department. The text in this section, popularly known as the Galton Report, is a highly critical examination of FitzRoy's forecasting approach and the overall management of the Board of Trade's Meteorological Department during his tenure (see Vol. II, 3.1 'Forecasts and warnings' and 3.2 'Galton report').[1]

Francis Galton (1822–1911) pioneered fields as diverse as meteorology, statistics, psychology and even eugenics.[2] His high scientific reputation in the 1860s was probably decisive for his nomination by the Royal Society and his influence on the other two members of the committee. The Galton Report, a lengthy 43-page document divided into three parts plus 18 appendices and presented to the British Parliament on 13 April 1866, was devastating for FitzRoy.[3] While paying symbolic tribute to him, it virtually disapproved of all his actions and achievements, from his statistical compilations and method of presentation to the cessation of his work on statistics and the Meteorological Department's publications, selected 'without any plan' (p. 14). The 'Recommendations' and 'Estimate of Cost' sections, reproduced here, precede the final 'Conclusion' section which transcribes the answers to a series of questions posed by the Board of Trade; both sections proved crucial. The two sections were instrumental in bringing the Kew Committee of the British Association for the Advancement of Science largely under the control of the Meteorological Committee for the next three decades. Galton is concerned with scientific rigour – a more systematic and theoretical rather than purely empirical approach – because it is a feature he does not find in FitzRoy's methods, despite the questionable accuracy of his storm forecasts. He therefore applies his academic and statistical expertise to recommend the immediate cessation of daily forecasts and the continuation of storm warnings as long as their principles are adequately defined and tested. The pioneering nature of FitzRoy's forecasts undermines Galton's efforts, but he nevertheless concludes that doubts about their accuracy and the processes used to formulate forecasts, including the interpretation of data, all point to serious methodological flaws. In the 'Resources' section, he estimates an annual cost considerably higher than the current expenditure of the Meteorological Department, allocating a significant part of the budget to the establishment of six stations equipped with self-recording instruments to investigate the laws governing weather changes.[4]

Notes

1 Aitor Anduaga, *Politics, Statistics, and Weather Forecasting, 1840–1910. Taming the Weather* (London and New York: Routledge, 2020), pp. 180–81, 310–14.

2 Derek Forrest, *Francis Galton: The Life of a Victorian Genius* (London: Paul Elek, 1974); Karl Pearson, *Life, Letters and Labours of Francis Galton* (Cambridge: Cambridge University Press, 1940–1930), 3 vols.; John C. Waller, 'Gentlemanly Men of Science: Sir Francis Galton and the Professionalization of the British Life-Sciences', *Journal of the History of Biology*, 34 (2001), pp. 83–114.

3 James Burton, 'The History of the British Meteorological Office to 1905' (The Open University, PhD diss., 1988), pp. 99–107.

4 J. Malcolm Walker, *History of the Meteorological Office* (Cambridge: Cambridge University Press, 2012), p. 73.

33

REPORT OF A COMMITTEE APPOINTED TO CONSIDER CERTAIN QUESTIONS RELATING TO THE METEOROLOGICAL DEPARTMENT OF THE BOARD OF TRADE [GALTON REPORT]

Source: 'Report of a Committee Appointed to Consider Certain Questions Relating to the Meteorological Department of the Board of Trade' [Galton Report], 1866 (3646), LXV, pp. 37–41.

41. *RESULTS TO BE LOOKED FOR FROM THE ABOVE.*

As the science of weather-changes is so little understood and excites so much interest, it would probably be desirable to publish charts and discussed observations more freely at first than would afterwards be necessary. It is very much to be hoped that by these means the subject would attract the attention of men eminent in science, who are now repelled by the impossibility of obtaining information in a form they can use, without previously undergoing an excessive amount of purely clerical labour.

If these steps are taken we may hope that at no distant time the laws which govern the changes of weather in the British Isles will be so far understood as to enable Meteorologists to place the practice of foretelling weather on a sound basis. If a considerable proportion of the various states of weather can be grouped under definite categories, and if each of these categories is found to change into other definite states, with more or less regularity, it will only be necessary to determine the category under which the prevailing type of weather should be classed, in order to arrive at a knowledge of coming changes. And this may probably be done by means of a limited amount of telegraphic communication. To take the least favourable view of the subject, the knowledge obtained by means of the observations we have recommended will furnish a complete check on such predictions as may be made, and will either enable us to reduce the practice of foretelling Weather into a certain system governed by clear and intelligible rules, or will enable us to conclude that no such system or rules are possible.

DOI: 10.4324/9781003650294-51

42. *RECOMMENDATIONS.*

The following are our recommendations on the subject of Weather Telegraphy, Daily Forecasts, and Storm Warnings, and upon Observations of Weather within or affecting the British Isles, viz.: –

1. That the system of telegraphing the weather from distant stations, as proposed by M. Le Verrier, and adopted by that officer and by Admiral FitzRoy, be continued.
2. That the places from which telegrams are to be received shall be those from which they are at present received, with power to add to or diminish their number as circumstances and advancing knowledge may require.
3. That these telegrams shall be published as at present, but arranged in geographical districts.
4. That the publication of daily forecasts of weather probable on the North, East, South, and West coasts shall be discontinued.

   That the "Remarks" or summary of the general results of the telegrams, such as M. Le Verrier publishes in his Weather Bulletin, and such as Mr. Babington has recently appended to the daily Forecasts, shall be continued, but that the office intrusted with the duty shall not hold itself bound to issue such remarks daily as a matter of course, but shall only do so when it has reason to believe that there is some general view or some conclusion of interest to be derived from the reports.
6. That the practice of issuing Storm Warnings shall be continued, but with the following modifications: –

   (*a.*) That the Signals shall for the present be confined to the indication of a probable gale, without attempting to indicate from what quarter.
   (*b.*) That they shall not be hoisted unless there is reason to expect the gale within 36 or at the outside 48 hours.
   (*c.*) That when hoisted, they shall continue up until all immediate expectation of further gales has ceased.
   (*d.*) That whilst the Signals indicating Direction are discontinued for the present, care shall be taken so to arrange the Signals for Force as to enable the Signals for Direction to be added hereafter.

7. That the officer of the Meteorological Department issuing the Storm Warning for Force should also at the same time, so far as he is able so to do, make, but not issue or publish, a prediction of the probable Direction of the coming gale, endeavouring in so doing to render it as specific as possible, *e.g.*, whether within any particular quarter of the circle.
8. That this officer shall note down at the time, and reduce into an exact shape afterwards, the maxims or principles which have guided him in making

the Signal of Force or Prediction of Direction; the facts to which those maxims are applied; the mode in which he has applied and combined them, the value he has attached to each of them, and the value of the probability which he has thus obtained, and which is indicated by the Signal or Prediction.

9. That the maxims so acted upon shall be reduced into a clear and definite shape, and kept in the office ready for reference.
10. That the present practice of collecting miscellaneous information concerning daily weather from newspapers and other sources be discontinued.
11. That a careful check upon the correctness of the predictions issued for Force, as well as upon those made, but not issued, for Direction shall be kept. If the recommendations we have made with regard to the collection of observations of the weather changes in the British Isles be adopted, such observations will provide this check; otherwise the observations and returns heretofore made to the Wreck Department of the Board of Trade should be continued, but with the addition, that a return should be made from every station of every gale felt there, whether a Storm Signal is hoisted there or not.
12. In whatever way these observations are made, the result should in each instance be carefully digested and compared with the Prediction, and with the reasons for making it. In case of error or omission, whether as regards Force, Direction, Time, or Place, it should be noted; and an endeavour should be made to ascertain how it occurred, and the maxims acted on should, when necessary, be modified accordingly.
13. If the observations are collected and digested by an office distinct from that which issues the Storm Signals, the Signals, with the reasons for them, and the results of the observations, should be mutually communicated by the officers to each other, so that the one may be a check on and assist the other. In this way the practice may be brought into the shape of a determinate system resting on a sound inductive basis.
14. In the meantime the returns already obtained by the Wreck Department of the Board of Trade, though not complete, inasmuch as they are only made when and where a Storm Signal is hoisted, afford valuable material for tracing the rise, progress, and direction of most, if not of all, the violent gales which have happened in the British Isles during the last five years. These returns should, if possible, be digested and utilized.
15. *Finally, we recommend that the Variations of the Weather in the British Isles and in the adjacent Ocean be carefully observed, charted, and discussed. We have entered so fully into the subject above that it is needless here to repeat our recommendations. We will only add that of all our recommendations on this part of the subject, it is in our opinion the most important.*

PART III.

ESTIMATE OF COST.

43. *COST OF EXISTING METEOROLOGICAL DEPARTMENT.*

We give in the Appendix a full account of the cost of the Department from its institution in 1856 to the present time. The aggregate amount to the end of the financial year 1865 will have been about 45,000*l.* The annual expenditure has increased from 3,240*l.* to, say, 5,500*l.*; but was in one year, 1863–4, as much as 7,100*l.* The sums expended on instruments and other expenses connected with Ocean Statistics have greatly diminished, viz., from an average of 2,215*l.* 19*s.* 6*d.* for the years 1856 to 1860, to an average of 1,613*l.* 6*s.* for the years 1860 to 1865; whilst the expenditure has, in the latter years, been increased by a sum averaging 2,011*l.* a year spent on Telegraphy and Storm Warnings. The expenditure for 1864–5 was 1,144*l.* 14*s.* 8*d.* on instruments 2,735*l.* 10*s.* on Telegraphy and Storm Warnings, and 1,134*l.* 17*s.* on salaries, making 5,460*l.* in all.

44. *RECAPITULATION OF WORK TO BE DONE HEREAFTER.*

We now proceed to consider in what manner and at what expense the work we have recommended can be done. That work is as follows: –

I. OCEAN STATISTICS.

1. Completion of work now in progress, viz.:

   a. *Winds.*

   Charts for South Pacific.
   Charts of Trade Winds for Indian and Pacific Oceans.
   Winds generally. Arrangement of existing "Collecting" and "Grouping" Papers.

   b. *Ocean Currents.*

   Arrangement of existing "Collecting" and "Grouping" Papers.

   c. *Sea Temperature.*

   South Atlantic. "Collecting" to be completed.
   Arrangement of existing "Collecting" Papers.

   d. *Temperature of the Air.*

   Tabulation and publication of results already obtained.
   Arrangement of existing "Collecting" and "Grouping" Papers.

   e. *Vapour Tension.*

   Publication of monthly means of wet bulb as already obtained, and making and publishing comparison of monthly means of wet and dry bulbs.
   Arrangement of existing "Collecting" and "Grouping" Papers.

f. *Barometer.*

Publication of monthly means already obtained.
Arrangement of "Collecting" and "Grouping" Papers.

2. Issuing instruments and registers to merchant ships.
3. Extracting the whole of the observations now in the office, and such other observations as may be hereafter obtained, to the number of (say) 1,650,000, in the manner explained above.
4. Reducing, digesting, and tabulating the observations so extracted.

II. Weather Telegraphy: Foretelling Weather: and Observations affecting Weather in the British Isles.

1. Telegraphing to and from out-stations.
2. Examining telegrams daily for the purpose of remarks and of Storm Warnings, and recording progress.
3. Establishment and maintenance of six stations for Meteorological Observations in the British Isles.
4. Collection of observations from lighthouses or other intermediate stations and from the Atlantic.
5. Digesting, tabulating, charting, and publishing the results.

45. *MEANS AND METHOD OF EXECUTING THIS WORK.*

It is not within our province to suggest alterations in a Government Office, still less to propose the establishment of a new Office. But we can hardly estimate the cost of what we recommend without forming an hypothesis as to the way in which it should be done, and in forming such an hypothesis we have adopted what appears to us to be the most efficient as well as the most economical plan.

The collection of Observations from the captains of ships is a function which can probably best be performed through the medium of such agencies as a Government Office can command, and which was in fact well performed by the Meteorological Department before its attention was devoted to the practice of foretelling weather. We assume, therefore, that this function will remain with the Board of Trade.

The Digesting and Tabulating Results of Observations is on the other hand a function which requires a large knowledge of what the state of the science for the time being requires, as well as exact scientific method. This function is one that has not been satisfactorily performed by the Meteorological Department. And we believe that it would be much better as well as more economically performed under the direction of a scientific body, – such as a Committee of the Royal Society or of the British Association, if furnished with the requisite funds by the Government, – than it will be if left to a Government Department. The establishment already existing at Kew might probably be easily developed so as to carry into effect such a purpose. It would in that case become a Meteorological centre to which all observations of value, whether made on land or at sea, and whether within the British Isles or not, would be sent for discussion and reduction. We have, therefore, in the following estimates assumed that all Meteorological Observations made on land, whether at the stations recommended by the Royal Society, or at the Lighthouses or Coast Guard Stations, as well as all observations at sea, shall be referred to and discussed under the direction of such a scientific body as we have mentioned; and we have also assumed that the aid afforded by Government would be in the shape of an annual vote so made as to leave the Royal Society, or other scientific body charged with the duty, perfectly free in their method and in their choice of labour, but upon the condition that an account shall be rendered to Parliament of the money spent and of the results effected in each year.

The completion of the work now in progress in the Meteorological Department may, on the above hypothesis, either be performed by that Department at once, or if the proposed change be made immediately, may be placed in the same hands in which the future discussion of Meteorological Observations is placed, and we have, in our estimates, dealt with it accordingly.

The procuring and sending of daily telegrams, and the issuing of Storm Warnings, is intimately connected with the discussion of Meteorological Observations in or near the British Isles, and ought, we think, to be placed under the same scientific body which superintends the discussion of those observations. For the convenience of telegraphing it will probably be necessary that part of the staff employed under this body, whilst in connexion with Kew, should occupy two or three rooms in London. But any expense in hiring such rooms will be less than the expense of the premises at present occupied by the Meteorological Department, which are, we understand, to be pulled down, whilst the persons employed for a part of the day on telegraphy, will be available during the greater part of the day for the discussion of observations.

The publication of results of Meteorological Observations at sea, which are of immediate utility to navigators, either in the shape of charts or otherwise, appears to be a function properly belonging to the Hydrographic Office of the Admiralty. We have accordingly assumed that it will be performed by that Office.

a. Estimated Cost of this Work.

On these assumptions we make the following estimate: –

Ocean Statistics:		£
Issue of Instruments and Registers, annually		1,500
Discussion and publication of results		1,700
	Total	3,200

This expenditure ought to terminate in about 15 years, as by that time a sufficient number of observations to determine the Meteorological Means will have been collected and discussed.

Weather Statistics in and near the British Isles:	
Six Stations with Self-Recording Instruments: Collecting Observations from intermediate Stations, Lighthouses, Ships, &c.; discussing Observations, Charting, and publishing Results, annually	4,250
Besides an outlay, to begin with, of 2,500*l.* and whatever sums may be needed for additions to the buildings at Kew.	
Telegraphy and Storm Warnings, annually	3,000
Grand total annually	10,450

As regards expenses to be incurred by the Hydrographic Department in getting out and publishing charts, &c., for the immediate use of navigators, we are unable to say with precision what it may be found necessary to do, and we are therefore unable to give an estimate.

46. *REASONS FOR PROPOSED INCREASE OF EXPENSE.*

The expense of what we propose is larger than the expense hitherto incurred. But this is unavoidable unless either the original object of the Meteorological Department or the system of Storm Warnings is to be abandoned. The Meteorology of the ocean is, as we have stated, as important an object now as it was in 1854; and we feel ourselves justified in believing (especially with such a promise of success as is held out by the Meteorological Registers already collected) that the Government and Parliament will not now abandon an object taken up by them after much consideration in 1854, and that they will not be satisfied to leave the matter in its present incomplete and useless condition. If the grant originally made had been steadily applied to this object and had not been diverted to other objects, the work would by this time have advanced far towards completion; and we do not doubt that it may be completed within the time, and for the sum we have mentioned above.

The prognostication of Storms is a branch of practical Meteorology which has been superadded to the original Functions of the Department, and to which a large part of the funds originally granted for the purpose of Meteorological Observations at Sea has been devoted. It is one far too important, too popular, and too full of promise of practical utility to be allowed to die. But the present treatment of it is, as we have shown, incomplete and unsatisfactory, and it cannot be made complete or satisfactory without the new system of observations, and consequent additional expense, which we have recommended.

These Observations are the foundation; the Telegraphy and Storm Warnings are the superstructure; and we have no hesitation in saying, in the interest of practical utility as well as of science, that if the expense we have recommended is thought to be too large, and any part of what we have proposed is to be postponed for the present on account of expense, the part to be postponed should be that part which recommends the present continuance of the attempts to prognosticate weather. To continue them in their present condition without an endeavour to determine the principles and rules on which they should be founded, would, in our opinion, be injurious to the fame of the eminent officer who has originated them and discreditable to the country.

For these reasons we have no hesitation in making the various proposals mentioned above, and in recommending the consequent increase of expenditure.

4.2

DAILY WEATHER CHARTS

Editorial Headnote

4.2 Daily Weather Charts

Although some nineteenth-century meteorologists prioritised physical research over weather forecasting mapping, there was an overwhelming majority from national weather institutes for whom research must be subordinated to public service. The mass production of weather maps and the processing of observational data therefore became the *goal* par excellence of these institutes. As one of America's leading meteorologists, Cleveland Abbe, stated, the 'daily routine of observing and forecasting is not in itself research'.[1] These sources illustrate the processes involved in the production and dissemination of weather maps at the British Meteorological Office and the scientific bureaucratisation required in the whole endeavour.[2]

The earliest, by Robert H. Scott (1833–1916), sets out why official meteorologists should, indeed must, produce weather maps and distribute them in newspapers: the visual representation of weather data and the timeliness of information make it easier for the public to understand weather patterns compared to textual descriptions, and help to anticipate changes. Scott advocates the social and educational value of weather information, emphasising its usefulness for safety and preparedness, especially in relation to severe weather events. Scott, a mineralogist and amateur geologist, had devoted himself almost exclusively to these two fields, except for sporadic forays into meteorology, such as his translation from German of Heinrich Dove's *Das Gesetz der Stürme* (*The law of storms*, 1862).[3] However, in 1865 he became director of the Meteorological Office, thanks to the influence of his close friend Edward Sabine, then president of the Royal Society. His ability as an administrator, rather than a forecaster or scientist, at the Office – where he was to remain as the executive head for thirty-three years – was reflected in his many articles and works, including his two books *Weather Charts and Storm Warnings* (1876) and *Elementary Meteorology* (1883), which went through nine editions.[4] 'Weather Charts in Newspapers' is an article that well reflects his concerns about the effectiveness of storm warnings and the challenges of weather forecasting and its dissemination to the public. By introducing weather maps, newspapers play a key role in raising public awareness and understanding of weather patterns.

The second text excerpted here is quite similar in tone and descriptive perspective, in the sense that it shows that the purpose of the Met Office was service-oriented meteorology. 'The Publication of Daily Weather Maps and Bulletins' is the paper read by Scott at the International Meteorological Congress held in Chicago on 21–24 August 1893. His extensive institutional experience as director of the Met Office and other positions in international organisations equipped him with first-hand knowledge of the operational challenges and successes in weather reporting. The paper is a description of the scientific bureaucratisation of the Met Office through its daily and weekly weather reports. It aims to highlight the importance of comprehensive weather reporting and the role of citizen contributions in enhancing meteorological data collection. Budgetary constraints are a major challenge insofar as they affect the scale and character of weather maps and bulletins. The level of funding allocated to weather services determines the resources available for the preparation and distribution of these maps. Consequently, the extent of the area covered and the level of detail included in the publications are directly related to the public funding received. Throughout the text, Scott articulates his conviction that meteorological services should have the autonomy to decide what to publish, rather than being dictated to by external bodies such as congresses. This leaves room for partnership with newspapers such as *The Times*, which published official weather data for over thirty years. Thus, Scott illustrates, and seeks to prompt, not the service to political or external dictate, but the autonomy of a bureaucratised service oriented to the public interest.

'The Times Weather Chart' is a short anonymous article published in the journal *Nature* on 15 April 1875. Although the London trade newspaper *Shipping Gazette* had been publishing daily wind charts provided by the Met Office since 1871, the *Times* weather chart is believed to be the first weather map to be published regularly in a British newspaper.[5] The daily weather report was the main activity of the Office's forecasters and was distributed daily to *The Times* and other media. The *Nature* article discusses the significance of *The Times* weather map, which aims to provide readers with a visual representation of weather patterns and forecasts. The article shows how, in its early stages, the Met Office did not adopt the isobaric map design devised by Hippolyte Marié-Davy in France in the 1860s, but the hybrid composition proposed by Francis Galton, consisting of dotted lines (rather than isobarometric lines) for the gradations of barometric pressure, figures (rather than isotherms) for temperature at major stations, and words for sea and sky conditions.

Notes

1 Cleveland Abbe, 'Meteorology and the Position of Science in America', *The North American Review*, 174:547 (1902), pp. 833–44, on p. 843.

2 Aitor Anduaga, *Politics, Statistics, and Weather Forecasting, 1840–1910. Taming the Weather* (London and New York: Routledge, 2020), pp. 327–34.

3 Patricia M. Byrne, *Robert Henry Scott*, in *The Dictionary of Irish Biography* (Dublin: Royal Irish Academy, 2024), available at https://doi.org/10.3318/dib.007951.v1

4 Jim Burton, 'Scott, Robert Henry (1833–1916)', in H.C.G. Matthew and Brian Harrison, *Oxford Dictionary of National Biography* (Oxford; Oxford University Press, 2004), vol. 49, pp. 465–66; Alastair D. Wood, 'Scott, Robert Henry (1833–1916)', in Bernard Lightman (ed.), *The Dictionary of Nineteenth-Century British Scientists* (Bristol: Thoemes Continuum, 2004), vol. 4, pp. 1790–91.

5 Mark Monmonier, *Air Apparent. How Meteorologists Learned to Map, Predict, and Dramatize Weather* (Chicago, IL: The University of Chicago Press, 1999), p. 273.

34

WEATHER CHARTS IN NEWSPAPERS

Robert Henry Scott

Source: Robert H. Scott, 'Weather Charts in Newspapers', *Journal of the Society of Arts*, 23, 1875, 776–782.

MISCELLANEOUS.

WEATHER CHARTS IN NEWSPAPERS.

By R. H. Scott, F.R.S.,
(Director of the Meteorological Office).

The recent appearance of weather charts in some of the daily papers has attracted the attention of many to whom such charts are new, and who, though they may have realised the fact that the weather all over the British Isles on the same day is not the same, do not see how the conditions prevailing in one district influence those in another, or how, by a comparison, *e.g.* of the weather in Ireland and Scotland, with that in England on any day, we can form some idea of the changes which will probably have taken place by next morning.

The honour of having been the first to render possible the discovery of such relations as those just indicated, by the publication of daily charts, is due to Le Verrier, who, in 1858, brought out his *Bulletin International* containing a lithographic weather chart for Europe, and the example set by him has been followed by this country, whose Daily Weather Charts date from March 11, 1872. and are now distributed at the rate of about 500 copies a-day to subscribers, and for exhibition at various public offices, seaports, and fishing stations. It is hardly necessary here to allude to the charts issued by the Signal-office, Washington, of which three appear for every day, or to Captain Hoffmeyer's charts for the whole of Europe, which, however, appear some months after date, and are only useful for the detailed scientific study of weather.

The advantage presented by such a graphic account of the weather, as compared with a mere tabular record of observations, need hardly be mentioned; it is almost like the difference between a picture and a description. But I may here

DOI: 10.4324/9781003650294-53

explain that the charts which accompany our own Daily Weather Reports are four in number, and give a representation of – (1.) The barometrical pressure. (2.) The temperature. (3.) The wind and sea disturbance; and (4.) The weather and state of the sky over the United Kingdom, and some of the adjacent parts of the Continent, for 8 a.m. every morning.

It is not, however, with the daily weather charts that I have to deal at present, but with the charts which have appeared since April 1 in the *Times,* and of late in the *Hour* also, in which it is endeavoured to show the leading features of the four charts just mentioned combined in one chart, and so I must confine my remarks to the amount of information which these convey.

It is obvious that a circulation of 500 copies daily is absolutely nothing as compared with the number of readers of any of our leading papers, and so for a long time the attention of the Meteorological-office has been directed to the undoubted desirability, and the questionable possibility, of providing in a short space of time a chart suitable for the columns of a newspaper. The credit of having persistently urged the matter till it met with final success is due to Mr. Francis Galton, F.R.S., while the practical details of the process have been devised and perfected by Messrs. Shanks and Johnson, of the Patent Type Founding Company.

It may be of interest to see how such a chart is produced, what it shows us about the weather of the day for which it is drawn, and finally what is its use as to forming a judgment of the weather probable for the next day, although this last cannot be given within the limits of an article such as the present, even were our knowledge sufficiently codified into regular laws and rules.

The production of the chart is a pretty simple matter. The observations at about 50 stations in these islands and on the Continent are taken at about 8 a.m., and telegraphed in cypher to London, where they arrive, when all goes well, before 10 o'clock. As fast as they come in the figures are put down in their proper places on a large chart, and as soon as this chart is finished, so as to afford a reasonably complete view of the general state of affairs meteorological for the day, the chart for the newspapers is commenced. The space which can be allotted to the chart in the columns of the *Times* is but small, and so it is necessary to condense into it as much information as possible without sacrificing clearness. What this information is wo shall explain later on. Once the chart is drawn, it is sent to the Patent Type Founding Company's works in Red Lion-square, and there copied mechanically, at a replaced scale, on a slab of a special composition, which has been formed in a mould bearing in bold relief the outline of the land on the map. This outline, of course, comes out in the impression as a deep groove. The engraving is done by a drill, the depth of the cut being regulated by the workman. The only speciality about this part of the process is that the composition cuts quite true, without risk of chipping. The curves are copied direct from the chart, while the words and letters are put on by means of templates, so as to ensure uniformity in the type.

The instant the engraving is completed the slab is ready to furnish a cast, which is, in the first instance, taken in rather fusible metal to save time in cooling. This first block being obtained, the ordinary process of taking an impression from it in paper ensues, and this being effected, it is a simple matter to produce from this impression any number of stereotype blocks type high, which are then sent to the newspaper office, set up with the type, and worked off in the usual way, just like a paragraph of letterpress.

The next point then is what this chart shows us. In the first instance we see the dotted lines; these are what are called "isobars," from two Greek words meaning "equal pressure." These lines are obtained by drawing lines on the chart to join together the places where the barometrical readings are the same, and, as we shall see presently, they are the most important lines we can give on a weather chart, to show us, when their indications are read aright, what the winds and weather are likely to be. The other entries on the chart are the temperature at certain prominent stations; the state of the weather and the sea disturbance, which are given in words, the latter being always in capital letters; and though last not least, the wind, which is shown by arrows flying *with it,* not *against it,* like the vane of a weathercock. The force of the wind is shown by five grades, which are as follows: – O calm, → light, → fresh to strong, ↣ a gale, ↣ a violent gale. The final result is a chart like those given below, and it can produced in about an hour after the completion of the original chart in the office.

With regard to the actual interpretation of the chart, I must pre-suppose in my readers some knowledge of the elements of meteorology, else it would be hopeless, within the limits at my disposal, to render what I have to say intelligible. The most important principle, of which sight must never be lost, is that of the relation of the wind, both in direction and force, to the distribution of barometrical pressure. Wind is produced by differences of pressure, for the air naturally flows from a place where pressure is in excess, to one where it is in defect, and this motion of the air is wind. The wind moves not in straight lines, but in great curves or sweeps, and its motion is ruled by the distribution of pressure, as we shall soon see. A glance at almost any of these charts will show that, even on the calmest day, there are appreciable differences between the barometrical readings taken at the same time over the United Kingdom. Meteorology is not at present able to explain what the causes of such differences of pressure are – if it were we should have made many steps towards the accurate prediction of weather – and we must for the present take these areas of high and low pressure respectively as existing, but we may assert that on the mutual action of these areas on each other is all our weather, with its manifold and perplexing changes, depends. The areas are called cyclonic and anticyclonic, from the Greek word *κύκλος* a circle; Piddington, the author of the "Seaman's Handbook of the Law of Storms," having been the first to propose the generic name of "cyclones" for the typhoons of the Eastern Seas and the hurricanes of the Atlantic, which are each connected with an area of very low pressure (*i.e.,* an area where the barometer has fallen very rapidly), round which the wind sweeps with terrific velocity. Practically, at present, any region of relatively low pressure is called cyclonic, and any region of relatively high pressure, anticyclonic. If at any place, or over any district, the barometrical reading is lower than at the places all round it, that place or district is the centre of a cyclonic area;

if, on the contrary, the reading at the place or district is higher than at places all round it, that place or district is the centre of an anticyclonic area.

A chart like those now under consideration is rarely sufficient to show the full extent of an area of either class, but we can gather from the general course of the isobars, and the directions of the several winds which are shown on it, where the respective areas lie. These areas or regions, as they are also called, are shown on a chart by the course of the isobars which enclose them. On such a chart there are several isobars corresponding to the various gradations of barometrical pressure. Take the chart for April 29, given below (p. 779). The isobar of the lowest readings is that of 29·9 ins., it is therefore the nearest to the centre of the cyclonic area, and accordingly it is taken as a sign of the existence of such an area to the northward of the North Sea, and that this is a correct supposition is proved by the direction of the winds in Norway, as will shortly be pointed out, and as is seen from the explanatory remarks on that chart. The isobar of 30·2 ins. is the highest on the chart, it is nearly closed, and is more or less surrounded by the isobar for 30·1 ins. These isobars show us that an anticyclonic area is lying over the entrance of the English Channel, and this statement is confirmed by the sweep of the winds at Scilly, Paris, and Biarritz, as will be at once explained. On this chart the isobar of 30 ins. marks, so to speak, the neutral area, but cyclones and anticyclones *are relative terms.* The reading may be above 30 ins. at the centre of a cyclonic area, or below 29·5 ins. at the centre of an anticyclonic area. The greatest care must be taken not to identify the isobars with the areas of atmospherical pressure which they mark out. The isobars are not cyclonic or anticyclonic, though the winds may be said to have a cyclonic or anticyclonic sweep. That the definition given above of a cyclonic area is fully borne out by the chart in question we may show by the fact that the lowest reading reported that morning was 29·8 at Skudesnaes, and *all round it,* as far as reports go, we had higher readings. Wick, 29·85; Aberdeen, 2989; Cuxhaven, 30·07; the Scaw, 29·93; Stockholm, 29·89; Christiansund, 29·87. The report from the Shetlands had not arrived when the chart was drawn. While speaking of the extent of the areas, we may say that while, as a general rule, the cyclonic areas are far more limited in superficial extent than the anticyclones, the former sometimes attain prodigious dimensions, one on the 19th of January, 1873, having covered the space from the Bay of Biscay to Iceland.

It may now be asked, how does the direction of the wind enable us to judge of the distribution of pressure? This is the great discovery of recent times in weather knowledge. The air either flows out of an anticyclonic or into a cyclonic area, but it moves in a curved course sweeping round the centre, where there is a calm in both cases. The direction of this motion in the northern hemisphere is *with* watch hands in the former, and *against* watch hands in the latter case; in other words, if we started from the south point of any area and drifted before the wind which was blowing there, we should in the former case move as an East wind, and afterwards change through S.E. and S. to S.W. and so on; in the latter case, we should begin to move as a West wind, and afterwards change through S.W. and S. to S.E., &c. If we consider either of these instances carefully, we shall see that when we are drifting with our back to the wind we have always a higher barometrical reading on the right hand than on our left; this is, in other words, that the direction of the wind is nearly parallel to the isobars. This is all stated very concisely by what is known as Buys Ballot's Law, which is as follows: – "Stand your back to the wind and the barometer will be lower on your left hand than on your right."

Hence we see that from the direction of the wind we can draw positive conclusions as to the distribution of pressure. If we have a south wind in the west of Ireland, we know that there must be an area of low pressure over the Atlantic outside our coasts to produce it. In the same way, westerly winds in Caithness and Shetland show positively that the barometer over the Faroes, and probably over Iceland too, is lower than it is with us.

It is needless to multiply examples like the above, and I must say something about the means which we possess of judging of the wind force from barometrical readings. The force, moreover, depends on "the difference of pressure over a definite distance. If the barometer in London is an inch higher than it is say in Edinburgh, there will be a strong wind from the westward over the north of England – in fact, a storm, while if the difference in readings between the same places is only a tenth of an inch there will be hardly any wind at all.[1]

This is all simply expressed by the idea of a "gradient," as a means of gauging these differences. The barometrical gradients cited in the remarks on the daily charts are in hundredths of an inch of mercury, and the definite distance is 60 geographical miles, and it is found that the wind will not reach such a force as to be too strong for yachts and small craft unless the gradient is at least six hundredths of an inch. The direction of the wind will always be nearly at right angles to the line along which the gradient is cited. These principles form the very foundation of our knowledge of weather, and they are nothing but a development of Buys Ballot's Law.

It need hardly be pointed out how the principles which have just been mentioned deal a death-blow at the old rules for weather which are expressed in the words still to be seen on barometer scales, "fair," "change," "rain," "stormy," &c., which are little less than utter nonsense. It is undeniable that there is more chance of strong wind when the barometer is low than when it is high, but this arises from the circumstance already mentioned, that cyclonic areas are usually much smaller than anticyclonic, so that when the barometer is low there is probability of a steep gradient, from adjacent higher readings, existing, than when the readings are high; although it sometimes happens that the barometer in these islands will remain for a day or two below 29 inches, *i.e.,* below "stormy," without much wind, as the gradients are slight.

I shall now proceed to give practical instances of the use of these charts, but in doing so I, as before stipulated, must give my readers credit for an acquaintance with a portion of the A B C of practical meteorology, the fact of the difference between the two great currents of air, the equatorial (S. and S.W.) winds which, are warm and moist, and the Polar (N. and N.E.) winds which are cold and dry. I must also premise that it is not possible to find in summer such telling examples of the various types of weather as are met with almost daily in winter, so that we must only do the best we can with such charts as have already appeared in the *Times.*

I shall take instances of the two classes of areas with their concomitant weathers, and strive to show something of what may be learnt from them. The period which I shall first describe is that from April 27–30, which was cyclonic, although not very markedly so; *i.e.*, there was a general deficiency of barometrical pressure near the United Kingdom, as is shown by the passage of several cyclonic areas over these islands; and see what it teaches us not only about cyclones or areas of barometrical depression, but about the laws of their motion.

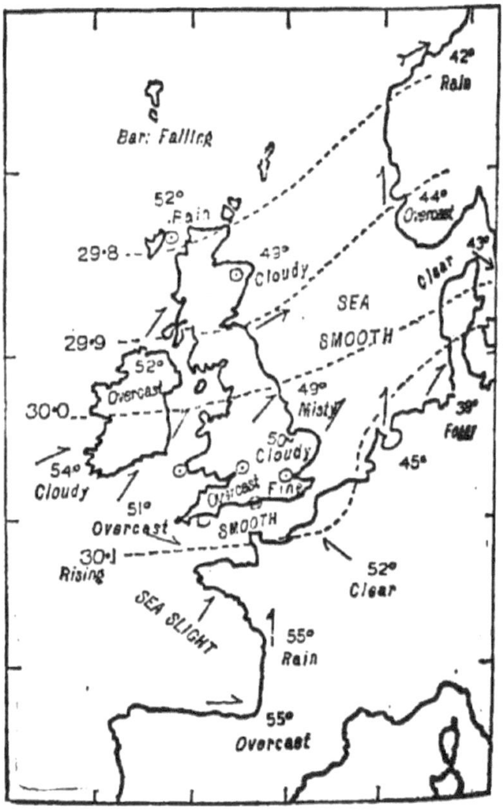

April 27, 8 a.m.

In the first instance we have four isobars shown, indicating that pressure is least off our northern coasts, and highest in the south. The winds are generally nearly parallel to the isobars, with a tendency of influx of air towards the region of least pressure, but to this there are are some exceptions. The south wind at Skadesnaes is affected by the coast line of Norway; the fresh north-west winds in Denmark belong to a system of depression which has just passed over that country, and the winds in France are irregular, being merely light local airs. The general wind is from W. to S.W., but while in the south of England, near the region of high pressure, the weather is warm and fine, though cloudy, mist is reported from Scarborough, and in the north of Ireland and Scotland the weather is showery, even with thunder storms (a frequent accompaniment of advancing depressions) on the eastern shores of the latter country, while rain is falling in Caithness and at Christiansund in Norway also, with a south-west gale. The probability from these conditions is that, by next morning, the area of low pressure will have advanced to the eastward, and that the winds in Scotland will have veered to west or north-west.

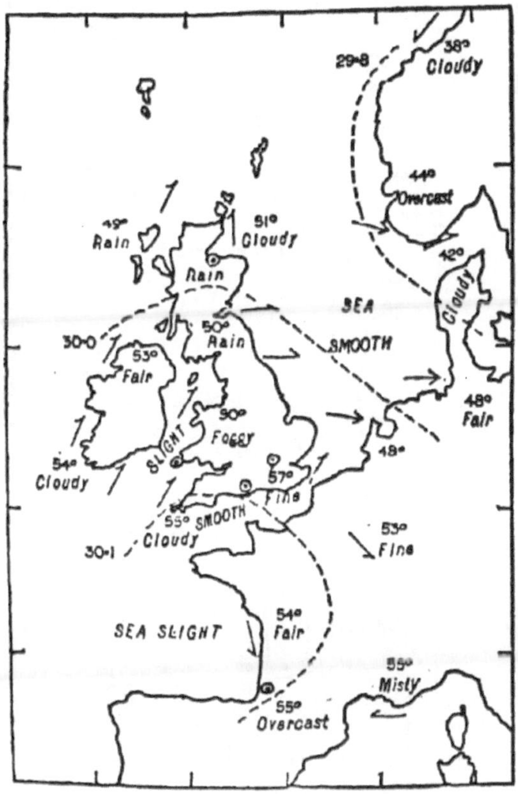

April 28, 8 a.m.

The chart for next morning shows us that the cyclonic area existing over the sea on the previous day has advanced over Norway, and the winds shown on the coast of that country belong to its rear, being N.N.E. to N.N.W., confirming the statement just made as to the position of the centre. The isobar of 29·8 ins., which on the 27th passed from the island of Harris to near Trondjem in Norway, now sweeps from just outside the Norwegian coast and across Denmark to near Copenhagen. The reduction of pressure over Germany had caused the area of high pressure to become contracted, so that the winds in France take an anticyclonic sweep round it, and the weather in the south of England remains fine. In the North, however, we have some signs of approaching disturbance in the shape of the advance of another cyclonic area with its system of winds. The barometer had risen briskly since the previous day, and is now, at 8 a.m., giving way again in the Hebrides, while cloud and rain have become more general in the north.

We see by the chart for next morning that the new depression is following the track of its predecessor pretty closely, but its centre is situated near the Shetlands,

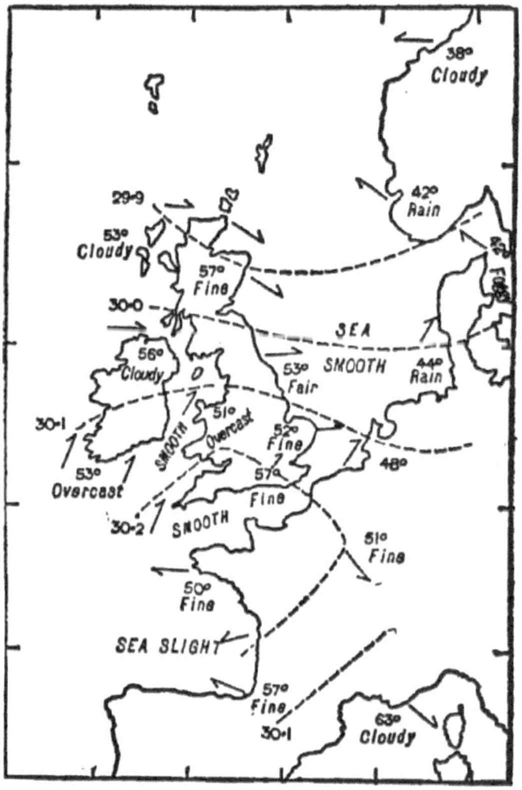

April 29, 8 a.m.

for the easterly winds in Norway *must* be on the northern side of the centre in accordance with Buys Ballot's Law. although from the absence of information we cannot draw the closed isobars which would mark it out more definitely. However, the direction of the winds in Norway from the eastward, shows us that the area is not a large one, and that it will probably cross the North Sea. The fog and mist have reached the eastern shores of that Sea, while in these islands the sky has cleared, and the anticyclonic system of the previous day in the south-west has not changed its appearance materially.

The next chart, for April 30, exhibits a very interesting state of affairs. The small depression, whose course we have traced, lay over Jutland and the Skager Rack, and could we see Sweden on the chart, we should find the cyclonic circulation round it complete, while the easterly winds in its rear have extended to Scotland. The area of high pressure has moved eastwards, and its outline has assumed a most irregular shape; but our weather, in England, remained almost unchanged. In Ireland, however, the wind has backed to south, at all stations, with a rough sea and cloudy sky, and of course, according to the principles so often quoted, the isobars run north and south, pointing to a low pressure over the Atlantic, the normal conditions accompanying our south-west wind at any time of year.

The sea which came in was not caused by the force of the wind on our own coasts, but by gales outside; in fact these gales grew to such violence on the 3rd of May, as to force the *Faraday* to buoy the telegraph cable she was laying on the banks of Newfoundland, and they sent to our coasts, two days later, the sea which broke up the ill-fated *Schiller* in a few short hours, though hardly a breath of wind was stirring to lift the fog which hid the deadly rocks of Scilly from the captain's view.

These disturbances which I have traced were the

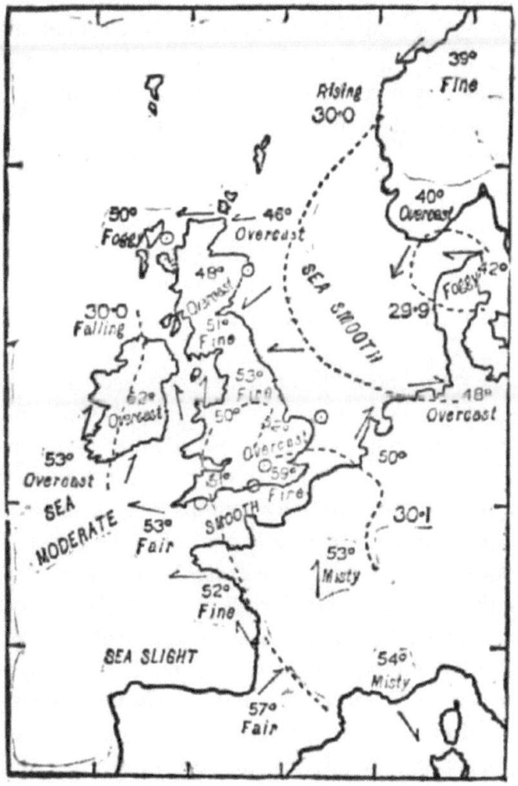

April 30, 3 a.m.

final break-up of our protracted winter, and the whole series of charts show that when we have westerly winds and fine weather, we may expect that to the northward of these islands the wind will be less steady, and will have a tendency to veer and back between S. and N.W., while the successive depressions pass by, or over, us to the eastward.

Such a direction of motion as this is general whenever the area of high pressure lies over France. The cyclones follow the main course of the stream of air, and as that is westerly on the northern side of the anticyclonic area, the cyclones move from W. to E., as was the case on the 27th. When the district of high readings lies more in the south-west, the systems of depression take a more south-easterly course, as we see from the instance of the 29th.

As an example of an anticyclonic period, and the changes which it brings with it, I shall take May 31st and the two following days. On the first of these days a belt of high pressure stretches from Holstein across England and Ireland, enclosed by the isobar of 30 inches. To the northward and southward of this the isobar of 29·8 inches appears, indicating the position of two areas of comparatively low pressure on each side of the central belt. The wind is not so strictly anticyclonic as might be expected, but still it draws from S.W. in the North of Scotland, through W, in Denmark, to N.E. at Cuxhaven, and E. in the Channel. The sharp change of direction of the wind round the pointed end of the region of high pressure is very striking, and it will be noticed that the general motion of the wind is *with* watch hands, or directly opposite to what it was in the instances previously treated. Within, the central area the breezes are light and variable, but over the rest of the chart we see the two great currents represented – the equatorial, with its S.W. winds, blowing over Scotland and the West of Norway, and bringing with it a warm temperature, loudy skies, and showery weather; while to the southward the true polar current appears, as a cold, dry, east wind, with a clear sky. Experience teaches us that with such conditions we need not expect any very sadden change of weather.

By next morning we find that the barometer has risen more than 0·2 in. over Scotland and the North Sea, and so the anticyclonic system has to some extent changed its place and appearance, having opened out on its eastern and closed in on its western sides, for from the fact of the readings at Stornoway and Valencia being both below 30·2 ins., we may venture to draw the isobar over the sea on our western coasts between those stations. The effect on the wind has been to displace the equatorial current, which now only shows at Stornoway and Christiansund, with its cloudy skies, while the easterly winds have gradually extended northwards over England, and the clear weather is reported even up to Caithness, with this remarkable peculiarity, that, as we are dealing with the conditions of summer, we find the warmth of the east wind increasing, inasmuch as the temperature of the north of Europe is much higher at that season than at the period of our usual east winds. Far to the southward, in Provence, we see dull weather and rain, where the atmosphere is no longer under the anticyclonic influence.

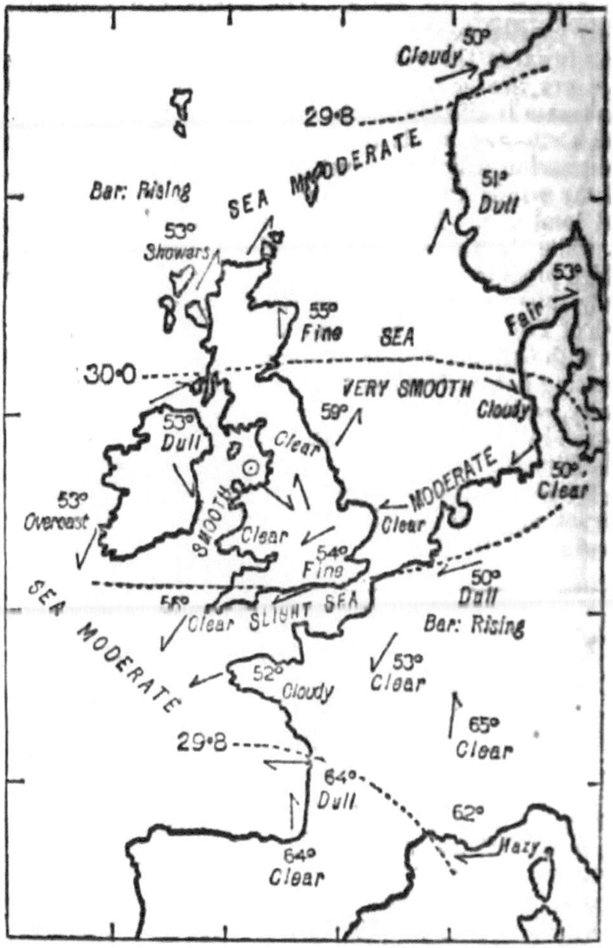

May 31, 8 a.m.

On the chart for the next day we see that the whole anticyclonic system of air circulation has moved away to the north-eastward, and the central area of high pressure lies over Norway. The easterly winds have become universal over the United Kingdom, and the fine weather has come with them, but the air is foggy at several stations. Over the Bay of Biscay we see the cyclonic curve of the wind, with rain setting in – the unfailing indication of the approach of an area of depression. This latter system appeared over France on the morning of the 3rd, and entirely obliterated the anticyclonic actions which we have just been tracing.

Here, then, we have an area of high pressure moving very slowly to the north-eastward, and drawing up after it the belt of easterly winds. The area is more

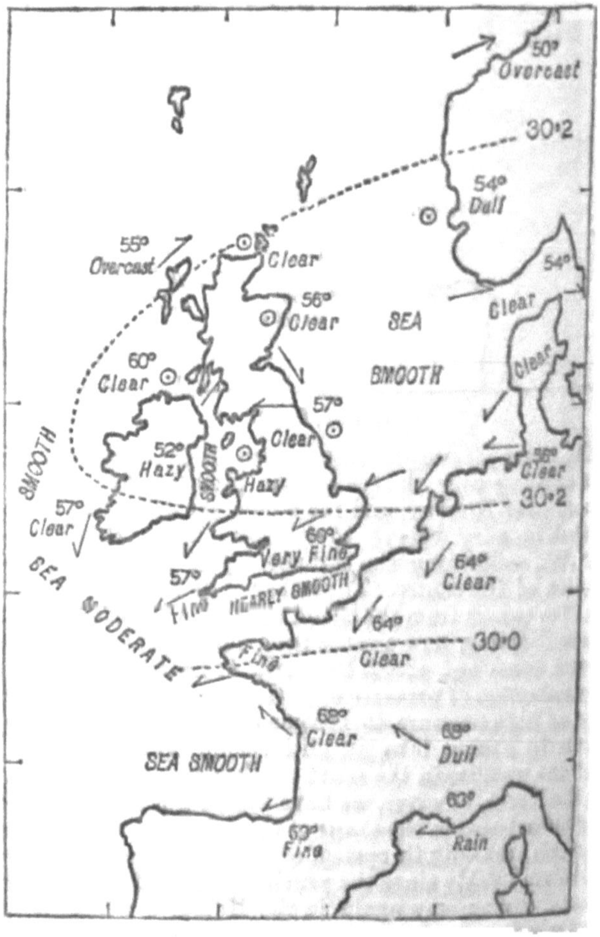

June 1, 8 a.m.

extensive than that of the cyclonic systems formerly noticed; its motion is more sluggish, and the atmospheric circulation it determines is less rapid.

The moral we may therefore draw from this latter series of charts is, that whenever we see an anticyclonic system developed on the chart we may calculate on its persistency for some days, and we need not, in general, expect high winds.

Space would not permit me, even if there were the charts to produce, to show what our chances of weather would have been if the anticyclone had lain over Ireland, when a succession of cyclonic disturbances might have been expected to take a southerly course over the North Sea, causing the northerly winds on the eastern edge of the anticyclone, *i.e.,* over Great Britain, to freshen, perhaps to the force of a gale, as each individual system passed us by.

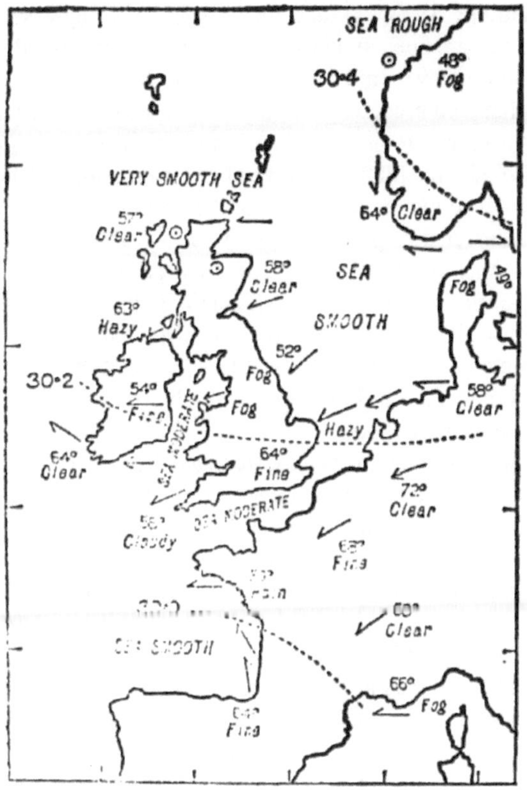

June 2, 8 a.m.

Such are the principles we have to guide us in judging of weather and in issuing warnings of storms. In winter if the characteristics of the different winds are more decided, the motion of the disturbances is often more rapid, and the whole condition of affairs is far more serious. Accordingly it must be confessed that our warnings are not unfrequently too late to be of much service to our exposed western coasts. One great reason for this is insufficiency of information. It is simply impossible that reports taken once a day can give an account of all the changes which supervene during the twenty-four hours. It is equally clear that to see at breakfast time a chart for the preceding morning is far less useful to us than a chart for the previous evening, not to say, even for midnight, would be. The attainment of improved rapidity of publication is merely a matter of £. s. d. Given unlimited money and the thing can be done.

A stereotyped chart can be delivered to a London newspaper within about four or five hours after the observations have been taken all round our coast, but the mere cost of telegraphy for a single set of say 25 reports is £ 1 5s. a-day, or about

£400 a-year (including Sundays), and the expense of a night service of reports from such stations as are open at a late hour, with the extra attendance at the London office entailed thereby, could not be less than £1,000 a-year. Had the Meteorological Office that sum at its disposal for the purpose, the work could be done, but it is hardly reasonable to ask the Government to pay such an amount for the ostensible purpose of furnishing recent weather information for the public press.

It is a very favourite matter with, critics of European meteorological work to contrast the weather charts, published thrice a-day at Washington, with those which are issued only once a-day here. In such strictures the circumstance is always ignored that the cost of weather telegraphy in the United States is £50,000 a-year, exclusive of all salaries, while our own Meteorological Office can only devote to the same object an annual sum of less than £3,000.

In conclusion, I may make a short digression with reference to the principles on which storm warnings are issued.

As soon as the telegraphic reports indicate that a disturbance of the atmosphere, probably accompanied by a gale, has either reached us or is near our shores, warning telegrams are issued to such parts of our coasts as are apparently threatened. The principles which we have to guide us in forming a judgment on these points cannot be stated in a paper like the present, but they are, in very general terms, as follows: –

The approach of a cyclonic disturbance is shown by the fall of the barometer and the changes of the winds at the different stations, in accordance with Buys Ballot's Law. Corroborative evidence is afforded by the variations of temperature and of weather, by the roughness of the sea, by the appearance of haloes, auroras, &c. The coasts threatened in the first instance are usually those where these symptoms first appear. The probable subsequent path of the storm, if storm there be, is shown by the general distribution of barometrical pressure, as already explained, and by the behaviour of the barometers at our different stations as the time wears on.

The degree of success which has attended our warnings in these islands, on the average of the last two years, has been that over 45 per cent. have been followed by severe gales, and over 88 per cent. in addition have been followed by winds too strong for fishing boats or yachts, though in themselves not severe gales; this gives a total percentage of success of nearly 80.

It cannot, however, be too emphatically stated, that the meteorologists who have had most experience in dealing with meteorological telegraphy in Western Europe are most decided in their opposition to any idea of their being able to foretell weather with certainty in the present condition of our knowledge.

Source Endnote

1 This very statement, almost *totidem verbis,* occurs in Webster's "Recurring Monthly Periods of Atmospheric Action," London, 1857, p. 201, but it unfortunately attracted no attention when first published.

35

THE PUBLICATION OF DAILY WEATHER MAPS AND BULLETINS

Robert Henry Scott

Source: Robert H. Scott, 'The Publication of Daily Weather Maps and Bulletins', in Oliver L. Fassig (ed.), *Report of the International Meteorological Congress Held at Chicago, Ill., August 21–24, 1893* (Washington D.C.: Weather Bureau, 1894), pp. 6–9.

The subject which has been placed before me is one which hardly admits of any very decided treatment, inasmuch as the scale and character of the maps to be published in each country must depend firstly on the amount of money which can be appropriated to the service of preparation and issue of these maps, and secondly on the extent of area which the maps are intended to cover.

It seems to me to be out of place and useless to prescribe for any gentleman, at the head of a meteorological service, what he ought to publish; of that he is a far better judge than any other person, or collection of persons, such as a congress possibly can be. And, as moreover, no congress can possess any executive power to enforce its resolutions, I fail to see the utility of proposing such.

To take a single example. One of the oldest meteorological bulletins in Europe is the "Bulletin Météorologique du Nord." This contains the observations from Denmark, Norway, and Sweden, but no maps of any description. The congress can hardly go so far as to recommend the three gentlemen under whose direction this publication appears, to change its character and adopt the system, say, of the Weather Bureau at Washington. Is there the slightest probability that their respective governments would increase their annual allowances so as to render such a scheme practicable? I say nothing of other offices whose bulletins are even less full than that which I have cited as an instance.

I shall, therefore, confine myself to an account of what the experience of thirty-two years in the preparation of weather reports and of twenty-four years in their issue to the public has shown this office as being well suited to the requirements of the population of the British Islands, some two hundred copies being issued daily to subscribers, in addition to the free issue, as described in our annual reports.

The daily weather report. – This consists of a large sheet of royal quarto size, which appears daily, and is accompanied monthly by a sheet containing corrections of occasional errors, and also reports which from any cause have arrived too

DOI: 10.4324/9781003650294-54

late for insertion in the daily issue. The bound volume of these reports for the last six months of 1891 contains also tables of mean values of the most important elements of the reports for a period of years – in most cases twenty, but in the case of rainfall for twenty-five years.

The information given in the report is that received by telegraph, and it is conveyed by the use of the International Code, recommended for introduction by the Permanent Committee of the Vienna Congress at its meeting at Utrecht in 1874, and finally adopted by the congress of Rome at its fifth meeting, April 22, 1879. This code provides for the transmission in the morning telegrams of information sufficient for the preparation of two maps, one for the morning of the day on which the telegram is dispatched and one for the previous evening.

The information conveyed in the telegram relates to pressure, temperature, humidity, wind direction and force, weather at the epoch of observation, amount, if any, of rain, or of snow (measured as water), and condition of the sea surface. (It need not be said that the last entry is blank for inland and for sheltered stations.)

From these observations there are two charts prepared for 8 a. m., one showing the barometer, wind, and sea disturbance, (the wind by arrows and the sea disturbance by hatching), the other showing the temperature by isotherms at 10° apart, and the rain in figures, where it exceeds 0.5 inch. Changes in pressure or temperature are printed in words across the face of the respective maps.

The chart for the previous evening is not published by the office except in its weekly weather report, which will be described presently, but a copy is supplied to "The Times" newspaper, and appears in its morning issue of the following day, and so secures a very extensive circulation.

A copy of the 8 a. m. chart is also forwarded to "The Times," and incorporated in the second edition, but the circulation of that edition is not very extensive.

Both of these copies are prepared expressly for "The Times," and at the sole cost of that journal, which for more than thirty years, ever since meteorological telegraphy was organized by Admiral Fitz Roy in 1860, has been conspicuous by the prominence it has given in its columns to meteorological information. In fact, for some years, the entire service for the preparation of these 6 p. m. charts was carried on at the sole cost of "The Times," a fact which affords strong evidence of the public interest in weather intelligence evinced in this country.

The weekly weather report. – This was commenced in 1878 at the suggestion of eminent agricultural authorities, in order to supply for the different agricultural districts statements of the temperature and amount of rain for the week, and of their differences from their respective averages. In 1884 this report was materially improved by the insertion of figures illustrating the weekly march of cumulative temperature, that is, of the number of "day degrees" of temperature above or below 42° F. (approximately 6° C.), which, according to the late Alphonse de Candolle, is the degree of temperature at which active vegetable growth may be assumed to commence. A popular explanation of this cumulative temperature will be found in a paper read by me before the International Health Commission in 1874. An explanation of the scientific principles on which the calculation of the values published weekly is based will be found in a paper by Lieut. Gen. R. Strachey, which appeared in the "Quarterly Weather Report" for 1878.

At the present date, 1893, this report contains on the first page, for each of twelve districts:

For temperature. – The average and the absolute maximum and minimum. The mean for the week, and its difference from the average for the week.

For accumulated heat. – The number of day degrees above and below 42° F. for the week, and their respective differences from the mean, with similar information for the interval elapsed from the beginning of the current year to the last day in the report.

For rain. – The number of wet days. The total fall for the week, and its difference from the average, and similar information, as before, for the interval from the beginning of the year.

For sunshine. – The number of hours recorded during the week, its percentage of the possible duration, and its difference from the average, with similar data for the interval since the commencement of the year, and general remarks on the weather for the week.

Page 2 gives information for each of the stations, as regards temperature, rain, and sunshine, with differences from averages for the week.

Then follow weather maps for the whole of Europe as far eastward as Odessa, Moscow, and Archangel, giving, respectively, pressure and wind for 8 a. m. and for 6 p. m., and temperature and weather for 8 a. m. only.

Remarks are given for each day, and the report concludes with a table of sunshine values for additional stations in the United Kingdom.

Appendices have, in successive years, appeared in connection with the "Weekly Weather Report," and *inter alia* these have contained figures giving, for each of the districts into which these islands have been divided, the weekly and progressive values of the different elements for each year as far back as 1879.

The daily and weekly weather reports are accompanied by monthly summaries, giving, for calendar months, a brief summary of the weather over the United Kingdom.

This is a brief account of the amount and character of the information which the experience of this office has led it to issue daily, weekly, and monthly, for the use of the public.

36

THE TIMES WEATHER CHART

Source: 'The Times Weather Chart', *Nature*, 15 April 1875, 11, 473–474.

MANY of our readers will have noticed the unusual appearance of illustrations in the *Times* in the shape of the small charts which have been appended to the Daily Weather Reports since the 1st inst. This measure has been the long-postponed carrying out of the line of action indicated by the Meteorological Committee in their Report for last year, and the chart in its present form differs but little from that printed as a specimen in that Report. We subjoin the chart for the 13th inst., 8 A.M., published in yesterday's *Times.*

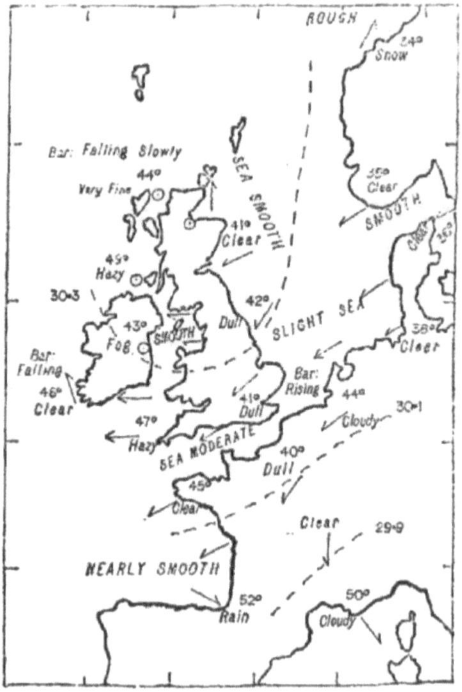

The dotted lines indicate the gradations of barometrical pressure, the figures at the end showing the height, with the words "Rising," "Falling," &c., as required. The temperature at the principal stations is marked by figures, the state of the sea and sky by words. The direction and force of the wind are shown by arrows, barbed and feathered according to its force. ⊙ denotes calm.

DOI: 10.4324/9781003650294-55

The method of preparation of the chart seems simple enough at present, but it has been the fruit of much thought, as the problem of producing, in the space of an hour, a stereotype fit for use in a Walter machine has not been solved without many and troublesome experiments.

In the first place, a material had to be provided which would admit of being engraved rapidly without burr or chipping, and would, without further preparation, serve as a mould for type metal. Secondly, drill pantagraphs had to be adapted to engrave the lines, and to be furnished with a gauge so as to vary their depth at pleasure.

The actual process is as follows: – The outline of the land is kept standing, and the composition is run in a mould bearing this outline on one face. The block, which is now an outline chart of the British Islands, is then placed under the pantagraph drill, which reduces the original drawing, furnished from the Meteorological Office, to one-fourth. The barograms and wind-arrows are put on direct from the drawing, the figures and words by means of templates, in order to ensure uniformity in the type.

The instant the block is engraved it is ready to be stereotyped, and then it is a simple matter to adapt it in the usual manner to the cylinder of the machine.

The initiative in this new method of weather illustration is due to Mr. Francis Galton, and the practical details have been carried out by Messrs. Shanks and Johnson, of the Patent Type Founding Company.

It is hardly necessary to allude to the value of such charts as these as a means of leading the public to gain some idea of the laws which govern our weather changes. As soon as they appear in our afternoon papers, we may hope for a more intelligent comprehension of the difficulties which beset any attempt to foretell the weather of these islands for the space of even twenty-four hours.

We may safely say that with these charts we have not seen the end of weather illustration, which was set on foot more than four years ago by Sir W. Mitchell in the *Shipping Gazette,* and has been continued daily; but whatever improvements may hereafter be introduced in the process, it must be remembered that the credit of breaking the egg is due to the gentleman we have named.

4.3

PUBLIC SERVICE

Editorial Headnote

4.3 Public Service

National meteorological services flourished in Europe and America in the second half of the nineteenth century, and with them came structures, staff and budgets allocated by governments to promote activities aimed at storm warning, weather forecasting and climatology. In Britain, *The Times* newspaper helped raise public awareness of weather conditions by regularly publishing storm warnings and forecasts from the Met Office (see 4.2 'Daily weather charts').[1] Routine operations of collecting data and forecasting were inextricably linked to public service and immediate results. A fundamental concern was understanding – and at the same time enhancing – the relationship between the service and the citizen, which some approached from the perspective of forecasting efficiency and others explored through formulas for voluntary citizen cooperation. Frederic Gaster (1842–1915) of the Met Office addresses the challenges of issuing warnings and the need to communicate them effectively to society, while Major Henry H.C. Dunwoody (1843–1933) of the US Signal Office seeks to combine the needs of state weather services with citizen participation.

Gaster published 'Weather Forecasts and Storm Warnings: How They Are Prepared and Disseminated' in the *Quarterly Journal of the Royal Meteorological Society*, as the result of a lecture delivered to the Society on 18 March 1896 at the invitation of the Meteorological Council. Gaster was a familiar figure at the Society's meetings, having been a member for more than fifty years. In 1879, he became head of the Met Office's Department of Storm-Warnings and Weather-Forecasts, a post created by Robert FitzRoy in 1861, the abolition of which (after FitzRoy's death in 1865) caused a strong public reaction (see 4.1 'Galton Report'). Gaster held the post until he retired in 1909.[2] Throughout this informative piece of writing, aimed at those unfamiliar with the subject, Gaster demonstrates his mastery of the art of weather forecasting: in his first paragraph he states that the purpose is to set out 'the manner in which weather forecasts and storm warnings are prepared and made known in this country' (p. 212). His ability to predict weather sequences is borne out by his regular collaboration with *The Times* newspaper, where he was responsible for writing weather articles for many years. At several points throughout the pages of the 1896 article, Gaster warns of the need for simultaneous barometric readings from various locations to improve the reliability of weather forecasts. He recognises that although predictive practice has advanced, significant challenges remain in issuing warnings due to potential errors in observation and communication. Gaster thus offers insights into the challenges faced by a public service in terms of the accuracy and effectiveness of weather forecasts and storm warnings.[3]

National meteorological institutes, such as the US Weather Bureau, coordinated stations operated by regular paid observers covering a small part of the national territory, not all of it. Therefore, interest in involving citizens as volunteer observers in fields such as agricultural meteorology was often a priority. Dunwoody is best known as an inventor who, after retiring in 1906, patented a radio wave detector using carborundum crystals as the detector element.[4] From the late 1860s, however, he was chief weather forecaster for the War Department's Signal Office, where he devised a system of storm warnings. His 1883 book *Weather Proverbs* and 1895 book *The Value of Forecasts* established him as an authoritative voice in official American meteorology. The article excerpted here, 'Functions of State Weather Services', is a paper he read at the International Meteorological Congress held in Chicago on 21–24 August 1893, the proceedings of which were edited by Oliver L. Fassig in Washington D.C. in 1894. By this time Dunwoody was president of the General Association of State Weather Services and had published the article 'State Weather Service Organizations' in the *Bulletin* of the Weather Bureau in 1892. Dunwoody argues that state weather services are vital to the collection and dissemination of climatological information, as this data supports agricultural practices and informs farmers about weather conditions that affect crop yields. In the organisation of these services, citizen participation is key: they help to collect weather data, their cooperation makes it possible to regularly publish the *Monthly Weather Review* and the *Weather Crop Bulletin* (both of which owe their existence to Dunwoody) and the willingness of the collaborators is more economical and sustainable in the long term than the remunerability of Weather Bureau officials.[5]

Notes

1 James Burton, 'The History of the British Meteorological Office to 1905' (The Open University, PhD diss., 1988).

2 R.H.C. 'Frederic Gaster, 1842–1915'. *Symon's Meteorological Magazine*, 50(600) (January 1916), p. 186.

3 J. Malcolm Walker, *History of the Meteorological Office* (Cambridge: Cambridge University Press, 2012), p. 123.

4 Bartholomew Lee, 'How Dunwoody's Chunk of 'Coal' Saved both de Forest and Marconi', *AWA Review*, 22 (2009), pp. 1–12.

5 Cleveland Abbe, 'The Meteorological Work of the U.S. Signal Service, 1870 to 189'. In Oliver L. Fassig ed., *Report of the International Meteorological Congress Held at Chicago, Ill., August 21–24, 1893* (Washington D.C.: Weather Bureau, 1984), vol. 1, pp. 232–84, on pp. 259–60.

37

WEATHER FORECASTS AND STORM WARNINGS

How They Are Prepared and Disseminated

Frederic Gaster

Source: Frederic Gaster, 'Weather Forecasts and Storm Warnings: How They Are Prepared and Disseminated', *Quarterly Journal of the Royal Meteorological Society*, 22, 1896, pp. 212–228.

A Lecture delivered before the Royal Meteorological Society, March 18, 1896.

HAVING been invited by the Council to bring before the Fellows a statement, in the form of a lecture, as to the manner in which weather forecasts and storm warnings are prepared and made known in this country, I have acceded with pleasure, and have to acknowledge with many thanks the honour which has been conferred on me by the request. In performing the task, however, I have to ask the indulgence of those who are already acquainted with the main features of the work, while I take them, and those who are not so familiar with the subject, through each stage of the proceedings, so that in the end we may all come to a clear and satisfactory perception of the principles on which the work is based.

Now, of the instruments employed in this work the barometer occupies a position so pre-eminently important that it is necessary to recall clearly its construction, and to consider what is indicated by its readings. We take a tube about 33 inches in length, closed at one end and open at the other; we fill this tube with mercury, and having expelled all the air, invert it in a vessel (or cistern) of mercury, so that the open end of the tube is below the level of the mercury in the vessel. The mercury in the tube at once sinks to a certain level, but the larger portion (about 30 inches in length) remains in the tube, leaving a vacuum in the upper part. Why does not the whole of the mercury leave the tube? Simply because it is maintained in its position by the pressure of the air on the mercury in the cistern. Remove that pressure by an air-pump, and all the mercury will come out of the tube; let in the air again and the mercury will at once reascend to the height previously marked, but will never *fill* the tube. As a consequence of this we see that, should the atmospheric pressure in the ordinary course of nature decrease,

DOI: 10.4324/9781003650294-57

a further portion of the mercury will flow out of the tube into the cistern, and we shall say "the barometer is falling"; while an increase of the atmospheric pressure will force a larger quantity of the mercury into the tube, and we shall say "the barometer is rising." Thus we come to two sets of terms –

(1) Barometer high = atmospheric pressure in excess of the normal, Barometer low = atmospheric pressure in defect of the normal; and
(2) Barometer falling = atmospheric pressure decreasing, Barometer rising = atmospheric pressure increasing,

the rate of change being described as slow, moderate, or rapid, according to circumstances.

This is all that a single, isolated instrument shows us; but there can be no doubt that by the careful consideration of its actual height and its rate of rising or falling, in connection with the simultaneous change in the direction and force of the wind, temperature, humidity of the air, and the increase or decrease in the quantity and density of the clouds and their height, isolated observers have been for years enabled to arrive at very shrewd conclusions as to the character of weather approaching their neighbourhood. Farmers, shepherds, and sailors, by constant practice, are among the most apt of such judges, especially when the physical features of their own locality at the time of observation form an element in leading them to their conclusions.

But the question long ago suggested itself to some, that if we could obtain simultaneous readings of the barometer, made at a considerable number of positions scattered over a wide area of country, we might by putting them on a map and looking at them in connection with the wind and weather, prevailing at the same time, arrive at some better conclusions as to the form and nature of the weather systems with which we have to do, and by comparing chart with chart learn something, perhaps much, more of the movements of such systems, and finally be prepared to deal with them for the good of mankind. The first endeavours in this direction were made by our forefathers in discussing the conditions prevalent during violent storms. The earlier investigators found that in these storms the wind tended to circulate round a central area in which the barometer was low, and thus it came about that such storms were called "cyclones." It was not till later that the cases of ordinary and lighter winds were similarly discussed.[1] The importance of such investigations and of the facts which they disclosed, gradually forced itself on the public, until in the present day hardly any civilised country is without its Meteorological Office or Weather Bureau, in which one of the duties is to establish a series of widely spread stations, which shall communicate by telegraph with a central office once, twice, or thrice daily, at certain fixed hours reporting the pressure and temperature of the air, the direction and force of the wind, the weather, etc., as soon as the observations had been made.

In the map, Fig. 1 (p. 214), we have the positions and names of the stations in the British Islands from which such information is received, together with many foreign stations,[2] the information from which is, by a courteous system of international exchange, available for publication in the *Daily Weather Report.* From this map it will be seen that the stations cover the whole of the British Islands and many parts of Western and North-western Europe. From all of these localities telegrams are received once, from several of the others twice, and from a considerable number of them three times a day, the hours of observation being 8 a.m., 2 p.m., and 6 p.m. Now let us see what the barometric readings taken at one of these epochs show us. In Fig. 2 (p. 215) we have a set plotted to the positions of the stations, and it is at once evident that not only are the readings unlike one another, but that all the lowest readings are gathered over or around one particular position – viz. the east of Scotland – while the highest are found in France and Germany, and that the values increase steadily from the former to the latter region. We want to know more definitely, however, the form of the area in which such readings are congregated, and to do this we find out the points at which the readings are precisely similar, and join them by lines. In the present case it is evident that the reading of 29·50 in. would lie between Aberdeen (29·46) and Leith (29·58), between Aberdeen and Ardrossan, (29·62) between Aberdeen and Nairn (29·51), between Aberdeen and Wick, and between Aberdeen and Skudesnaes – in fact, between the two last named stations the readings of 29·5, 29·6, 29·7, and 29·8 will all be found at certain positions along a straight line joining the stations. We place a dot on the map where each reading of 29·5 in is to be found, and we then join them by a curved line, and at once it becomes evident that the line takes a somewhat oval form. On drawing in a similar way the lines joining the dots which mark the values of 29·6 in. and 29·7 in. we shall find that they run in a course somewhat parallel to that of 29·5 in., and enclose an oval-shaped area in which the barometer stands lower than in the regions adjacent. We therefore call this a region of "low barometer" or "low pressure," and the lines are called "isobars." On continuing the process, however, it is found that the line for 29·8 in. shows a tendency at its extremities, to draw away from this low pressure area; those of 29·9 in. and 30·0 in. show a still greater tendency in this direction, and when we come to those for 30·1 in. and 30·2 in. we find them enclosing another somewhat oval area, but in this the barometric readings are higher than those in the adjacent regions. This area is termed an area of "high barometer" or "high pressure," as opposed to those of low pressure previously described,[3] and if the whole of the earth's surface were so treated several of these areas, both of high and low pressure, would be discovered.

The next question which suggests itself is this. Is there any connection between these high and low pressure systems with which we have become acquainted, and the winds and weather observed in their neighbourhood? Undoubtedly there is; and to make that connection evident we must give up the hypothetical diagram in Fig. 2, and refer to Figs. 3 and 4 (p. 216), which are drawn from actual observation. Both of them are compiled from observations made in winter, and are therefore intercomparable. Fig. 3 shows a large area of low pressure, having its central part (in which the barometer is below 28·5 in.) lying over and off the west

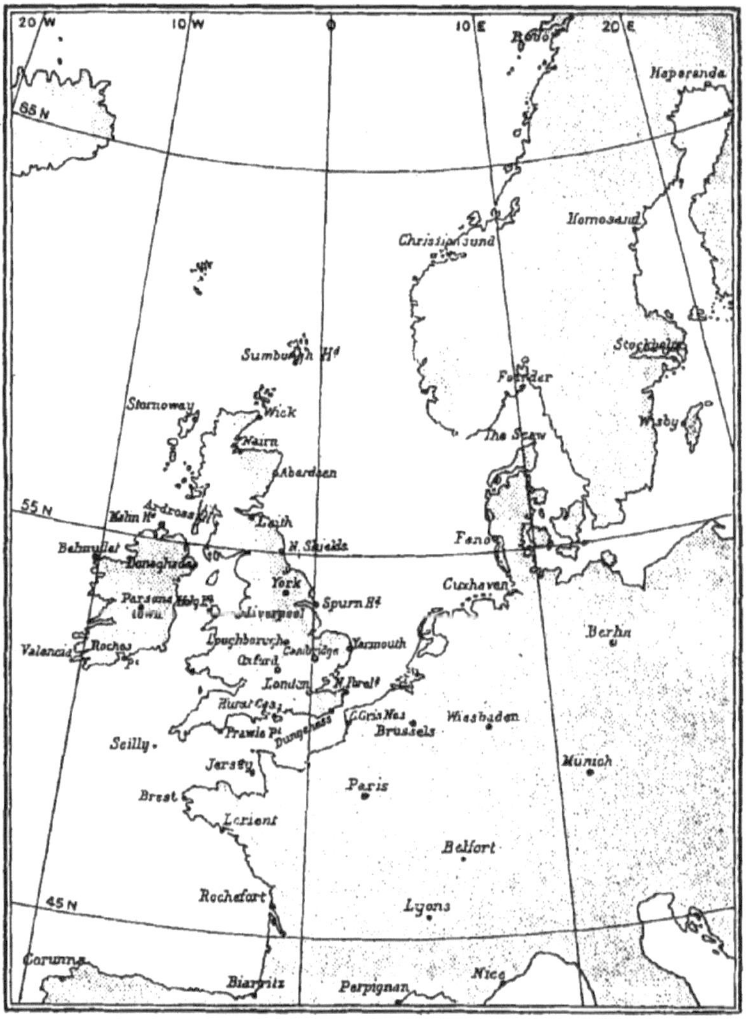

Fig. 1.

of Scotland. The cylonic system covers the whole of the United Kingdom, and the barometric "gradients" (or differences of pressure for a given unit of distance) are steep. The arrows on the diagram show the winds prevailing at the time, and of these an explanation is given in the footnote.[4] What do they show? Westerly and South-westerly winds strong to gale in force on all our western coasts, strong West-southwesterly winds in the channel, but a moderate Southerly breeze off the east of Scotland, a South-easterly breeze at Sumburgh Head, and an Easterly wind in the Minch. In Norway the wind is Westerly, influenced by another system which had previously passed over us, and was disappearing over the north of

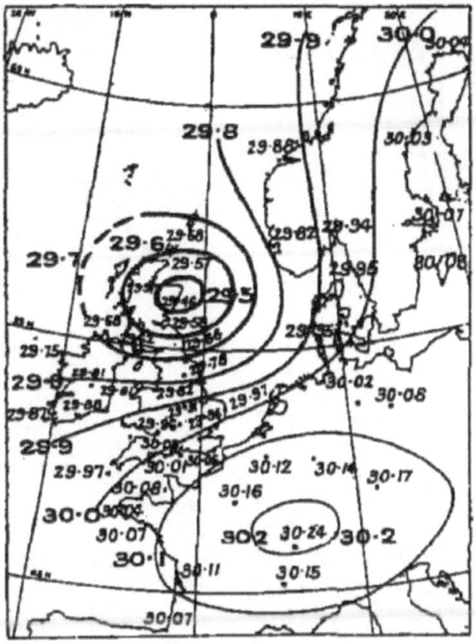

Fig. 2.

Russia. Now look at the temperatures; they range between 51° and 49° over all the southern half of the area, but are as low as 30° to 37° in the north, where either the wind was Easterly, or the pressure distribution was approaching more nearly to that of the "high pressure" systems than to the low-pressure system with which we have to do. Mark, too, the weather – squally and rainy on almost all our coasts, with snow in the south-west of Norway and in the north of Scotland. Even as far south as Rochefort the sky is overcast; and the morning observations of the following day show that gales and rains had been very general.

We come now to Fig. 4. Here we have a high-pressure system, the central area of which lies over the Christiania Fjord, where the readings exceed 31·0 in., while the system extends southwards to the middle of France, and westwards over the whole of the United Kingdom. The winds are North-easterly over Denmark and North Germany, Easterly over England, Easterly and South-easterly in Ireland, and South-easterly to Southerly at the Scotch stations. Here also we have a circulation of wind round the system, but its tendency is to draw *outwards*, *from* the centre, whereas in the case of the low pressure system it tended to flow *inwards*, *towards* the centre. Moreover, in the high-pressure system the winds are moderate to light in force, as opposed to the gales and strong winds of the low-pressure system, and had the specimen chosen for the high-pressure system been, as is often the case, a system of larger area and slighter intensity, the winds would have

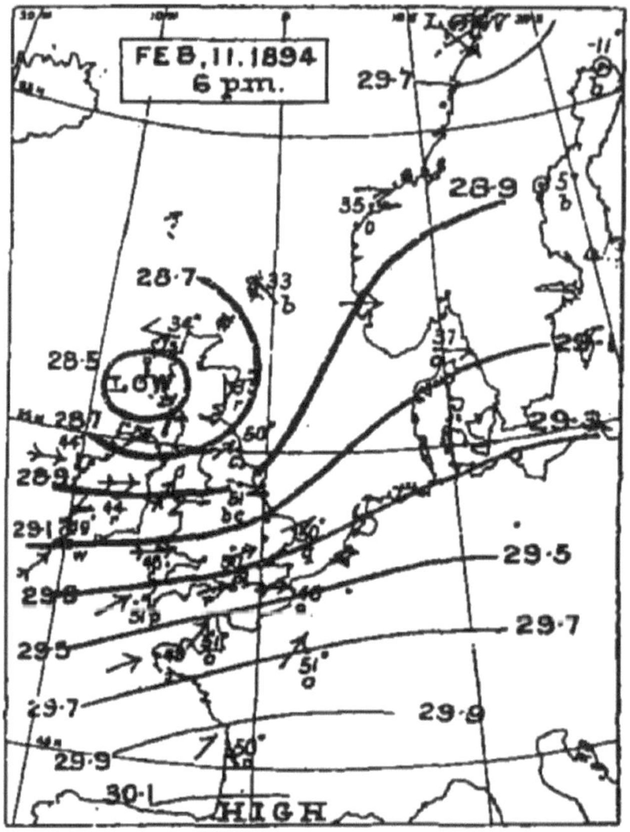

Fig. 3.

been lighter still and in many regions calm. In fact, variable airs and calms usually prevail over the central parts of the high-pressure areas, while strong winds mark those of low pressure. Let us now examine the temperatures in this system – they are all very low, the only readings which exceed the freezing-point being those in the west of Ireland, while those reported from Germany and France are as low as 18° or 19°. The weather, too, is fine and dry, as opposed to the squally rainy conditions shown in Fig. 3.

We learn, therefore, from these illustrations, that the two systems, the high and the low pressure, are diametrically the opposite of one another in pressure, temperature, wind, and weather; and, as the wind draws *out from* the high-pressure centres, and *in towards* the low, it is evident that the former systems "feed" the latter at the surface of the earth; and it is only reasonable to suppose that the reverse of this operation takes place in the higher regions of the atmosphere.

One or two more facts must be noted before we tabulate the important facts at which we have arrived. The first is that in summer time the anticyclone is the

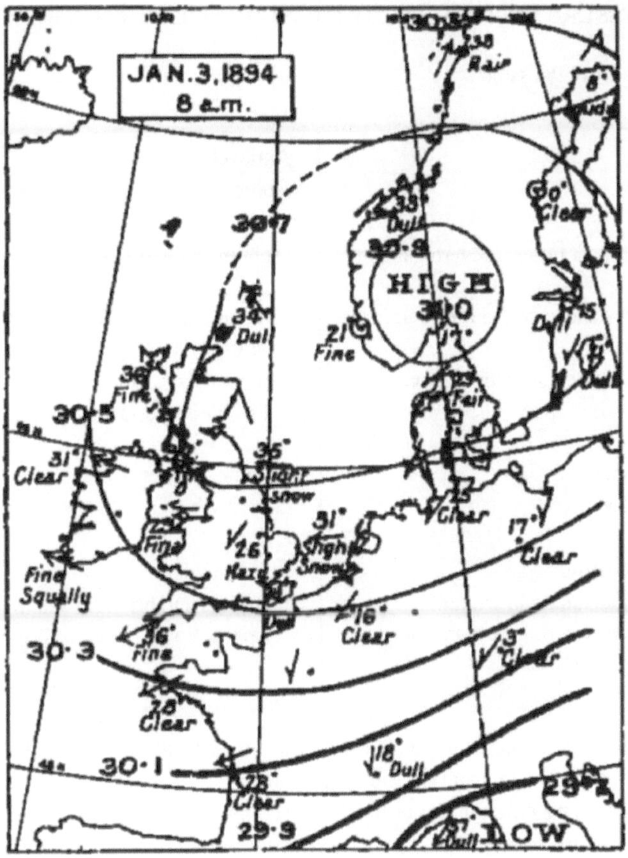

Fig. 4.

warm weather system, as owing to the fineness of its weather, and the many hours of sunshine the solar rays are felt very strongly in heating the earth and air, while the low-pressure systems, owing to the density of their clouds, and the effect of their rains, become the cold-weather systems of this season by cutting off the solar rays. As the apparently circular movement of the air round the low-pressure system had led to their being called "cyclones" (from *kuklos,* a circle), the opposite circulation round the high-pressure systems gave rise to the term "anticyclones" as applied to them. Again, from the fact that in cyclones the barometric column is depressed below the level which it maintains in adjacent regions, the systems have come to be called "depressions"; and it has been suggested that for the opposite reasons the anticyclones might with equal propriety be termed "appressions."

In the following table there will be found all these normal characteristics of the primary systems of weather to which we are subject – a table, the value of which will be apparent to every one who takes any interest in the matter.

Pressure Systems and their Normal Characteristics.

Low Pressure.	High Pressure.
Names.	
Cyclones.	Anticyclones.
Depressions.	Appressions.
Winds.	
Strong in force; at times a severe gale, or hurricane.	Light in force, often calm,
Circulate left-handed round the centre of the system.	Circulate right-handed round the centre of the system.
Draw in spirally *towards* the centre.	Draw out *from* the centre towards the neighbouring cyclones.
Temperature.	
Low in summer.	High in summer.
High in winter.	Low in winter.
Weather.	
Rough and squally.	Quiet and dry.
Rain in summer, snow in winter, thunderstorms in both seasons, but especially in summer.	Cloudless and bright in summer, with haze at times; foggy or bright in winter.

We come next to the consideration of the movements of these systems – are they at all dependent the one on the other? and if so in what way? With regard to the anticyclones no rule whatever has been recognised as to their movements; but the movements of the cyclones appear to be in a limited degree dependent on their position with regard to the anticyclones – it being found generally that cyclones tend to move round the anticyclones from left to right – so that if pressure be highest in an anticyclone lying over Continental Europe and having its central area over Germany and the Baltic, then a depression arriving off the west of Ireland from the Atlantic will move north-eastward, outside our west and northwest coasts, and travel away over Northern Europe towards Siberia. On the contrary, with pressure highest over Northern Europe and stretching out over Scandinavia towards Scotland, any depression which may appear over North Germany (and these are usually very slight) will tend to move westward over the Netherlands, and the north of France. As a rule these latter systems are secondary to a larger, and more general (primary) system lying at that time over the Mediterranean Sea or Southern Europe. Before going more fully into this question, however, we must get some further insight into what are termed "secondary" systems.

Secondary systems may be either of the low or high pressure type, and perhaps it will be well in considering them to confine our attention in the first place to the former. They are observed sometimes as *small cyclones*, tending to move round their primaries from left to right – as if carried on by the general movement of the air-currents in which they are developed – at other times as *hemicyclones*, *i.e.* systems of which only one half of a cyclone (the half remote from their primary) is to be traced; they obey the same law of movement as is observed with their more fully developed brethren. Other secondaries assume the form of the letter V, and are termed "V-shaped" systems: they are ordinarily developed between two anticyclones as a primary depression passes by the mouth of the hollow which separates them. A chart showing both the cyclonic and hemicyclonic secondary systems will be found in Fig. 5 (p. 218) for 8 a.m. on February 10, 1894, where the primary depression is to the northward of the Shetland Isles; the secondary cyclone is shown lying over Kent, while a hemi-cyclone is approaching the west of Scotland from the Atlantic. The effect of such secondary systems is to throw the bad weather which would have prevailed closely round the primary system to a greater distance from its centre – the space intervening between the secondary and primary being marked by a very temporary spell of bright weather to be

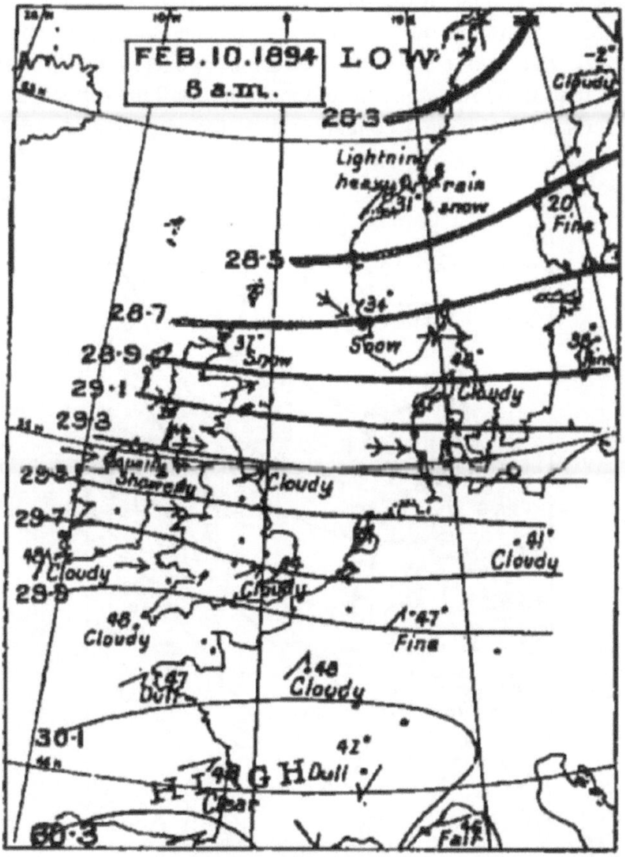

Fig. 5.

succeeded by the disturbed conditions normal to the primary when the secondary has passed. "Y-shaped" disturbances are marked, as far as the weather is concerned, by most of the features which maintain in secondary cyclones or hemi-cyclones, but the winds in their front and rear are set with regard to one another at an angle so acute, that when their central line (or diameter) passes over any station, the winds shift abruptly, sometimes momentarily, through a considerable angle, in a sharp squall, and if the gradients on the rear side be steep, the change is often productive of serious damage to shipping, owing to the unexpected change, and the severity of the gale which is developed. A good instance of this is shown in the chart for 8 a.m. on January 26, 1894 (Fig 6, p. 218). It is not very uncommon for small cyclonic systems of great intensity, commonly called "tornados," to be developed in the apex of the "V," and the evil qualities of these small, but most destructive, storms are sufficiently well known, although happily in this country we know very little of their ravages.

Let us now turn to the secondary high-pressure systems. Here we find varieties, almost precisely the reverse of the secondaries which belong to the low-pressure class. The similarity in their outline, as shown by the isobars, is simply remarkable, but the fact that in one class the pressure is in excess, while in the other it is in defect of that in the surrounding regions, makes all the difference as to their other characteristics. We have secondary anticyclones, pure and simple, such as

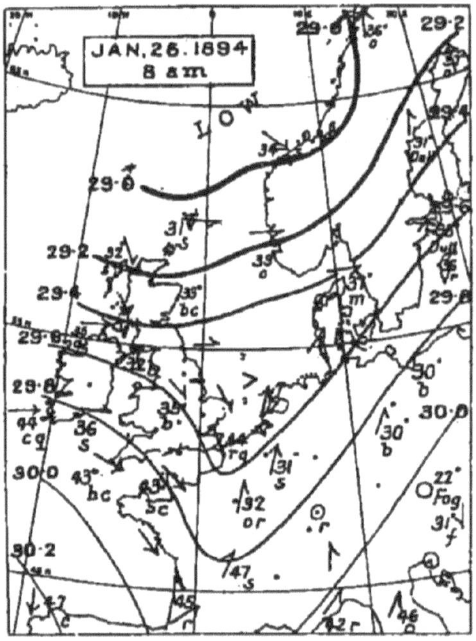

Fig. 6.

is shown in Fig. 7 (p. 219) for 8 a.m. on March 10, 1896, in which the primary anticyclone covers the Bay of Biscay and neighbouring parts of the Atlantic, while the secondary lies over the midland and southern counties of England. The effect of this secondary is (as in the case of the secondary cyclone) to throw out the conditions observed in the primary much farther to the northward (*i.e.* away from the centre of the system) than would have been the case had the secondary not existed. The next form of secondary high-pressure system is that called the "ridge," a feature shown to some extent in the Fig. last referred to, but much more strongly in Fig. 8 (p. 219), which represents the conditions prevailing at 6 p.m. on February 14, 1894. On this occasion a large anticyclone lay over France, Western Germany, and the south of our islands, but the "ridge" spread northward (like a spur from a mountain chain) covering the whole of the British Islands, and almost touching the Faro Isles. The winds and weather in this kind of system are exactly the opposite of those in the "V-shaped" depression, and spread out the anticyclonic condition of the primary high-pressure area much farther to the north-westward than would have occurred had the ridge not been present. This form is found extending from anticyclones between two depressions as they move along the margin of the high-pressure area, and may perhaps be considered as the natural result of such an arrangement rather than an independent system of itself – it being hard to conceive that two somewhat circular depressions could

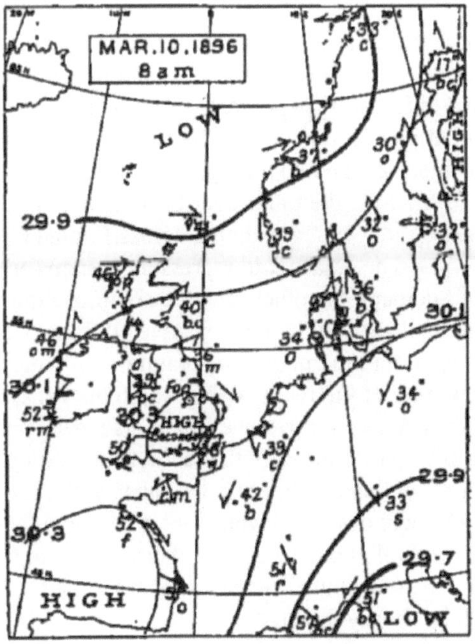

Fig. 7.

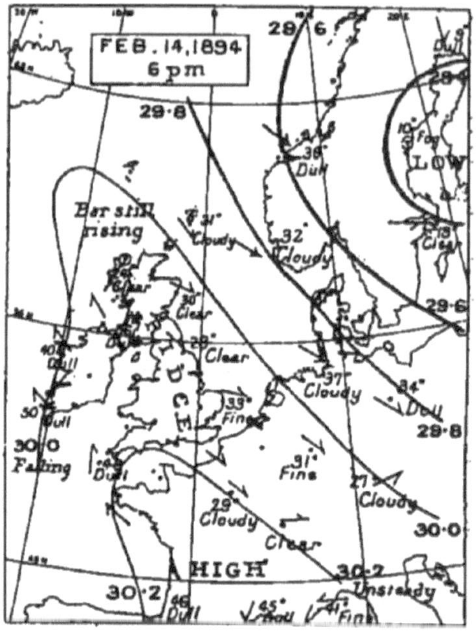

Fig. 8.

follow one another round an appression without such a ridge intervening. A further development of this form of secondary is found in the union of two "ridges" emanating from two different high-pressure systems, and uniting in the form of a "col." Such a formation is shown in the chart for 6 p.m. on February 26, 1896 (Fig. 9, p. 221), when one of the two high-pressure areas lay over Ireland and the adjacent parts of the Atlantic, the other over Scandinavia and Northern Europe, the two being united by the "col" stretching across England, the south of Scotland, and the North Sea. The adjacent low-pressure systems are shown lying one to the northward of Scotland, the other over the south of France and the Gulf of Genoa, and are well marked. Within the area covered by the "col" the weather is fine, the winds are light, and the night temperatures (in winter) are low; in fact, the anticyclonic condition of the two systems which the col unites are continued throughout its length. The two anticyclones, of course, need not occupy the positions shown in the illustration in order that a "col" should unite them, but it is necessary that each should extend towards the other in the form of a "ridge," or the weather conditions here described would not hold true. Should the major axes of the anticyclones be parallel one to another there would, in fact, be produced a complex condition which has hitherto been known by the term "trough," and to which attention will be shortly directed.

It is evident, from what has already been stated, that if the distribution of pressure were always in accordance with one of the types hitherto described, and if

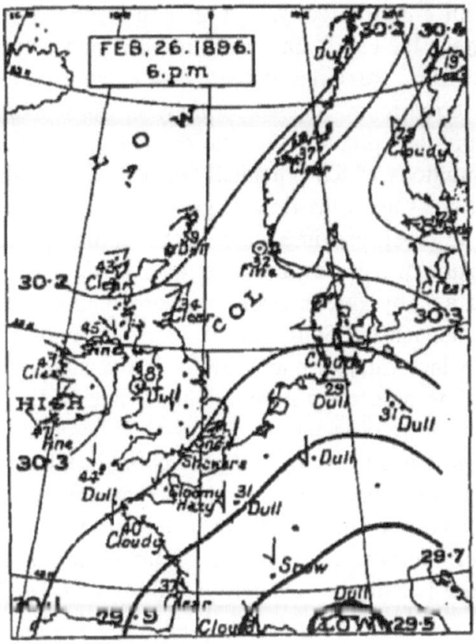

Fig. 9.

the changes were of a simple and regular character, the forecasting of weather and the issue of warnings for storms would be matters comparatively easy of execution; but this is not so. Both the low and the high-pressure areas undergo at times important alterations, both of intensity and of form, as they approach or pass over our area. Sometimes a depression which, on advancing towards us, is large and deep, fills up almost suddenly, and the warnings issued turn out to be wholly or partially unnecessary; at other rare times a disturbance apparently of slight importance develops an energy in no way foreshadowed by the barometrical indications, as far as our present knowledge enables us to understand them, and storms are then experienced where none were looked for. More commonly the advancing depression extends over a larger area than it appeared likely to cover, and consequently a rather larger part of our coasts than had been warned experiences a gale. On other occasions a sudden increase of pressure far outside our area of observation will develop steep gradients where only slight ones were shown before, and gales result on coasts which had not appeared to be threatened. All such matters as these are causes of inquiry and careful consideration, and it is cheering to be able to say they are gradually coming more and more within the grasp of those whose duty it is to perform the very trying and often thankless task of issuing the warnings.

With regard to mere "weather" conditions, however, there is a distribution of pressure which produces rains, so heavy and so fitful, that it can at present be forecasted only with great difficulty, and at times baffles entirely all attempts to deal with it. I mean the condition which in the official publications has been termed a "trough" of low pressure. This must not be confused with the "trough" of a depression referred to by Mr. Abercromby in his works, and which, as a meteorological entity, has probably no real existence. An instance of a trough is found in the conditions which prevailed between May 11th and 14th, 1886. The distribution of pressure at the time of greatest intensity is shown in Fig. 10 for 8 a.m on May 13th. At that time there were two high-pressure systems, one over the northern parts of our area, the other over Spain and the Mediterranean, but they were separated by an elongated area (or trough) of low pressure lying over the English Channel, the Scilly Islands, and the south of England. In this there were three distinct minima, one over the Scilly Islands, a second over the northern part of Devonshire, and a third over the Straits of Dover; and, as a rule, it will be found that such a distribution of pressure is favourable for the development of local depressions which, while they may not produce much wind, cause heavy falls of rain in many places. Torrents of rain had prevailed over our midland and north-western counties, and in the east of Ireland, with Easterly and North-easterly winds. A very similar condition has occurred on the morning of the day on which this lecture is delivered (March 18, 1896), and we all remember how steadily the rain has poured down for many hours together in London. There are

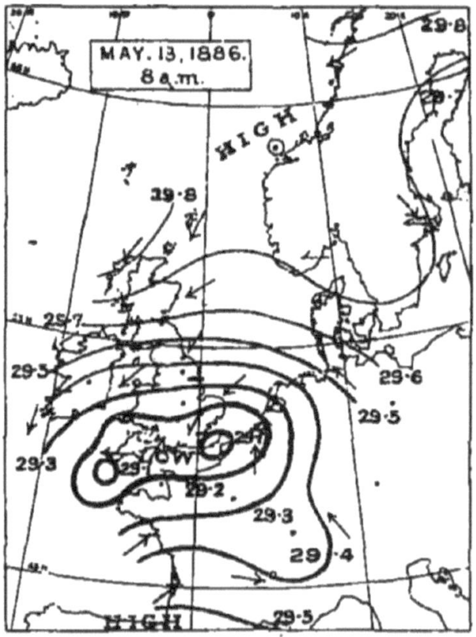

Fig. 10.

other minor disturbances to which attention might be drawn, *e.g.* the thunderstorm disturbances of summer, and other local systems, most of them having the accompaniment of heavy rain – the bulk of it coming with an Easterly wind – but time forbids their detailed consideration at present.

Let us now see what we can gather from the facts which have been adduced so far. We have considered the main features of high and low pressure systems and their bearing on one another, and have noticed the principal forms of primary and secondary systems, and the laws which appear to govern their movements. If these features be as has been stated, it is obvious that in preparing forecasts, or issuing storm warnings, our first consideration must be the positions in which the pressure is highest and where lowest, then the steepness of the gradients, and the indications (if any) of the approach of any depressions, and whether they appear to be large or small, deep or shallow, primary or secondary. Then come the questions as to what coasts are likely to be affected, and the further question whether, by the rate of the barometric fall, and the manner in which it is spreading, whether the high or the low pressure system is obtaining the mastery, and whether one system is spreading decidedly over one neighbourhood to the exclusion of the other.

Let us take a few typical maps and see what they show. We have, first of all, in the map for 8 a.m. on December 25, 1895 (Fig. 11, p. 223), a high-pressure area lying over Northern Europe, a primary depression off our south-western coasts,

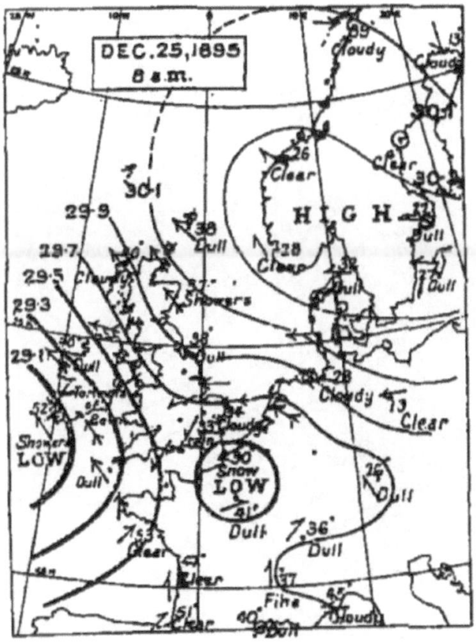

Fig. 11.

and a shallow secondary over the Channel. South-easterly winds were then general over Ireland, and Scotland, and were blowing hard; torrents of cold rain were falling in the south-east of Ireland, some rain and snow over our south-eastern counties, and very slight snow on the north-east of Scotland; as the depression filled up these conditions gave way to finer weather. Take next the conditions prevalent on February 10, 1894 (Fig. 5, p. 218), when with high pressure over France and very low pressure over the north of Scandinavia, the wind was generally Westerly, moderate to fresh at the southern stations, and strong to a gale in the north, and the little rain which was reported fell almost entirely over the northern parts of the kingdom. Now let us take the map for January 16, 1896 (Fig. 12, p. 223). In this case the barometer is highest over the Bay of Biscay, lowest in a depression over the southern parts of Scandinavia, the gradients being steepest over the North Sea. We notice that in this case the winds are Westerly to Northerly, and are blowing hard over the Skagerrack, and strongly over the North Sea and Scotland, in all of which regions squalls of rain, graupel, and snow are prevalent, while very fine weather is being experienced on the south-west coast of the British Islands and the west coast of France. In the next, and last case, that of January 3, 1894, and already referred to as Fig. 4, p. 216, we have the highest pressures over Scandinavia, and the lowest over the Mediterranean, but with gradients slight except in the south and southeast. The winds we find are Easterly (NE to SE), and blow hard on the Dutch coast, through the Straits of Dover, and

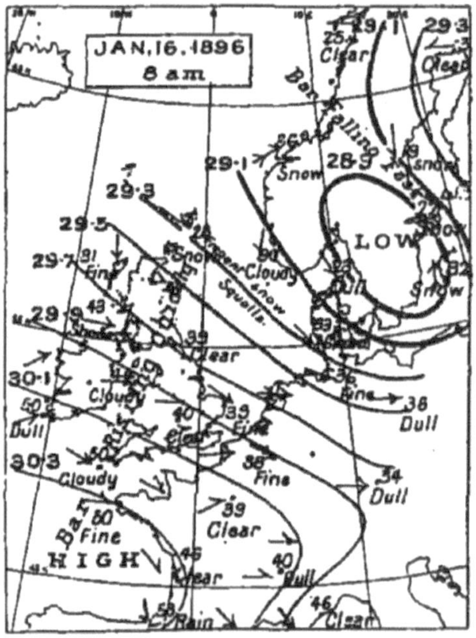

Fig. 12.

at the mouth of the Thames. The weather is extremely cold, and some very shallow secondary disturbances over the southern half of our area are causing snow squalls, in the south-east and east, while in the west and north the weather is fine and bright. From these four types of pressure distribution it is evident that the main condition on which the direction and force of the wind depends is the distribution of pressure over the area with which we are dealing; the strength of the wind varies with the steepness of the gradients, the parts of the area affected by disturbed weather depend on their proximity to the region of lowest pressure, or to their being affected by the secondary systems which move round the primary in its passage. Given as the statistical condition on which our judgment is based, the distribution of pressure at the time of observation – we have to consider the rate of fall in the barometer in the different sections, and the tendency of the wind to draw cyclonically round any region where the fall is most decided – these details point to the direction from which new disturbances, if any, may be expected. Temperature plays a most important part as an aid to the other observations, as the thermometer, no matter how low, almost invariably rises in front of an advancing depression, and in some instances the rise is very rapid. Again, if two depressions are following one another across the kingdom – say from west to east – and the ridge which separates them should pass over the country in the night – then, in the winter time, no matter how warm and wet the winds of the first disturbance may have been, the fall of the thermometer as it passes off is very rapid, and the subsequent rise as the new one comes on is equally striking. In this way it is recognised that if after a South-westerly gale and rain the sky clears very quickly, and there is great terrestrial radiation of heat at night, perhaps causing sharp frost on the grass, this occurrence is indicative of the early advance of a new depression with warm winds and rains, as before. Even in conditions when cold North-westerly winds and snow squalls are prevalent over our northern and eastern districts, owing to some depression lying over Scandinavia, or the eastern shores of the North Sea, the approach of any new disturbance from the north-westward or northward is heralded by a decided rise of the thermometer, first at the northern, and later on at the eastern and southern stations, accompanied by a temporary backing of the wind to West or South-west.

In Fig. 13 we have a diagram giving a hypothetical instance of a distribution of pressure over the Atlantic, showing certain anticyclones with depressions moving along their northern side, and showing every form of secondary system referred to in this paper. It is a modification of a diagram published by Mr. Abercromby some years ago in his book entitled *Weather,* and is very instructive.

From the facts that have been all too briefly touched upon it must be evident that we still require a clearer and more certain knowledge of the nature of the atmospheric circulation – of the manner in which the air rises in depressions, and comes down to earth again in the anticyclones. We require some means of ascertaining the movements in the higher regions of the atmosphere, and to accomplish this various means have been tried. Ballooning has been attempted, but this is unreliable, and is of course impossible just at the time when the indications would

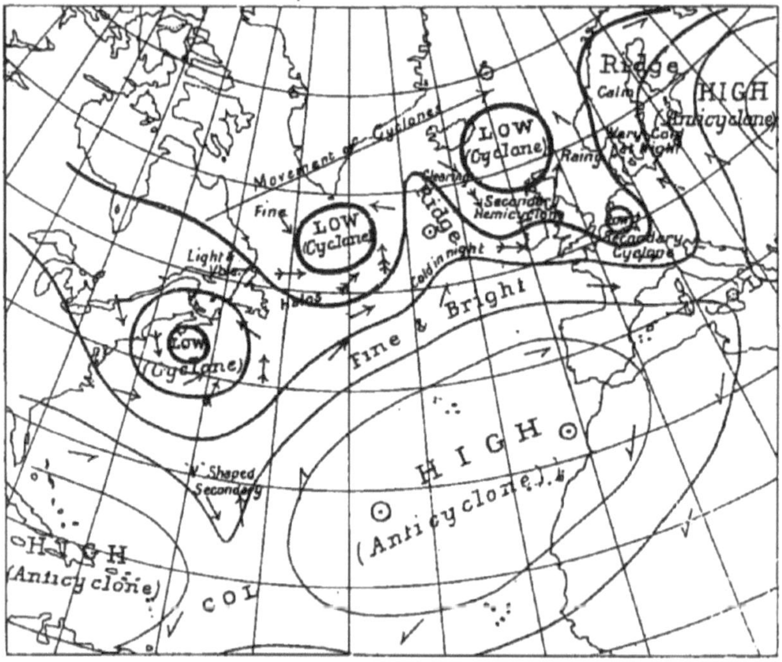

Fig. 13.

be most valuable – viz. immediately within the swirl of storms or in various positions around them. Guns have been employed for firing shells high into the air, where, on exploding, they produce a cloud of dense smoke, the movements of which can be watched. A more usual and permanent plan has been to build observatories on the peak of some mountain – such as that on the top of Ben Nevis in our islands, on the Puy-de-Dôme and the Pic du Midi in France, and several others in various countries. The observations of pressure, temperature, and vapour, made simultaneously at the top and at the base of such mountains, are yielding much, valuable information to meteorology and other branches of physical science; but up to the present they have not been of so much service, as aids to the forecasting of weather, as it was hoped they might be – probably owing chiefly to the obstruction which the hill, or range of hills, offers to the wind currents, so that the direction and force of the winds recorded are often of a purely local character, and do not indicate accurately the general movement of the air current in which they are situated. We are consequently thrown back upon observations of clouds and their movements. The value of such observations is steadily increasing, because much has been done to settle the question as to the altitude above the sea at which certain well-defined forms usually appear. Take in the first case the *Cumulus*, or heap cloud, in which the particles of water are found to be drawn together in large masses, which rise from an almost horizontal base, and often tower up, like masses of wool, to a great height. The variations in the size and density of such clouds are of themselves of great value in judging of coming weather, but the altitude of their base being approximately known, their movements show the direction in which the air current is travelling in which they (among the lowest forms) are floating. Then there are some of the higher forms of *Sheet* (or *Stratus)* clouds, which sometimes appear as a pall (*pallium*), and, when broken into detached portions, are easily recognised by the form of the cloudlets so developed. These, too, show by their movements the direction and velocity in which their air current is travelling, and their height can be approximately estimated. Still higher we have clouds of the *Cirrus* type, floating at about 25,000 to 30,000 feet, or even more, above the sea. Such clouds are readily known by their very delicate and filamentary structure, appearing sometimes as feathers, sometimes like locks of hair, sometimes like bunches of seaweed, and at others in great white masses or sheets, through which the sun and moon can be seen, often surrounded by a halo. Much has already been done to show the connection between the movements of these clouds and those of depressions in their neighbourhood, and much more will undoubtedly be ascertained ere long, when an accurate knowledge of the different forms of clouds and their classification has become more general among observers than it is at present. There is, however, a serious drawback even to this branch of the work. It is that thick lower clouds often prevent us from observing those at the higher levels, especially within the area covered by a storm, and we are then unable to obtain the very information we require. Observations on or outside the margins of such storms have, however, proved of enormous value already, and such observations should, if possible, be made more generally. In my opinion they are the most valuable of all the "tallies" on the movements of the upper currents which have ever been proposed; and there can be no doubt that in a short time, when the methods of observing them are improved, their importance will be more fully recognised.

It has been frequently suggested that the accuracy of the forecasts and storm warnings might be materially improved if we received information daily by wire from the United States, as to the general distribution of pressure over North America, and the storms which had left their shores moving eastwards, together with reports of the weather experienced by large steam vessels on their way from this country to the United States. The system was tried a few years ago for several months. It was shown then that a certain number of storms did probably come across the Atlantic and affect the weather on our coasts to a serious extent. More, however, passed away so far to the northward that their influence on our weather was not appreciable; others broke up in their journey, and the plan was abandoned, because not only was the information useless as a direct warning, but because the knowledge that a storm was prevailing over the Western Atlantic often caused those on this side to issue warnings prematurely. To cry "Wolf! wolf!" too often, is sufficient to ruin the reputation of any system of warning.

The proprietors of the *New York Herald* endeavoured for some years to warn our shores directly from the United States, and any one who wishes to judge of their success should read Mr. Scott's report of an examination into their value published while the system was in progress. The warnings were apparently abandoned for some years, but during the past few months there has been a resuscitation of them in the *Times* and some other papers. A more lamentable failure it is not possible to imagine.

Others have proposed that a vessel moored some few hundreds of miles to the westward of Ireland, and connected by an electric cable with this country, could telegraph information of advancing storms which would make the storm warnings almost perfect. From experience of the value of reports from an isolated station I must say I have no faith in such an experiment, even if the maintenance of the vessel in the position assigned to it, and the continuity of the electric communication could be kept up. Of the cost I say nothing. I think that at least three such vessels would be required, moored say in long. 28° to 30° W., and in lat. 50°, 55°, and 60° N., on a line between the Azores and Iceland; The cost of such a system, if it could be maintained, would be enormously in excess of its value to mariners.

This brings us to the close of the first part of our subject. We have now to consider briefly in what manner the forecasts and storm warnings are made known.

The forecasts which are issued to the public generally are drawn up twice daily – viz. at 11 a.m., from the 8 a.m. observations, and at 8.30 p.m., from the 6 p.m. observations. The former appear in many of the afternoon and evening newspapers; they are posted up soon after noon at the Mansion House, and several other places both in the City and at the West End, and are supplied for a small annual fee to such of the clubs as subscribe for them. The evening forecasts are supplied to all the principal newspapers, either directly or through the various news agencies, throughout the kingdom, and thus appear on the following morning in all parts of the British Islands. Forecasts, prepared at 3.30 p.m., to cover the same period of time as those prepared at 8.30 p.m., are issued gratuitously during the hay-harvest season to certain selected agriculturists in various parts of the kingdom; and any of them (11 a.m., 3.30 p.m., or 8.30 p.m.) are issued either by wire, or letter, or personally to applicants, who pay a small fee for the information in addition to the cost of transmission. The success of such forecasts is shown by the fact that year after year the same applicants are desirous of receiving and publishing (and in many cases of paying for) them, for weeks together. But it is to be understood the forecasts are prepared for large districts or areas of country, and that at times, when quite right for such districts as a whole, they are somewhat faulty in certain localities, where some peculiarity in the position of a station, with regard to ranges of hills, modifies the weather in that region, and renders the result less satisfactory. The districts for which the forecasts are prepared are shown in Fig. 14.

The storm warnings are made known by being telegraphed to a large number of stations on our coasts, and in these places the fact that such a warning has been received is made known to the shipping authorities and others by the hoisting of a black "cone," which indicates the probability of rough weather – commencing from some southerly direction, when the apex of the cone is downward, and from some northerly direction, when it is upward. The telegram itself is exhibited, in a

Fig. 14.

frame, in some prominent and well-known position at each port, and explanatory pamphlets are distributed from time to time among the seafaring community at the positions referred to, and are published in many nautical works as well. In addition to this the information is communicated to the daily newspapers in a manner similar to that adopted in the case of the forecasts. The results of these storm warnings are very satisfactory, the coast gales being less liable to local influences than the ordinary winds and weather at inland stations. Moreover, the fact that warnings are circulated by telegraph increases their practical value, by annihilating so much of the time which is necessary ere the forecasts can appear in print.

The winds observed not only at the telegraphic reporting stations of the Meteorological Office, but also by the observations made at many lighthouses and light-vessels round our coasts, are used in checking the accuracy for these storm signals, with the result that in 1894–95 no less than 93 per cent of the warnings were justified by the occurrence either of gales or strong squally winds. In Fig. 15 (p. 228) are two curves, taken from the last annual Report of the Meteorological Office, and showing the improvement which has been observed in this work during the past ten years.

It is sufficient to point out the gradual increase in the percentage of the accurate warnings, and the decrease in the percentage of the unsuccessful warnings, in order to show that the work is improving greatly. Of course in some years the gales experienced are less strong than in other years; and as the warnings have to

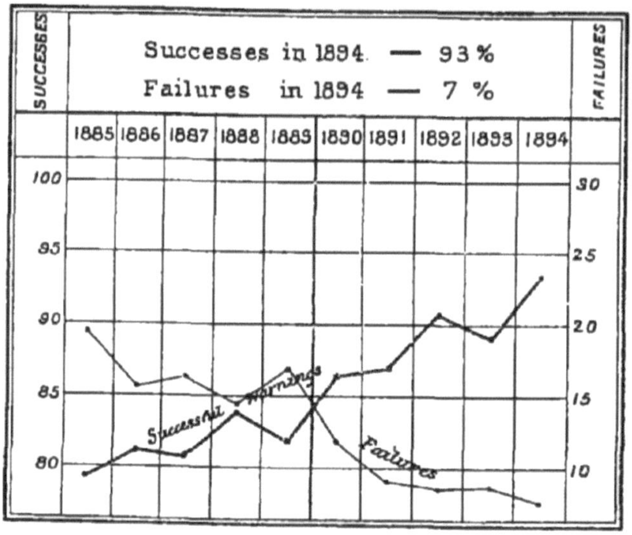

Fig. 15.

be issued on very slender information, if they are to reach the coasts in time to be of practical value, it happens that in some years the percentage of success will be smaller, and that the proportion of warnings justified by strong squally winds to those justified by strong gales will be increased somewhat. These are defects to be remedied only by constant study, by a gradual and judicious increase of the area covered by the observing stations, and of accuracy both in the observations themselves and in their subsequent transmission by wire. On several occasions whole coasts have been needlessly warned, either because a wrong barometric reading has been taken at some important station, or a telegraphic error has crept into the message in course of transmission; and the only thing to do was to act promptly and warn the coasts, or to risk the advent of a bad gale without the coasts having been warned.

Here I must close. Much has been left unsaid which might have been said if time permitted; but all that has been attempted has been to explain the broad principles which are acted on in carrying out the work of forecasting the weather and warning the shipping of the British Islands.

Source Endnotes

1 It is in no way proposed to go minutely into the history of this work, but merely to mention it with the object of showing how certain terms came into use in dealing with weather systems.

2 The map is not large enough to give the positions of Lisbon and P. Delgada (Azores), information from which is supplied by the Portuguese authorities.

3 The isobars prevailing are as a rule not so regularly formed as those in the diagram, which is a typical case drawn merely to aid those who read this lecture to understand thoroughly the main principles observed in drawing isobaric maps.
4 The force of the wind is shown on the various diagrams by the number of barbs and feathers on the arrows. Thus two barbs and four feathers indicate a violent gale; two barbs and two feathers a fresh gale; two barbs and no feathers a moderate gale or fresh wind; one barb only, a moderate to light wind. A small circle indicates a calm.

38

FUNCTIONS OF STATE WEATHER SERVICES

Henry Harrison Chase Dunwoody

Source: H.H.C. Dunwoody, 'Functions of State Weather Services', in Oliver L. Fassig (ed.), *Report of the International Meteorological Congress Held at Chicago, Ill., August 21–24, 1893* (Washington D.C.: Weather Bureau, 1894), pp. 9–13.

State Weather Services are organizations for the collection and dissemination of climatological and other information. They depend almost wholly upon the voluntary co-operation of intelligent and public-spirited citizens, whose individual reports collected at the several central stations form the basis of their publications. These publications are reviews of the prevailing weather conditions published monthly, and bulletins issued weekly during the season of planting, cultivating, and harvesting of crops, giving the more important weather features and their effect upon growing crops from week to week. Through State weather service organizations, the daily weather forecasts and special warnings of the National Bureau are distributed to large numbers of stations throughout the country.

There are three independent lines of work, each dependent upon its special class of contributors who serve in the capacity of (1) meteorological observers, taking observations of temperature, rainfall and miscellaneous data; (2) crop correspondents, who, during the crop season, render weekly reports of farming operations, the growth, maturing, and harvesting of crops, and the effects of the prevailing weather conditions thereon; (3) the forecast displaymen, who display flags or sound whistle signals representing the weather forecasts of the National Weather Service. It not infrequently happens that one person serves in more than one capacity and sometimes co-operates in all the three distinct lines of work.

In the United States there are less than 175 meteorological stations conducted by the regular paid observers of the Weather Bureau, or about one station for each 22,000 square miles of territory. The utter inadequacy of the data supplied by these stations for purposes of detailed investigation of special localities is therefore plainly apparent, making the State weather service an absolute necessity for the prosecution of such work.

Although the work of collecting voluntary meteorological observations and publishing the results was begun in Iowa as early as 1875, and in Missouri in

 DOI: 10.4324/9781003650294-58

1878, the organization of State weather services for the active prosecution of work on the lines previously referred to may be said to have begun in 1881 and 1882, since which time the number of meteorological stations has steadily increased, there being now about 3,000 stations taking and recording meteorological observations daily. With this extensive system it is possible to determine the special climatic features of every section of the country to an extent that would be entirely impossible were it not for the existence of local weather services.

All State weather services issue monthly reviews of the prevailing weather conditions, and many of these publications are issued in elaborate and attractive form, rendering them valuable and interesting. In many of these monthly reviews, besides giving a general discussion of the daily temperature and precipitation, observations are published in detail. While it would be difficult to correctly estimate the great value of this particular line of State weather service work, a more popular feature is the weather-crop service. From the beginning of the crop season until its close, weekly reports of the weather conditions and the effects of the same upon farming operations, the growth of crops, etc., are collected at the several State weather service centers. These weather-crop reports are mailed by the correspondents so as to reach the central station on Tuesday morning, and, as far as possible, cover the week ending with Monday. Upon receipt they are carefully summarized and a brief discussion of the general conditions prepared, which, with the detailed reports from the several correspondents, forms the State crop bulletin. The official in charge of each State service on Tuesday morning sends a telegraphic summary of the more important features of the week to the National Weather Bureau in Washington.

The entire territory of the country being covered by local services, complete information as to weather and crop conditions is had from every section of the United States. These telegraphic reports are published in full in the National Weather-Crop Bulletin, and, with the charts of temperature and precipitation departures, form the basis of a general discussion of the weather and crop conditions for the whole country.

The charts of temperature and precipitation departures are prepared from the data collected principally from U. S. Weather Bureau stations and serve in a general way to show how the temperature and rainfall of each week compares with the normal of the corresponding period.

This weather-crop service is, with the exception of the general weather forecasts, the most valuable work being done by the National Bureau, and is the most popular feature of State weather service work, being of greatest interest to agriculturists, although the bulletins are eagerly looked for by those interested in other pursuits. To the intelligent farmer it affords a means of supplying accurate and important information as to the condition of crops, enabling him to form reliable estimates as to supply and demand. In some States the editions of the local weather-crop bulletin have already grown to very large proportions, and the demand for the bulletin is constantly increasing. More than 11,000 copies of the Ohio weather-crop bulletin are printed and distributed weekly. As an illustration of the importance of this work, it may be stated that a material change in the condition of the cotton crop in the State of Texas influences the cotton markets of the world; and it is the work of the State weather service that presents weekly impartial and reliable information as to the actual weather and crop conditions prevailing throughout each season.

The publicity given the State and National weather-crop bulletins through the press of the country is so extensive that an accurate estimate of the combined bulletin and newspaper circulation would be difficult of computation. The full text of the National bulletin, including the special telegraphic reports from the various States, is telegraphed each week by the press associations and printed on Wednesday in the large dailies. The agricultural press make a specialty of the bulletin, and some reproduce in their columns the charts of precipitation and temperature. The patent-sheet papers also find the bulletin an attractive item, and they extensively print the bulletins of the States covered by their circulation. The Missouri bulletin is printed in nearly one hundred patent-sheet papers issued by the Kellogg and Western newspaper companies.

The late Prof. George H. Cook, for several years Director of the New Jersey Weather Service, in the work of organizing the New Jersey service, summarized the importance of the State service as follows:

> It will be the means of soon securing better predictions of weather changes and storms.
>
> It will bring the benefits of the National Weather Bureau of the United States into every county participating in the State local organization.
>
> It will soon prepare the State for a system of storm signals displayed from railroad trains that will be widely beneficial to agricultural interests.
>
> It will give to every county the Government standards for temperature, rainfall, wind velocity, humidity, etc., which are sources of useful public information.
>
> It will put within reach of local agricultural societies means of accurate observations which, in the course of years, must be valuable to any locality in the study and adaptation of cereals.
>
> It will bring the science and methods of the National Weather Bureau within the reach of the high schools of the State, offering teachers and pupils alike excellent opportunity to study a wide range of the application of science to foster and protect agricultural industry.
>
> It will lead to the collection of rainfall statistics to enable engineers to better estimate the supply of canals, also the sudden downpours to guard against in laying out sewers in cities. It will lead to a correct knowledge of rainfall over the different watersheds of the State, for the purpose of giving data for supplying the water works of cities, towns, and villages.
>
> It will lead to the forming of reliable meteorological records for use in legal cases.

It will lead to publishing the temperature of summer resorts, drawing attention of outside parties to their desirability as summer residence.

It will lead to a better practice of medicine, when physicians throughout the State can study disease with reliable and accurate meteorological facts by their side – and for sanitary purposes correct meteorological statistics are invaluable to the practitioner in applying preventive remedies for the public good.

The growth and popularity of these services were such that in November, 1885, Gen. W. B. Hazen, Chief Signal Officer, invited the directors of all State weather services to assemble in Washington for the purpose of mutual conference and discussion. Arrangements were accordingly made for a convention of the directors, which met February 24 and 25, 1886. At this conference many important subjects bearing upon State services were discussed looking to improved methods of taking and recording observations, and a general interchange of views regarding State service work was had. Much good resulted from this conference, and a report of its proceedings was published with the Annual Report of the Chief Signal Officer.

A second and more largely attended convention met in Rochester in August, 1892, at which the "American Association of State Weather Services" was formed, the constitution of which provides for annual meetings, and the convention of 1893 will be held in Chicago in August during the time of the meeting of the Meteorological Congress, August 21–24.

During the past year there have been prepared by many State weather services, for exhibit at the World's Fair, valuable and interesting charts illustrating graphically the special climatic features of the several States. Some of these exhibits have been prepared at much expense of labor and considerable pecuniary cost, and have been very favorably commented upon.

Part 5

WEATHER CARTOGRAPHY

5.1

DIAGNOSTIC

Editorial Headnote

5.1 Diagnostic

The second decade of the nineteenth century saw the emergence of cartographic representations in meteorology. The purpose of these early maps was not to foretell the course of atmospheric phenomena, but rather to create retrospective weather maps that could be used to ascertain the laws of nature and their physical causes. The sources in this section illustrate different ways in which these new and early cartographic representations of meteorology preceded the synoptic techniques later used by national meteorological services. While Heinrich Wilhelm Brandes (1777–1834) attempted to analyse the weather retrospectively, Alexander von Humboldt (1769–1859) sought to derive the main laws governing the distribution of atmospheric temperature from his representations.[1]

A professor of mathematics at the University of Breslau and the author of numerous textbooks, Brandes is considered the founder of synoptic meteorology.[2] He formulated his first ideas on the importance of synoptic observations in two letters to the *Annalen der Physik* in 1816 and 1819. In both letters, his ideas differed from traditional meteorology in privileging weather maps over traditional weather charts and wondered about their advantages 'if one could draw maps of Europe according to the weather for all the 365 days of the year'.[3] His method par excellence was the *synopsis*, or simultaneous and comparative review, of observations made at the same time over a wide area. In the letter of 1819, he identified two major storms on 6 and 12 March 1783 by supposedly mapping the weather from the recorded and preserved data of these two events. His fullest expression can be found in his book *Beiträge zur Witterungskunde* (1820) (Contributions to weather science), which was hardly understood at the time. He offers a synoptic interpretation of the atmospheric conditions during 1783, which is now considered as the first synoptic meteorology.

Brandes' extract re-evaluates the series of weather observations in terms of their usefulness as a tool for retrospective weather mapping.[4] The series of weather observations recorded by the Societas Meteorologica Palatina in Mannheim in 1783 – upon which Brandes draws for his research – is a 'buried treasure' that has remained 'without any use for science' (p. 26). From these series of observations, he derives appropriate comparisons and investigates the relationship between winds and air pressure. He proposes a relationship that is supposedly derived from the plotting of isolines and pressure centres. Here we are talking about assumptions or hypotheses, since neither his 1820 book nor his two letters contain any meteorological maps. The lack of cartographic evidence has caused some controversy among his biographers and historians,[5] with the majority accepting that Brandes reconstructed the weather maps and a few interpreting Brandes himself as using the conditional verb when referring to the maps.[6]

A polymath, naturalist, explorer and travel writer, Alexander von Humboldt was a celebrity in his native Prussia and throughout Europe. His numerous voyages of exploration rivalled those of Charles Darwin in prestige and influence. In 1807, Humboldt was sent on a diplomatic mission to Paris, where he remained for twenty years, as it was there that he was able to have his researches heard and properly evaluated by the world's leading scientists.[7] In 1817, he published several notable papers in the volumes of the *Mémoires de Physique et de Chimie de la Société d'Arcueil,* a scientific society founded near Paris in 1806 by Claude Louis Berthollet and Pierre-Simon Laplace. These works were the result of his extensive research on physical geography, meteorology and chemistry conducted during his famous expedition to the Americas (1799–1804). In addition to a study on the moisture content of the atmosphere ('Mémoires sur l'hygrométrie'), this Parisian society published two studies on temperature data collected by Humboldt during his travels: 'Sur les lignes isothermes', which introduces the concept of isotherm; and a more extensive version of this, 'Des lignes isothermes et de la distribution de la chaleur sur le globe', which identifies patterns in the global temperature distribution. An abridged English version of this work – excerpted here – was published in the *Annals of Philosophy* in 1818, and another, translated by David Brewster, appeared in the *Edinburgh Philosophical Journal* from 1820 to 1821 (see Vol. IV, 2.2 'Isotherm and mean temperature'). The fact that the original French work had a very limited circulation, because a large part of the volume perished in a fire, encouraged Humboldt to publish a German edition in 1853, *Von den Isothermen Linien und der Verteilung der Wärme auf dem Erdkörper* (Kleinere Schriften Bd.1, Stuttgart and Tübingen).

'Of Isothermal Lines, and the Distribution of Heat over the Globe' offers an interpretation of heat distribution that challenges the dominant view that climate depends only on latitude, showing how other factors such as altitude and ocean currents (in addition to latitude) influence temperature patterns. The work also defines a new concept, the 'isotherm', as the 'lines of equal annual heat' that can be plotted on a map. These lines do not run parallel to the equator, but cut across geographical latitudes at different angles, just like magnetic lines. Humboldt's isotherms make clear the power of precise observation whilst helping to eliminate errors and establish the laws of climate, in line with earlier projects such as the astronomical method of Tobias Mayer, who in 1775 formulated a theory of heat distribution over the Earth's surface in terms of trigonometric equations, or the magnetic declination charts or isogonic maps of Edmund Halley – both of which inspired Humboldt in this work through their precise observation and error analysis. Defined as an "empirical" concept, it also has an "idealistic quality", the embodiment of precise measurement that no theory or mathematical contrivance could draw and which deemed instrumental precision as the ideal of geodesy.[8]

Humboldt rewrote and disseminated his ideas on isothermal lines in other contemporary texts, such as his article 'Sur les lignes isothermes' published in the *Annales de chimie et de physique* in 1817. The peculiarity of this text with respect to the previous work is that it contains the only isothermal map that Humboldt ever published during his lifetime, which is extracted here; in fact, it was appended to this short text by François Arago and Louis Joseph Gay-Lussac (pp. 102–103) and given to the reader as a kind of *jeu d'esprit*: 'une carte que nous devons à l'amitié de M. de Humboldt, et dans laquelle il a representé graphiquement quelques uns de ses curieux résultats sur la forme et la situation des lignes isothermes, ou d'égale chaleur'. This fact reveals that at the time it was by no means obvious that tangible but non-geometric variables such as temperature or climate should be mapped in scientific writings, as was the case with geographical features.

Notes

1 Aitor Anduaga, *Politics, Statistics, and Weather Forecasting, 1840–1910. Taming the Weather* (London and New York: Routledge, 2020), pp. 42–7.

2 Bernhard Sticker, 'Brandes, Heinrich Wilhelm', in Charles Coulston Gillispie (ed.), *Dictionary of Scientific Biography* (New York: Charles Scribner's Sons, 1970), vol. 2, p. 420–21.

3 Quoted by Karl Schneider-Carius, *Weather Science, Weather Research: History of Their Problems and Findings from Documents during Three Thousand Years* (New Delhi: Indian National Scientific Documentation Centre, 1975); originally published as *Wetterkunde, Wetterforschung: Geschichte ihrer Probleme und Erkenntnisse in Dokumenten aus drei Jahrtausenden* (Freiburg: Verlag Karl Albert, 1955), p. 179.

4 Arthur H. Robinson, *Early Thematic Mapping in the History of Cartography* (Chicago, IL: The University of Chicago Press, 1982), pp. 73–4; John A. Kington, *The Weather of the 1780s over Europe* (Cambridge: Cambridge University Press, 1988), pp. 18–9.

5 Mark Monmonier, *Air Apparent. How Meteorologists Learned to Map, Predict, and Dramatize Weather* (Chicago, IL: The University of Chicago Press, 1999), pp. 18–22.

6 Aleksandr Khristoforovich Khrgian, *Meteorology: A Historical Survey*, 2nd ed. (Jerusalem: Israel Program for Scientific Translations, 1970), p. 138.

7 Malcolm Nicolson, 'Alexander von Humboldt and the Geography of Vegetation,' in Nicholas Jardine and Andrew Cunningham (eds.) *Romanticism and the Sciences* (Cambridge: Cambridge University Press, 1996), pp. 169–88; M. J. Bowen, *Empiricism and Geographical Thought: from Francis Bacon to Alexander von Humboldt* (London: Cambridge University Press, 1981); Michael Dettelbach, 'Romanticism and Resistance: Humboldt and "German" Natural Philosophy in Natural Philosophy in Napoleonic France', in Robert M. Brain, Robert S. Cohen, Ole Knudsen (eds.), *Hans Christian Ørsted and the Romantic Legacy in Science: Ideas, Disciplines, Practices* (Dordrecht, The Netherlands: Springer, 2007), pp. 247–58.

8 Michael Dettelbach, 'Humboldtian Science', in Nicholas Jardine, James A. Secord, and Emma C. Spary (eds.), *Cultures of Natural History* (Cambridge: Cambridge University Press, 1996), pp. 287–304, on pp. 481–85.

39

BEITRÄGE ZUR WITTERUNGSKUNDE

Heinrich Wilhelm Brandes

Source: Heinrich Wilhelm Brandes, *Beiträge zur Witterungskunde* (Leipzig: Johann Ambrosius Barth, 1820) pp. 26, 69, 73, 213, 270.

HISTORY OF THE WEATHER OF THE YEAR 1783

We have an almost incalculable series of weather observations, most of which, like a buried treasure, are without an use for science, for no one will take the trouble – to be sure very great trouble – to derive from the thousands of observations appropriate comparisons and thus to test whether we can get results from them. The results which one is wont to deal from the whole annual series – how many bright days, cloud days, rainy days there were in the year, how often the easterly or the northerly wind blew, etc. – are of course not to be despised; but the information obtained from these is, in fact, only very little, and we must necessarily attempt other computations if we want progress in weather science.

I report here one such experiment, which is not yet complete and in which I combine the weather observations carried out in 1783 all over Europe and in some other regions of the earth, in order to survey the whole course of the weather and the simultaneous changes in nearby and distant regions.

The question as to how this unequal pressure originated and how it is possible for the inequality of pressure to increase gradually without the neighboring air rushing in and producing an equalization before the inequality of pressure attains such a high value is more difficult. In fact, it often looks as if the air pressure is reduced for a long time at some places, and there would be no unusually stormy weather and no inrush in the neighboring areas; but when the inequality of pressure has reached a certain magnitude, then this flow of air, often accompanied by thunderstorms, suddenly breaks out – it is if the opposing forces, which till then had held back the neighboring air in spite of the low pressure, had vanished suddenly and only now the air of the type required by the last equilibrium endeavored to restore equilibrium. The origin of the unequal pressure can be ascribed, of course not entirely but partially, to the heavy rain which usually falls on such. Also during the period of time with which we are now dealing the rain was exceedingly heavy in North Germany and although the water which fell from the air might not

DOI: 10.4324/9781003650294-61

have been sufficient to reduce the masses which exerted the pressure by so much as 8 lines of mercury, yet a reduction of the pressure-exerted masses is affected by it. One cannot ascribe the reduction in pressure completely to this rain; for, the 8 lines of mercury exert as great a pressure as 9 inches of water, so that to reduce pressure by 8 lines a 9-inch deep rain must have covered the entire area, and in our case this amount of rain hardly falls in three or four months and so we cannot assume that such a great mass of air was converted into water. Hence one can really regard this rain itself only as an effect of the cause, which as the principal reason brings about the reduction in pressure, i.e.,; an effect which then, of course, contributes again towards further increasing this reduction.

In order to better represent the inequality of the air pressure which was most different during these days, I will mention from numerous observations that we possess, how far below the mean height the barometer stood. From this it is seen that the barometer was exceptionally low in Central England and one can specify fairly accurately the boundaries within which barometric readings of 14 lines below the mean height, 13 lines below the mean height, etc., occurred.

These observations give rise to several considerations. Obviously the region of lowest barometric reading moved away from the canal above North Germany; but together with this shift the depression of the barometric reading diminished or the air pressure in the areas where later the lowest reading occurred was not so very unusually low as in the areas where we first noticed the low barometric reading. Further, as soon as the low pressure area of the atmosphere had moved away from a place, the barometer rose very rapidly, so that if we could accurately draw the limits of equal pressure at any point of time, undoubtedly these, when drawn for every difference of pressure of one line, would be packed much closer on the western side than on the eastern side.

We can conclude from several of the occurrences mentioned above that there are events which move, as it were, from place to place over Europe. Among these phenomena, the shifting of the area of lowest barometric pressure seems to be of special importance and to merit special attention; for, with a sufficient number of simultaneous observations it cannot really be difficult to obtain quickly a series of empirical data on this point, which surely must yield definite results, especially if we were so fortunate as to have observations not only from the whole of Europe but also from the north coast of Africa, Asiatic Russia, Iceland and even from several regions of North America.

40

OF ISOTHERMAL LINES, AND THE DISTRIBUTION OF HEAT OVER THE GLOBE[1]

Alexander Von Humboldt

Source: Alexander von Humboldt, 'Of Isothermal Lines, and the Distribution of Heat over the Globe', *Annals of Philosophy; or, Magazine of Chemistry, Mineralogy, Mechanics, Natural History, Agriculture, and the Arts* 11:63/3, 1818, pp. 177–181.

THE unequal distribution of heat over the globe is one of those phenomena which has been long known as a general fact, but which cannot be exactly ascertained, with respect to its particular laws, until we have more correct data furnished by observation and experiment. To furnish these data is the immediate object of this paper; they are deduced from a great number of facts which have not been published; and if they are not sufficient to enable us to form a correct theory, they may, at least, lay a foundation for it, and will be useful in pointing out to travellers those objects to which they ought especially to direct their attention.

The distribution of vegetables and of organized beings in general depends upon the circumstances that are connected with latitude, longitude, and elevation; and of these one of the most important is atmospherical temperature. The means which the author enjoyed of making observations, during his residence in South America, have enabled him to establish some very valuable data, which could not have been obtained in any other situation, more especially those which require to be made at great heights above the level of the ocean. The highest point in Europe where any observations have been made, is the Hospice de St. Gothard, at about 6,400 feet above the sea; but in South America the town of Quito is about 9,000 feet, Huancavelica about 11,670 feet, and the mine of Santa Barbara as much as 14,400 feet above the sea, or more than double that of St. Gothard. In order to compare the results which were obtained in the equinoxial regions with the mean heat of temperate climates, it was necessary to find out different situations, at intervals of 10° of latitude, but on different meridians, the mean temperature of which had been accurately ascertained. These will form so many fixed points, through which the *isothermal lines*, or the lines of equal temperature, may pass. In collecting facts for ascertaining these stations, comparatively few of the numerous thermometrical observations that have been published could be employed.

DOI: 10.4324/9781003650294-62

Many of the observations contradict each other; in many cases we do not know under what circumstances they were made; and we are frequently obliged to reject such as, in other respects, appear correct, because we are not acquainted with the absolute height of the place at which they were made. This is remarkably the case with almost the whole of Asia; and it is not a little singular that, while there are more than 500 stations in the equinoxial district of America, many of them mere villages, or even hamlets, the altitude of which has been exactly determined, we are ignorant of the height of Bagdat, Aleppo, Ispahan, Delhi, and many other large and ancient cities in the Old World, In comparing the temperatures of different places, it is, however, quite necessary either that they should be upon the same level, or that a proper allowance should be made for any difference which there may be in this respect.

In the old continent the only good observations which can be employed to form our calculations are limited by the parallels of 30° and 70° of latitude, and by the meridians of 30° east and 20° west longitude; the extreme points of which are the island of Madeira, Cairo, and Cape North; it comprises about $^1/_7$ the circumference of the globe from east to west. There are many circumstances connected with Europe, partly depending upon its natural form and situation, and partly upon the state of its civilization, which have given a peculiar character to its climate, different from that of other regions in the same latitude. But as this has been the abode of men of science, they have considered the laws which regulate the temperature of that part of the world as what are applicable to all the others. In this, however, they have fallen into some considerable errors, the causes and the amount of which we must endeavour to discover and appreciate. It is not with the temperature of the atmosphere, and with the magnetism of the globe, as with those phenomana which are determined by a single cause, and may be considered as distinct from all disturbing causes. From their nature, they both depend upon many local circumstances, such as the constitution of the soil, and the disposition of the radiating surface of the globe; and it frequently requires much judgment and discrimination to decide what circumstances are to be taken into account, and what are not connected with the inquiry. The object is to ascertain the quantity of heat which every part of the globe receives annually, and which is of actual use to agriculture and the well-being of the inhabitants; not what depends solely upon the action of the sun, its height above the horizon, or the extent of its semi-diurnal arcs. The temperature of a climate depends both upon the action of the sun, and of various extrinsic causes; among these may be enumerated the mixture of the temperatures of different latitudes, produced by winds, the vicinity of the sea, the nature of the soil, the presence and peculiar form of mountains, and the existence of large tracts of snow or masses of ice.

In distinguishing between the *solar* and the *real* climate, we must remember that the local causes which modify the sun's action are themselves only secondary causes, effects which the motion of this luminary produces in the atmosphere. Many of the local or disturbing causes are necessarily connected with the nature of heat, and are felt over every part of the globe. The ocean tends to equalize the temperature of all the different regions by the mobility which necessarily belongs to it, and the currents of warm air which always flow from the equator to the poles, tend to diminish the rigour of the countries of the north. In estimating the action of the sun alone on the earth, we must have recourse to theory: this will not express the actual height of the thermometer in different situations, but it will show the relations between the mean temperature of different regions. By comparing the results of the calculation, not to the mean deduced from observations made in different longitudes, but to the mean temperature of a single point taken at the surface of the earth, we may proceed in our examination of what depends upon the sun, and upon all the other influences, solar or not solar, local, or such as extend to considerable distances. In this way we shall be able to form an interesting comparison between theory and experience.

For the first philosophical ideas upon the subject of solar heat, we are indebted to Halley. Marian afterwards extended our knowledge of solar action; but he fell into some considerable errors, which were rectified by Lambert, who instituted some important calculations, which, however, do not always accord with actual experience, and which, indeed, depend almost entirely upon mathematical principles. After him, the subject was taken up by Moyer and by Kirwan; and they proceeded more, upon observation, at least upon the method of endeavouring to approximate to a true system by collecting observations, and employing them to correct the theory. In speaking of the authors who have added to our knowledge on the temperature of the different parts of the earth, we must not omit to mention Cotte, who has collected a great number of documents; but they are not reduced into any kind of method, and are not all of equal authority.

These considerations lead us to the conclusion that, in investigating the distribution of temperature, it is important to distinguish between the results which are deduced from observation, and those which are derived from theory. We must collect all the authentic facts that can be obtained; and, after arranging them into regular order, we must submit them to what may be termed empirical laws. After having correctly ascertained the mean temperature of certain places, we may trace on the globe *isothermal lines,* which thus exhibit to the eye the relation of these places to each other. In determining the mean temperature of a particular spot, the old plan was to take the maximum and minimum temperature of the year, and to consider the middle number between them as the mean temperature; but this plan is obviously incorrect. De la Hire seems to have been the first who attempted to pursue a different method, and one founded on more just principles: observing the uniformity of the temperature of the vaults that are attached to the observatory at Paris, he proposed as a general fact, that the temperature of vaults was the mean temperature of the climate.[2] The other method, that of the maxima and minima, continued, however, to be generally adopted; and, by multiplying the number of the observations, it was rendered more correct, but still liable to inaccuracy. Some of the latest observers have noted the thermometer three times each day, and then taken the mean of these as denoting the mean temperature of the day; others have adopted the plan of observing the thermometer at two periods in the day, which are considered as indicating the maximum and minimum – sun-rise, and two hours after noon; while others, again, have satisfied themselves with observing the temperature at one period only, which has been found, by previous experience, to denote the mean temperature.

By comparing a great number of observations made between 46° and 48° N. latitude, we find that at the hour of sun-set the temperature is very nearly the mean of that at sun-rise and two hours after noon. When, besides noting the maximum and minimum, we take a middle observation, we shall fall into an error, if we simply divide the sum of the observations by three, without attending to the duration of the particular temperatures, and the place which the middle observation occupies between the extreme terms of the series. The middle observation should be at least four or five hours from either of the others; but, upon the whole, the two observations of the extreme temperatures will give us more correct results. Some very valuable observations have been made by noticing the thermometer from hour to hour, at different seasons of the year, and in different latitudes, until we are able to fix upon a number which may indicate the mean of the day. Serene and calm weather has been chosen on these occasions, and the thermometer has been carefully examined in this way, both in the observatory at Paris, and under the equator. These observations have tended to confirm the opinion that has been mentioned above, that the temperature of the earth corresponds with the mean temperature of the atmosphere, the disturbing causes nearly counteracting each other. In expressing the results of the observations on mean temperatures that have been made in various situations, it is convenient not to employ the numbers that are derived from any particular scale, but to consider the equator as the standard to which all the rest are to be referred, and to denote them all by numbers which have an arithmetical relation to it.

Having now ascertained the method of taking mean temperatures, and of reducing them to a general expression, we may proceed to examine the form of some of the isothermal lines. It has been long known that the temperatures are not the same in the same parallels, especially those in Europe and in America; but, from the facts that will be stated, we shall find that this difference is not so great as has been imagined. By constructing a table, in which we compare the mean temperature and the latitude of different places in the continents of Europe and America, we learn the amount of this difference, and we deduce from this the number of degrees of latitude which we must go northward in Europe in order to arrive at the same annual mean. From this train of observations, we find that the isothermal line, or band, which is considered as 32°,[3] passes between Ulea, in Lapland, latitude 66° 68′, and Table Bay, in Labrador, latitude 54°. The isothermal line, or band, of 41°, passes near Stockholm, latitude 60°, and St. George's Bay, in Newfoundland, latitude 48°. The isothermal line, or band, of 50°, passes through Belgium, latitude 51°, and near Boston, latitude 42° 30′. The isothermal line, or band, of 59°, passes between Rome and Florence, latitude 43°, and near Raleigh, in South Carolina, latitude 36°.

Source Endnotes

1 Abridged from the third volume of the Memoirs of the Society of Arcueil.
2 Mem. Acad. Sciences, 1719, p. 4.
3 The thermometrical numbers are all reduced from the centigrade scale to that of Fahrenheit. In the original, the numbers on the centigrade scale are, 0°, 5°, 10°. 15°, respectively.

41

SUR LES LIGNES ISOTHERMES

Alexander Von Humboldt

Source: Alexander von Humboldt, 'Sur les lignes isothermes', *Annales de chimie et de physique*, 5, 1817.

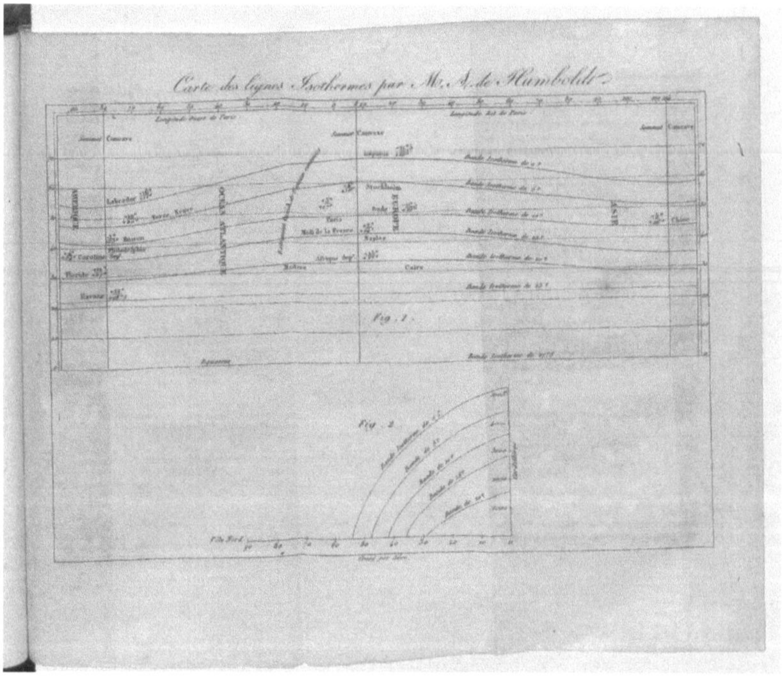

 DOI: 10.4324/9781003650294-63

5.2

FITZROY'S CHARTS

Editorial Headnote

5.2 FitzRoy's charts

The safety and predictability of shipping routes were a matter of intense interest in Britain and elsewhere in the mid-nineteenth century. Since its inception in 1854, the Board of Trade's Meteorological Department had been engaged in producing charts for navigators. The two texts in this section, written by British Royal Naval Officer Robert FitzRoy (1805–1865), show new graphic ways of meeting these challenges to maritime navigation. Drawing on his extensive nautical experience, FitzRoy uses new forms of empirical thinking to propose new wind charts and representations that go beyond those of his predecessors and are shaped by and transcend nationality and culture.[1]

From an influential aristocratic family, FitzRoy was appointed head of the new department created in 1854 under the title of 'Meteorological Statist to the Board of Trade' to direct the collection of weather data at sea (see 2.3 'Lagrangian Approach'). Like many of his contemporaries, he was initially attracted to the measures proposed by the American Captain Matthew Maury to improve the safety and efficiency of sea travel. From 1848 to 1873, Maury published a series of maritime charts known as *Pilot Charts*, which provided valuable information on prevailing winds, ocean currents and other navigational data (see Vol. II, 2.3 'Sailing directions'). The aim of the charts was to reduce the length of voyages and improve the predictability of sea routes. Capital for this purpose were Maury's famous 'wind roses', circular numerical tables showing wind frequencies, or more precisely, the average frequency and strength of the wind in each five-degree square during a given month. With these tables, the navigator could determine the optimum route, i.e. the statistically fastest route for a sailing vessel between two ports.[2] Although useful and effective, FitzRoy realised that Maury's *Pilot Charts* posed certain problems for users and that it was necessary to convert the data from these charts into a graphical form which, while retaining the information about wind distributions, could be more easily managed and interpreted. The outcome was the well-known Board of Trade Wind Charts, the first edition of which was published in August 1855, and by March 1856 there were eleven charts. The rivalry illustrates the variety of views on how the distribution of wind directions should be shown and the role that cartography could play in meeting the perceived challenges of maritime navigation.

FitzRoy had first put forward his idea in an earlier paper on 'Wind charts of the Atlantic, compiled from Maury's Pilot Charts', which he read at the twenty-fifth meeting of the British Association for the Advancement of Science, held in Glasgow in September 1855. He proposed that marine data should be grouped geographically using a series of numbered 10° latitude-longitude squares – a system usually attributed to William Marsden, Secretary of the Admiralty, but which, according to historian Duncan Carr Agnew, was actually invented by FitzRoy around 1854. By connecting the ends of the radial lines in the directions of the sixteen cardinal points, FitzRoy obtained a polygonal figure showing the distribution of the winds, which he called the 'wind star'.[3] In his opinion, Maury's Pilot Charts, which provide information by means of numbers, 'require more mental operation in their use than these graphical figures' (p. 39). To conclude, and following in Maury's footsteps, he proposes to compile wind charts for all known parts of the world, for smaller spaces and for every month of the year.

FitzRoy summarised these and other ideas eight years later in his *The Weather Book: A Manual of Practical Meteorology* (1863). Published in London by Longman and Green, it was intended as a treatise for the general reader with an interest in the weather.[4] Conceived as 'a holiday task, hastily performed', he admitted in a letter to the secretary of the Royal Society on 5 March 1863 that it was 'a so ill-digested and obscurely-written work' that he hoped to present him with a second, 'somewhat less obscure' edition.[5] But even so, it was this work that opened the door to the European academic world when, shortly afterwards, Hippolyte Marié-Davy of the Paris Imperial Observatory asked him for permission to translate his work into French, which was granted around 1865 with the publication of *Le livre du temps: Manuel pratique de météorologie* (Paris, Arthus Bertrand).[6] *The Weather Book* describes in detail his methods of forecasting, along with Heinrich W. Dove's theory of opposing currents, which was so influential on his concept of wind flow patterns (two schematic charts included in the appendix show these patterns around the British Isles).[7] The text and figures reproduced here are from the 'wind star', a type of diagram he devised to show the direction and strength of winds at a glance. The wind star consists of a series of lines radiating from a central circle, marked with latitude and longitude. While the lines represent wind directions, the longitude of each radial line indicates the frequency or proportion of time the wind blows from that direction. FitzRoy's wind stars proved to be particularly useful for route planning, as by examining them closely, sailors could quickly determine the prevailing wind directions and their relative frequencies, thus identifying the prevailing wind patterns at a particular location.

Notes

1 Derek Barlow, 'From Wind Stars to Weather Forecasts: The Last Voyage of Admiral Robert FitzRoy'. *Weather*, 49:4 (1994), pp. 123–32; Aitor Anduaga, *Politics, Statistics, and Weather Forecasting, 1840–1910. Taming the Weather* (London and New York: Routledge, 2020), pp. 362–64.

2 Fabien Locher, *Le savant et la tempête. Étudier l'atmosphère et prévoir le temps au XIX siècle* (Rennes: Presses Universitaires de Rennes, 2008), pp. 150–51.

3 Duncan Carr Agnew, 'Robert FitzRoy and the Myth of the 'Marsden Square': Transatlantic Rivalries in Early Marine Meteorology', *Notes and Records of the Royal Society of London*, 58:1 (2004), pp. 21–46, on pp. 33–4.

4 James Burton, 'The History of the British Meteorological Office to 1905' (The Open University, PhD diss., 1988), pp. 48–9.

5 J. Malcolm Walker, *History of the Meteorological Office* (Cambridge: Cambridge University Press, 2012), p. 50.

6 Pauline Halford, *Storm Warning. The Origins of the Weather Forecast* (Stroud: Sutton Publishing, 2004), p. 202.

7 Katharine Anderson, *Predicting the Weather: Victorians and the Science of Meteorology* (Chicago: The University of Chicago Press, 2005), pp. 118, 190.

42

WIND CHARTS OF THE ATLANTIC, COMPILED FROM MAURY'S PILOT CHARTS

Robert FitzRoy

Source: Robert FitzRoy, 'Wind Charts of the Atlantic, Compiled from Maury's Pilot Charts', *Report of the Twenty-Fifth Meeting of the British Association for the Advancement of Science Held at Glasgow in September 1855* (London: John Murray, 1856), vol. 2, pp. 39–40.

These diagrams are intended to show what winds prevail, at the four quarters of the year, in the Atlantic.

Each figure should be considered by itself alone, as the scales are generally very different, depending on the number of observations from which the respective diagrams are constructed.

Relative prevalence of wind (or calm) is shown in each square of ten degrees; but in no case is *absolute* amount given; nor is *strength* of wind exhibited, as it may be hereafter.

The navigator may be influenced, in shaping a course, by the probability of finding certain winds more or less favourable in certain localities.

To a sailing ship such considerations are most important; and a glance at these charts shows a seaman how the wind blows (usually) during a *season,* as readily as his "dog-vane" indicates the (apparent) direction at any *moment* of observation.

The diagrams illustrate Maury's *Pilot* Charts, in which similar information is offered by numbers, which require more mental operation in their use than these graphical figures.

In each square the numerical data contained in four of Lieut. Maury's five-degree squares are combined in the following manner.

Of a circle inscribed in any such square, the radius is taken as a measure of the sum of the greatest number of observations of the most prevalent wind; and other lines, likewise drawn (to leeward respectively) from the centre, and on the same scale, indicate the *relative* duration or prevalence of other winds (each observation referring to a period of *eight hours*), and through the extremities of these lines a boundary is traced.

As a circle is said to be generated by the revolution of a line around a point, so the figure representing successive directions of wind may be supposed to be

DOI: 10.4324/9781003650294-65

generated by the motions of a wind-vane, and the lines or points may extend from the centre (like the growth of crystals) in proportion to the *persistence* (or continuance) of the vane in their respective directions.

The *relative* amount or duration of calms is shown by a circle, of which the radius equals (according to the scale of the diagram) the number of (eight-hour) periods in which there was little or no wind.

The direction of wind is corrected, approximately, for variation of the compass.

The larger area of each figure is to leeward of the centre of the square (or inscribed circle).

The calendar quarters of the year are adopted advisedly, because the consideration of seasons in all quarters of the globe, and the examination of Maury's charts (including those of the trade-winds), induce the belief that extreme periodical changes of wind follow at a certain interval, rather than accompany the extremes of temperature or climate.

The small figures at the lower left-hand corner indicate the total number of (eight-hour) observations of calm, as well as of wind, recorded as having been made in that square; and the figures at the lower right-hand corner show, in decimals of an inch, the *unit of scale* employed in constructing the diagram *in that square*.

The force of wind is not shown, because it was not noted in the records from which these charts were compiled; but at a future time it may be given so combined and arranged as to indicate average strength as well as direction.

Nothing more is thus shown, in *a graphical manner,* than has been exhibited *numerically* in Maury's original *Pilot* Charts, whence solely the data for these were obtained.

For the few squares still blank, sufficient data have not yet been collected.

The number of observations used in constructing each diagram affords a scale of value for the figure, which may be augmented from time to time by fresh material, but need not be diminished, unless by a reduction of the *scale* (should the figure much outgrow its square).

The star-like form of the figures ("Wind-star") is merely a consequence of *grouping* observations under *principal* points of the compass.

It is proposed to compile wind-charts for all known parts of the world, for smaller spaces or squares, and for each month of the year, as soon as sufficient observations can be collected and employed.

43

THE WEATHER BOOK

A Manual of Practical Meteorology

Robert FitzRoy

Source: Robert FitzRoy, *The Weather Book: A Manual of Practical Meteorology* (London: Longman and Green, 1863), pp. 413–418.

E

WIND CHART DETAILS

SQUARE 375.

Subdivided into *a, b, c, d,* which subdivision may be continued by quartering and lettering *a, b, c, d,* as *e, f, g, h,* &c.

SOUTH ATLANTIC.

Brazilian Coast (near Rio do Janeiro).

For Three Months – January, February, March.

Four Windroses are condensed into this Diagram; namely, those for –

	a	*b*	*c*	*d*
Lats. and Longs.	20° + 25° S. 30° + 35° W.	20° + 25° S. 35° + 40° W.	25° + 30° S. 30° + 35° W.	25° + 30° S. 35° + 40° W.

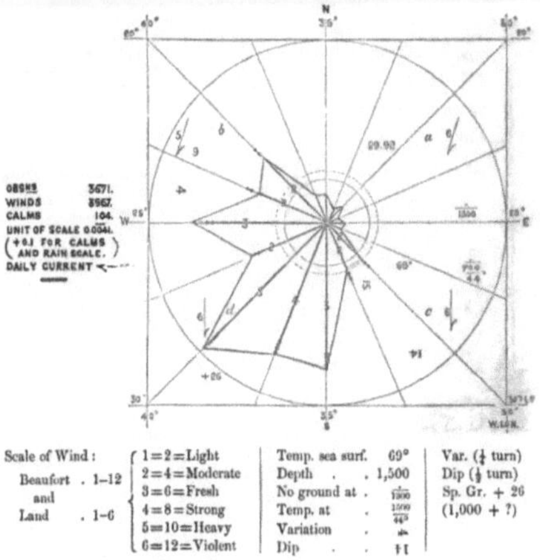

Scale of Wind: Beaufort . 1–12 and Land . 1–6	1 = 2 = Light 2 = 4 = Moderate 3 = 6 = Fresh 4 = 8 = Strong 5 = 10 = Heavy 6 = 12 = Violent	Temp. sea surf. 69° Depth . . 1,500 No ground at . $\frac{\cdot}{1500}$ Temp. at . $\frac{1500}{44°}$ Variation . Dip . .	Var. ($\frac{1}{4}$ turn) Dip ($\frac{1}{2}$ turn) Sp. Gr. + 26 (1,000 + ?)

SQUARE 338.

SOUTH ATLANTIC.
Brazilian Coast (near Bahia).
For Three Months – January, February, March.

Five-inch Square: half one side (= radius of inscribed circle) has 2,500-thousandths of an inch, in which measure the *unit for scale* is taken.

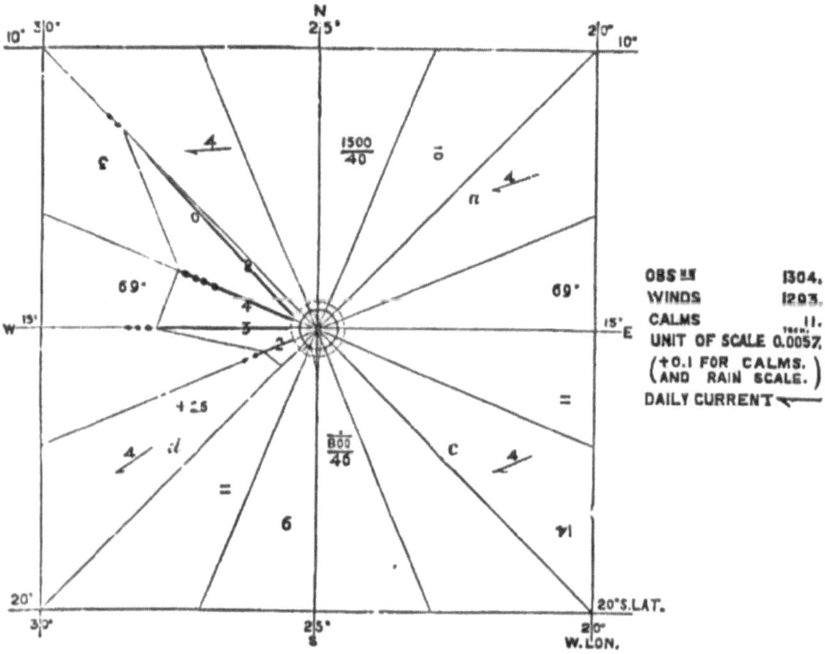

Let n = number of parts in *radius* (= 2,500)
n' = „ „ longest point (*radius*)
l = length of *one* part of radius = (·001)
l' = length of unit of scale;
then $n' : n :: l : l'$

$$\therefore \frac{n \times l}{n'} = l' = \textit{unit of scale}$$

Log. n (2,500)	= 3·3979
Log. l (·001)	= $\bar{3}$·0000
	0·3979
Log. n' = (439)	2·6425
Unit of Scale = l' = ·0057	= $\bar{3}$·7554

SPECIMEN OF METHOD ADOPTED IN SUBDIVISION OF SQUARES OF TEN DEGREES INTO WHICH THE SURFACE OF THE GLOBE IS SUPPOSED TO BE DIVIDED.

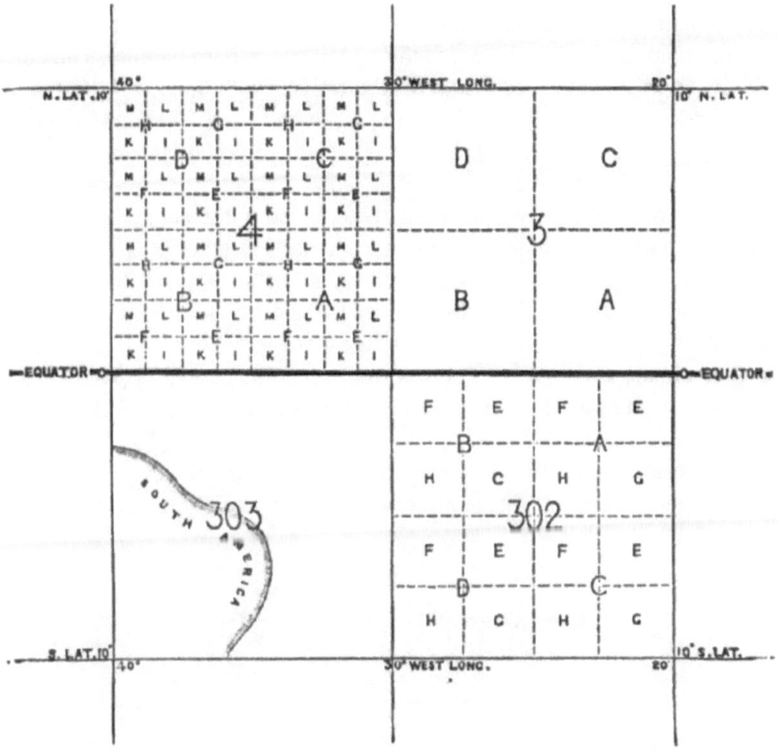

N.B. The letters in the above diagram may be of any character.

DIVISIONS OF THE GLOBE AND CONSTRUCTION OF WIND CHARTS.

Divisions of the Globe.

THE mode of exhibiting barometric and thermometric oscillations by diagrams is familiar to most persons; but, as the method here adopted to show the results of observations of wind, weather, and some other matters graphically, is new, it may require explanation.

The surface of the globe is supposed to be divided into squares of ten degrees each. Beginning at the meridian of Greenwich, on the Equator, the numbers go westward until the same meridian is regained; then, on the next circle, northward, between the parallels of 10 and 20 degrees of latitude; and so on, omitting the 10 degrees' space of latitude around the pole.

From the first meridian the squares S. of the Equator are numbered from 300 in a similar manner, but *southward* to the eightieth southern parallel. Thus distinguished by numbers not exceeding 600, all those below 300 being N. of the Equator, the locality of each frequented square may soon become fixed in the memory of a navigator, and serve (like provinces on land) to recall *spaces* to the mind, rather than points indicated only by latitude and longitude.

Observations made in any square may be referred to the centre of that square, as to an Observatory at which observations are made regularly.

In such spaces it is supposed that meteorologic occurrences will not generally be very dissimilar; and that, if all are referred to the centre of that square, or subdivision of it, in which they occur, the mean result of a great number of observations will give a reliable and approximately correct general average for practical use.

If between each time of observation eight hours elapse, three winds (or calms) may be noted for one day.

During a certain interval of weeks, months, seasons, or years, a number of observations are collected. These are classified and totalled under points of the

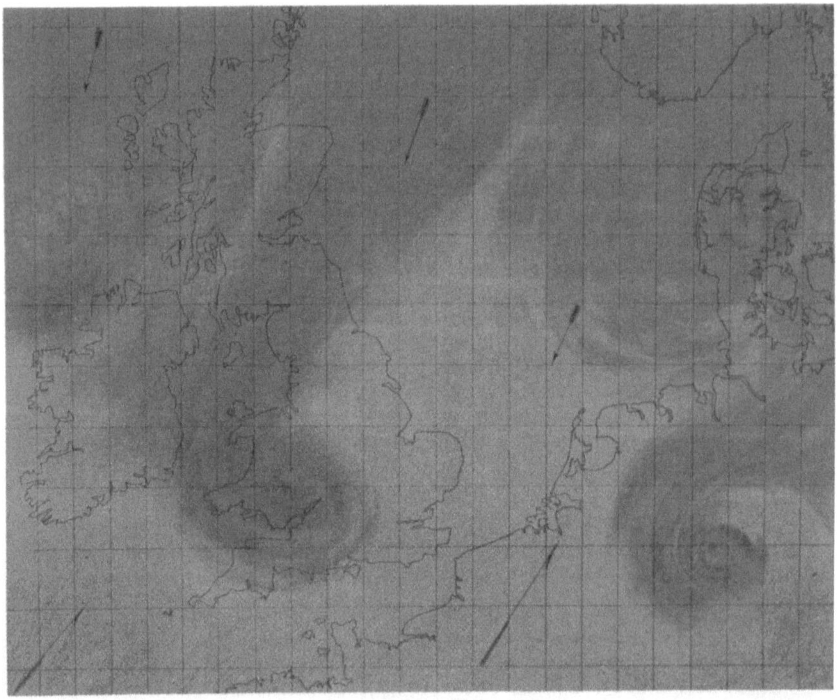

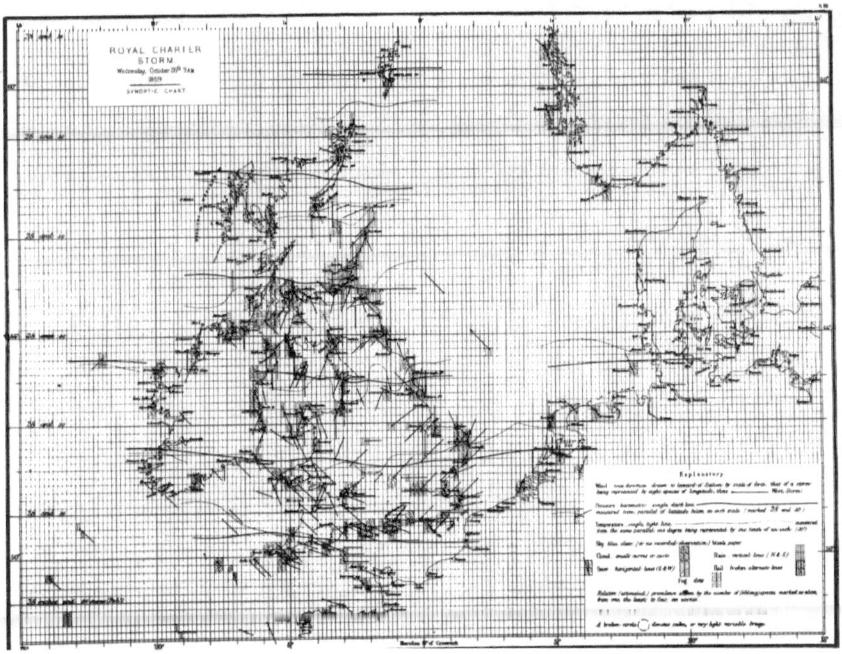

compass. A circle being inscribed in a square of so many degrees, the greatest number of observations under any point governs the scale of equal parts according to which the numbers of winds under each point are laid down, from the centre of the circle *leewardly* toward its circumference, the longest point being always equal to the radius. Through the extremities of these lines, on points or alternate points of the compass, a line is drawn, as a curve is drawn through ordinates, and the resulting diagram (a wind star) shows the *proportion* of winds.

The greater area of the figure being to leeward of the centre indicates, at a glance, the *relative* prevalence of wind from particular points, and its relative duration.

As a circle is said to be generated by the revolution of its radius, so a wind star may be said to result from the motion of a vane; and persistence in any direction may be supposed to increase the length of the corresponding line or point (like the growth of crystals).

The average strength of wind may be shown by parallel lines, or numerically, or by dots; the per-centage of calms by a circle of which the radius equals their number (according to the scale of the diagram); the per-centage of rain by a dotted circle, on a similar principle; and oceanic currents by the usual arrows and numbers. In addition may be given, without overloading the paper, confusing the eye, or using colours,[1] the per-centage of gales or storms (by marks on the point lines), deep-sea soundings, temperature, atmospheric pressure, and the specific gravity of sea water; besides variation of the magnetic needle, and perhaps other information.

Many wind charts for the principal oceans have been published, but they show part only of what may be given at a future time.

Source Endnote

1 Colours increase the expense of printing.

5.3

METEOROGRAPHICA

Editorial Headnote

5.3 Meteorographica

The Eulerian and Lagrangian approaches shaped weather mapping in the nineteenth century, with each side claiming the superiority of its method. These two alternative ways of conceiving changes in atmospheric variables adopted different cartographic techniques: while the former focused on variables fixed at a particular point in space with the barometric gradient as the dominant field, the latter focused mainly on tracking the trajectories of air currents. Between these two methods, however, a middle way emerged, the semi-Lagrangian method, whose clearest exponent was Francis Galton (1822–1911) and his work *Meteorographica, or, Methods of Mapping the Weather: Illustrated by Upwards of 600 Printed and Lithographed Diagrams Referring to the Weather of a Large Part of Europe, During the Month of December 1861.*[1]

The semi-Lagrangian approach entails the construction of iconic maps and schematic micro-maps that represented the flow field by streamlines, while showing changes in temperature and barometric pressure. This approach combines the virtues of its two predecessors: while using observations taken at fixed points close to the ground, it aimed to show an instantaneous picture of the overall flow field and the other meteorological elements. As an intermediate method, it came to prominence with Francis Galton in the 1860s (see 2.4 'Semi-Lagrangian'). The origins of *Meteorographica* date back to July 1861, when Galton issued a circular letter with a map to meteorologists across Europe, inviting them to send him meteorological data for December 1861.[2] By extracting a wealth of information from the simultaneous observations submitted by hundreds of meteorologists, Galton was able to publish a series of ninety-three weather maps of Britain and parts of Europe, illustrated with over six hundred printed and lithographed diagrams in black and white and colour, with texts in two languages, English and German.[3]

Meteorographica uses a fundamentally geometric and visual language and is based on the presentation of numerical and graphical summaries of complex and spatial data. The illustrations reproduced below show different levels of abstraction. The weather maps of 1 and 2 December 1861 are maps with weather glyphs showing rainfall, cloud cover and wind direction and strength by means of a series of templates or stamps, as well as letters indicating the wind direction (N, NW, etc.). The small, iconic three-dimensional maps on the top right show the deviations of barometric pressure from an average using red and black symbols to indicate the degree of divergence. The schematic micromaps, or miniature abstract panels, show the barometric pressure, temperature, wind direction and rainfall conditions, morning, afternoon and evening for each day of the month. The last set of micromaps shows the streamlines, which are the tangents to the wind speed vector. These illustrations allow the reader to follow the movements of the air currents and to observe the replacement of one air current for another based on the analysis of the schematic micromaps.

Weather mapping took many distinctive geometric and visual forms throughout the nineteenth century, and the semi-Lagrangian method of glyph maps and schematic micromaps never became widely adopted in Britain, let alone Europe. Nevertheless, these examples and those in the previous section illustrate how British meteorologists grappled with the search for the trajectories of air currents. Through the writings of FitzRoy before and Galton now, we can trace a growing interest in analysing atmospheric changes in terms of trajectories and streamlines, rather than relying solely on barometric distributions.

Notes

1 Crispin Tickell, 'Meteorographica and Weather', in Milo Keynes (ed.), *Sir Francis Galton, FRS. The Legacy of His Ideas* (New York: Palgrave Macmillan, 1993), pp. 140–51; Aitor Anduaga, *Politics, Statistics, and Weather Forecasting, 1840–1910. Taming the Weather* (London and New York: Routledge, 2020), pp. 186–92.

2 Nicholas Wright Gillham, *A Life of Sir Francis Galton: From African Exploration to the Birth of Eugenics* (Oxford: Oxford University Press, 2001), pp. 141–42.

3 Michael Friendly, 'The Golden Age of Statistical Graphics', *Statistical Science*, 29:4 (2008), pp. 502–35, on pp. 514–16; Katharine Anderson, *Predicting the Weather: Victorians and the Science of Meteorology* (Chicago: The University of Chicago Press, 2005), pp. 78–80.

44

METEOROGRAPHICA, OR, METHODS OF MAPPING THE WEATHER

Francis Galton

Source: Francis Galton, *Meteorographica, or, Methods of Mapping the Weather: Illustrated by Upwards of 600 Printed and Lithographed Diagrams Referring to the Weather of a Large Part of Europe, During the Month of December 1861* (Cambridge: Macmillan, 1863).

INDEX MAPS.

On the opposite page are four Index Maps, which show the names and other data of the meteorological stations selected.

A 1 and A 2 refer to the General Maps, which represent the weather on the morning, afternoon, and evening of each of the 31 days of December 1861.

B and C refer severally to enlarged maps of Britain and the Continent, one or other of which occupies the lowest corner of each right-hand page, and is intended to illustrate some interesting occurrence during one of the six adjacent epochs.

As an example of the meaning of the symbols in A 2, we will take those corresponding to North Unst Light. They are printed $\overset{235\,-\,1^\circ}{9\,.\,3\,.\,9}$; that is to say, 1st, the elevation of that station above the sea is 235 feet; on that account, 2dly, 1° Fahrenheit has to be added to the thermometrical record, if it be desired to reduce the registered temperature to the sea level; and 3dly, 9 A.M., 3 P.M., and 9 P.M. are the hours of observation, in approximate local time. The symbol  that is printed over the figures is merely another way of expressing the 1° and is a rude measure of the height of the station; 0°, 1°, 2°, 3°, 4°, and 5° being represented respectively by the symbols . My object was to give, at a glance, a general notion of the elevation of the area under observation, by the intensity of the shade employed.

WEATHER CHARTS.

On opening the book at any place, we see the morning, evening, and afternoon observations of two consecutive days, making six epochs in all. They occupy the whole of the left-hand page, and the left half of the right-hand one. On the other

DOI: 10.4324/9781003650294-68

half are printed six small symbolic charts, each of which corresponds to one of the foregoing epochs, and affords a convenient summary of its barometrical peculiarities. Red signifies *higher,* and Black *lower* than the average barometrical height (assumed at 29.95 inches). The symbols have been so arranged that the intensity of the shade is proportionate to the amount of deviation from the average.

Symbols in Barometrical Charts.

	Inches. Inches.	Inches. Inches.	Inches. Inches.	Inches.
Black	29.95 to 29.71	29.70 to 29.46	29.45 to 29.21	29.20 and below.
	○	⊙	*	•
Red	29.96 to 30.20	30.21 to 30.45	30.46 to 30.70	30.71 and above.

Below the barometric tables, is printed one of the enlarged maps, as mentioned above.

EXPLANATION OF THE SYMBOLS USED IN THE WEATHER CHARTS.

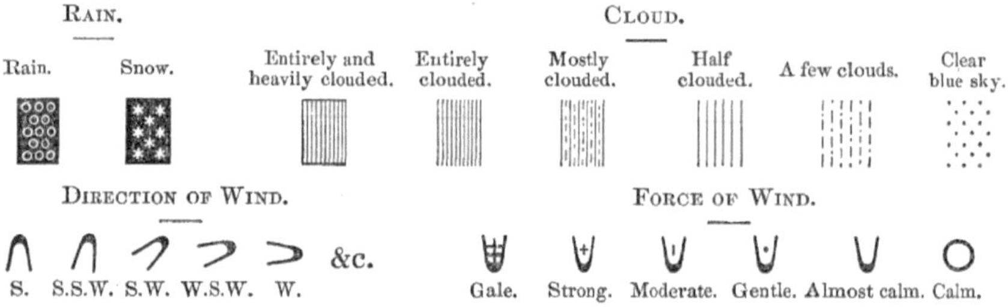

In each compound symbol the upper figures, as $\left\{\begin{matrix}30\ 39\\29\ 72\end{matrix}\right\}$ respectively signify $\left\{\begin{matrix}30.39\\29.72\end{matrix}\right\}$ to be the height of the barometer in English inches.

The lower figures 34–1 mean 34° is the height of Fahrenheit thermometer, and 1° is its excess over the thermometer with a moistened bulb.

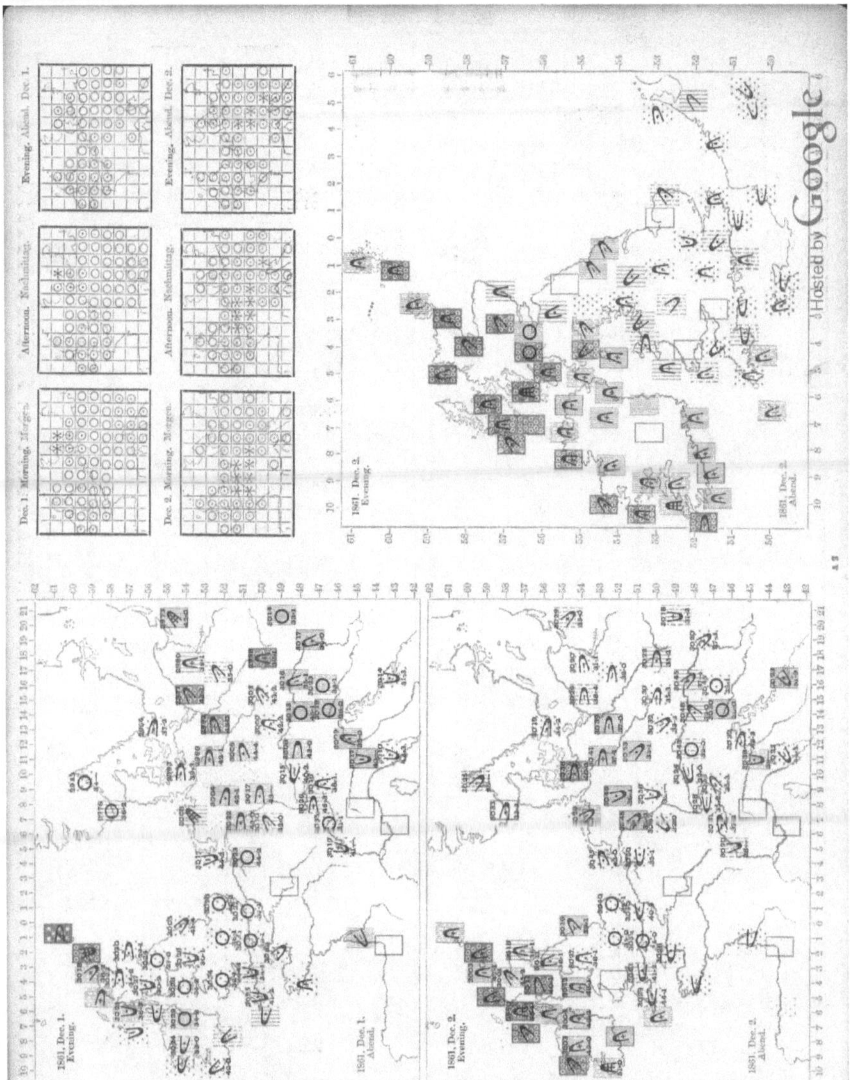

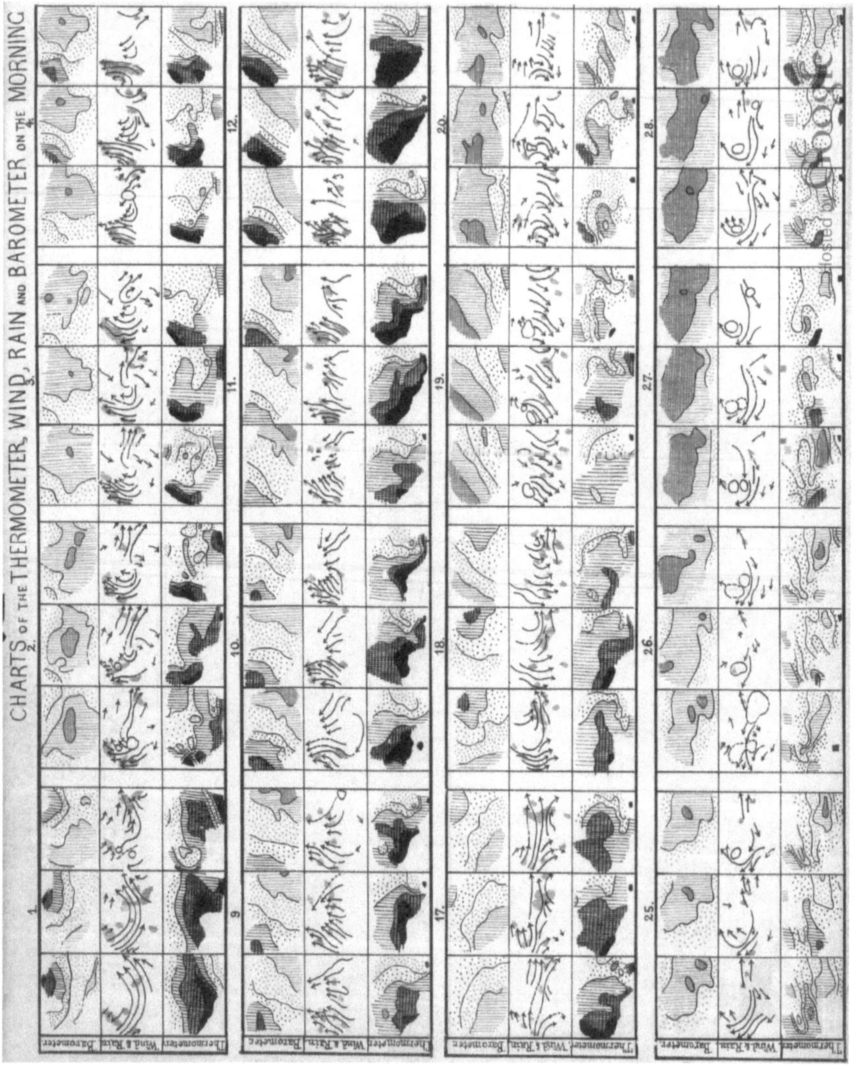
CHARTS OF THE THERMOMETER, WIND, RAIN AND BAROMETER ON THE MORNING
Barometer
Wind & Rain
Thermometer

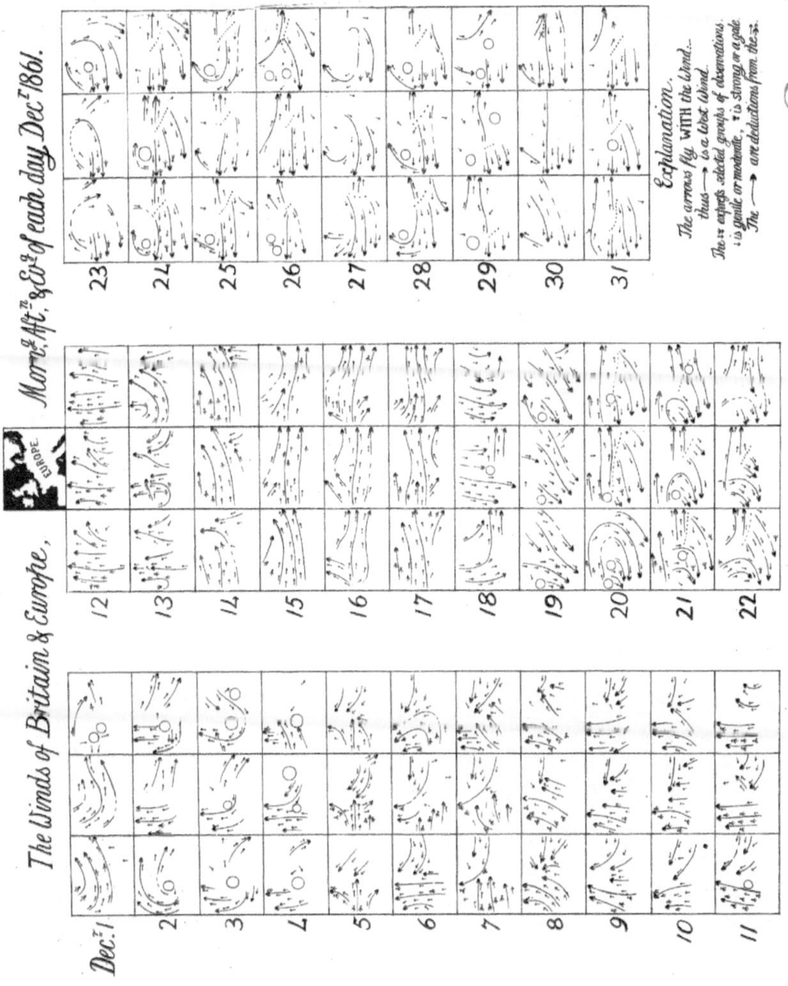
The Winds of Britain & Europe, Morn.g Aft.n & Ev.g of each day, Dec.r 1861.
EUROPE
Dec.r 1
Explanation.

5.4

STANDARDISING

Editorial Headnote

5.4 Standardising

The national meteorological institutes conceived of new standardised forms of weather maps. A key concern was to establish standards and symbols to be used on weather maps, in an attempt to impose an official character on state-sponsored weather maps in their military and agricultural services.[1] By the late 1870s, most Western countries with telegraphic services had established national meteorological institutes. The three texts in this section describe what official weather maps were, or should be, as agreed at international meteorological meetings. The first two (an anonymous text and that of Hugo von Schoder) are official reports of the International Meteorological Congress of 1873, while the third is a text by Mark Harrington (1848–1926) showing the standard form of isolinear weather maps in official institutions.[2]

The first two texts report on the agreements reached at the First International Meteorological Congress held in Vienna in 1873, although only the second deals exclusively with the issue of meteorological mapping. International associationism in meteorology took its first steps in the early 1870s. In the spring of 1873, the Austrian government invited governments with national meteorological services to a congress of directors to be held in Vienna from 2 to 16 September of that year. Most of the twenty-eight items discussed in Vienna had already been discussed at a preliminary conference held in Leipzig in 1872. These items, recorded in the *Proceedings* of the Congress, were translated from the official report and published in *Nature* on 7 May 1874. The Vienna congress, chaired by Christophorus Buys Ballot and attended by thirty-two representatives of twenty governments (all European governments except France and Prussia, plus the USA and China), was a serious attempt to standardise procedures, replacing national modes of notation and graphic forms with a unified international symbology for official meteorological maps and climatological tables. The aim was to develop a new cartographic and graphical culture in meteorology that would benefit from the advantageous aspects of common standards and community meetings. Although the resolutions and standards adopted served only as recommendations – national representatives were free to reject them – many representatives engaged with the resolutions on symbols and graphic forms adopted in Vienna. The standards gradually spread throughout the participating countries so that common features began to appear in the weather maps of all of them.

Hugo von Schoder was a professor at the Polytechnic School in Stuttgart and a delegate of the German Empire who had been appointed to the sub-committees on 'Rain and Evaporation' and 'Observation Hours' in Leipzig. His 'Appendix 1. to the Protocol of the Fourth Meeting' is included in the *Report of the Proceedings of the Meteorological Congress at Vienna*, published in London in 1874 by G.E. Eyre and W. Spottiswoode for H.M. Stationery Office, and translated into English from the original official report. It contains the list of symbols to be used internationally to indicate precipitation types and other phenomena on weather maps, as agreed in Vienna. The instructions on how and where to place them in the publications were so meticulous that, for example, it was recommended that the monthly rainfall *résumé* should show the sum of rain and snow days separately. The appendix details a total of twenty-one symbols, their shape, meaning and use, ranging from a black circle for rain and a star for snow to a black triangle for hail and three parallel lines for fog.

Mark Walrod Harrington, the author of the 1895 article 'History of the Weather Map', seemed to be well acquainted with the dynamics of international meteorological associationism, when two years earlier, as head of the American Weather Bureau, he had requested copies of synchronous weather maps for the dates 1–5 January 1892 from eighteen national meteorological institutes around the world.[3] Harrington was, at the time, a highly respected authority in meteorological circles: founder of *The American Meteorological Journal* in 1884, of which he edited the first seven volumes, he became the first civilian head of the Weather Bureau, having been professor of astronomy and director of the Detroit Observatory at the University of Michigan in Ann Arbor from 1879 to 1891.[4] Copies of the weather maps received were sent to the Weather Bureau library for further examination and comparison. Harrington drew upon a world forum, the International Meteorological Congress held in Chicago from 21–24 August 1893, to present the results of his research, 'History of the Weather Map', which appeared in Volume 2 of the Congress *Report* published by the Weather Bureau in 1895. Harrington compares the content, size and symbols of the various maps, noting that while some countries sought a balance between international standards and their own cartographic techniques (the case of the USA is paradigmatic in this respect), there was generally a uniformity of standard symbols, notations and forms on official weather maps.[5]

In 'History of the Weather Map', Harrington identifies the points of convergence and divergence in the mapping conventions of different countries. He shows that national meteorological services unanimously opt for isolinear mapping, with one map devoted to isobars and winds and another to isotherms and precipitation. Almost all of them also include forecasts or probabilities on their maps, although only France and the USA are able to foretell the occurrence of floods. As evidence of the adoption of uniform formats, all maps have one element in common: a synopsis, which gives an account of the prevailing weather conditions, showing changes over the last twenty-four hours. There are also remarkable similarities in terms of colour and synchrony: virtually all maps are 'black lines on blue water and white land' and maps covering a single country are constructed using synchronous observations (p. 331). Copies of weather maps received at the Weather Bureau enabled Harrington to determine the extent to which the weather maps of the national meteorological institutes reproduced the standard form of isoline maps and the degree to which they combined it with their own cartographic techniques.

Notes

1 Hendrik Gerrit Cannegieter, 'The History of the International Meteorological Organization, 1872–1951', *Annalen der Meteorologie*, 1 (1963), pp. 7–280, on pp. 10–11; Aleksandr Khristoforovich Khrgian, *Meteorology: A Historical Survey*, 2nd ed. (Jerusalem: Israel Program for Scientific Translations, 1970), pp. 133–34.

2 Aitor Anduaga, *Politics, Statistics, and Weather Forecasting, 1840–1910. Taming the Weather* (London and New York: Routledge, 2020), pp. 230–39.

3 Donald R. Whitnah, *A History of the United States Weather Bureau* (Urbana: University of the Illinois Press, 1961), pp. 61–81.

4 John D. Cox, *Storm Watchers: The Turbulent History of Weather Prediction from Franklin's Kite to El Niño* (Hoboken, NJ: John Wiley & Sons, Inc., 2002), pp. 109–16.

5 Willis Luther Moore, *The New Air World: The Science of Meteorology Simplified* (Boston, MA: Little, Brown, 1922), pp. 153–55.

45

REPORT OF THE PROCEEDINGS OF THE METEOROLOGICAL CONGRESS AT VIENNA. PROTOCOLS AND APPENDICES

Source: *Report of the Proceedings of the Meteorological Congress at Vienna. Protocols and Appendices* (London: G.E. Eyre and W. Spottiswoode for H.M. Stationery Off., 1874), translated from the official report – reproduced in 'The Meteorological Congress at Vienna', *Nature*, May 7, 1874, 17–18.

THE METEOROLOGICAL CONGRESS AT VIENNA[1]

THE Meteorological Congress, which held its meetings in Vienna from the 2nd to the 16th of September last, had its origin in a wide-spread conviction that since meteorology can be prosecuted with success only when it is treated internationally, uniformity of procedure among different nations is indispensable; and it was to bring about this uniformity that the Congress was convened. A preliminary Conference was held at Leipsig in August 1872, for the purpose of preparing the programme for the Congress. The Austrian Government issued invitations to other Governments to send delegates to the Congress. To these invitations every European country, except France, responded, and the United States and China were also represented.

The questions which were discussed, and the names of the delegates, have already appeared in NATURE.[2] The following is the deliverance of the Congress on these questions: –

1. A decision regarding the best mercurial barometer for stations of the second order was postponed to a future Congress. Aneroids should not be employed at stations where there is no other barometer, but they may be used as interpolation instruments alongside the barometer.
2. It was considered impossible to lay down fixed rules for general adoption in the protection of thermometers, on the ground that regard must be had to local conditions, and that the mode of exposure which is most to be recommended, in a space which is open and accessible to all winds, and at a height of 4½ to 6 ft., cannot be used everywhere.
3. Casella's minimum, and Hermann and Pfister's metallic thermometer, since they are found to become frequently deranged, cannot be recommended

DOI: 10.4324/9781003650294-70

for stations at which they cannot be kept in proper order and their errors ascertained, for minimum thermometers, amyl-alcohol is to be preferred to ordinary alcohol, as being less liable to distillation. It is recommended that maximum and minimum thermometers be read at the last observation of the evening, and entered on the day on which they are taken.

4. Reference having been made to the experiments on radiation by Symons, Stow, and Soret, further experiments were recommended to be undertaken by physicists, so that the subject might be brought into the sphere of the regular observations.
5. Lamont's method of observing earth temperatures, which consists of a wooden tube, to the bottom, of which the thermometer is let down, and up which it is drawn in order to be read, was recommended as giving more trustworthy results than thermometers with long tubes fixed in the ground. New experiments should be made in different countries, in order to decide the question at what depths observations should be taken.
6. The use of the wet- and dry-bulb hygrometer is in the meantime recommended, and the attention of physicists is drawn to the invention of some new apparatus by which the humidity of the air may be more accurately determined. Hair hygrometers can only be used with safety where care is taken to have their indications compared with those of the wet- and dry-bulb hygrometer, so as to determine their corrections, especially near the point of saturation, where the readings are often too low.
7. It was agreed to introduce the English designations of the directions of the wind: – N, = North, E. = East, S. = South, and W. = West, and to give only sixteen directions of the wind; and in the case of intermediate directions being observed, it is proposed to count them alternately to the one or the other, Lambert's formula is not to be recommended in deducing the mean direction of the wind; but, on the other hand, the frequency and mean force of the winds which correspond to the different directions should be given in numbers. In the distribution in the windrose, those winds whose velocity is less than ½ metre per second, or 2½ English miles per hour, are not to be regarded, but counted as calms. The direction of the cloud-drift should be observed and noted.
8. No general scale for the estimation of wind-force is yet recommended, but it is desirable that a gradual advance be made towards giving the velocity of the wind in metres per second.
9. Wild's apparatus for measuring the force of the wind, already in use in Switzerland, Baden, and Russia, was recommended for introduction at stations of the second order. The velocity of the wind obtained by anemometers should be expressed in metres per second, and tables should be prepared for the mutual conversion of metres per second, kilometres per hour, and English miles per hour.
10. The best form for the receiver of the rain-gauge is a circular one with the area of one-tenth of a square metre, that is, having a diameter of about 14 in. The

receiver of the rain-gauge should be placed at a height of not less than 1, and better, of 1½ metres above the ground, or at a height of from 3 to 4½ ft. In the published results the height above the ground should be stated. Where it can be done, the measurement of the rainfall should be at the end of the fall; in other cases the first observing hour of the day is recommended, in which case the amount is to be put down to the previous day. It is recommended that the duration of the fall be stated in hours.

11. It was agreed to introduce symbols for the character of the precipitation in the "Remarks" column, and to give in the monthly *résumé* the sum of the days of rain and snow separately; to have two columns, one for the quantity fallen, and one for the depth of the unmelted snow; and to give, in the yearly *résumé,* the maximum fall in twenty-four hours for each month. It was further recommended to state the number of days when the fall is less than 0·04 in. and 0·01 in.
12. Hail is defined to be as a precipitation of frozen water, in which the stones attain such a size that they may be expected to do damage to agricultural products.
13. (*a*) In order to obtain data regarding thunderstorms which admit better of comparison, it is recommended only to count the days of thunderstorms, but this is not intended to prevent individual observers from inserting in the column of "remarks," in addition, the number of the storms, the time of their commencement, their duration, direction of motion, &c.

    (*b*) As days of thunderstorm, only those are to be noted on which both lightning and thunder have been observed. If only lightning without thunder has been noticed, the entry for the day should be sheet lightning.
14. As regards evaporation, the evaporating dish should not be less than seven inches in diameter, and it is indispensable that it be absolutely identical as regards diameter and depth at all stations, if comparability is aimed at. The level of the water in the dish must remain constant, for the obvious reason that the evaporation is less the deeper the surface of the water stands under the edge of the vessel. Provision must be made for reading off the quantity evaporated with accuracy. The measurement of evaporation by means of floating apparatus on large surfaces of water should be introduced wherever possible.
15. (*a*) The degree of cloudiness is to be given by the figures 0–10, in which 0 represents a sky quite free from cloud, and 10 an entirely overcast sky. These figures refer only to the extension and not to the thickness of the cloud, the latter being indicated by accompanying expressions, such as "slight," "great," &c.

    (*b*) Arbitrary symbols representing rain, snow, fog, &c., were adopted.
16. It was resolved that the institution of observations on atmospheric electricity be recommended only for head observatories. As regards ozone, the existing methods of determining its amount in the atmosphere are insufficient, and the Congress therefore recommended investigations for the discovery of better methods.

17. It was agreed that for observations as well as for publications, the use of the same units of measure is desirable; that among all existing systems of measure the metric has the best prospect of universal adoption; that it is most desirable, if it be not possible to introduce uniform measures at present, to use henceforth only metric and English measures (with Celsius and Fahrenheit scales), and that all action is to be supported which tends to the introduction of the uniform metric system. It was also agreed that the Jesuits of observations, or the means, should be published in the metric scale as well as in the original scales.
18. The hours of observation should be chosen which give a close approximation to the true mean temperature of the day.
    The following are the suitable combinations: –

h.	h.	h.	h.	h.	h.		h.	h.
6	2	1	8	2	8	with min. temp.	8	8
7	2	1	9	3	9		9	9
7	1	9	10	4	10		10	10
7	2	9						

    Observations should be set on foot at a number of normal stations, especially in Turkey, East Indies, Australia, Southern States, and Brazil, in order to ascertain the corrections for the most important meteorological elements, such as temperature, pressure, and humidity.
19. As units of time should be chosen (1) the mean solar day of the place of observation, reckoned from midnight to midnight; (2) the civil year; (3) the civil months everywhere, the calculation of the monthly means being simply arithmetical; and (4) Dove's 5-day means (73 in the year) for a selected number of stations of each country. It is proposed to count the first 12 hours oi the day, from 1 to 12, as forenoon; and the following 12 hours, from I to 12, as afternoon; thus counting 12 o'clock midnight as the end of the day, and 12 o'clock noon as the close of the forenoon.
20. It is resolved to choose, as the periods for calculation of normal values, intervals of five years to be called *Lustra,* so that the next Lustrum will begin with January 1, 1876; and that as regards the more important data, old observations should be calculated in accordance with this proposal.
21. The existence of a system of weather telegraphy is, for all countries, considered to be a necessity; in addition to the direction and force of the wind, the barometric gradients at the time of observation should also be added. For purposes of storm warnings, the reduction of the barometer readings to mean sea-level for places not above 1,000 feet in height is admissible. For greater heights, the gradients are to be referred to the mean normal heights of the barometer at the stations. The relations of temperature, moisture, rain, cloud, and state of the sea and tides to storms, are recommended for investigation.

As regards storm warnings, each director should give his opinion on the probable course of atmospheric disturbances which are expected, or have already commenced, not as prophecies, but as *probabilities.* Only wind-force of 8, and upwards, of Beaufort's scale should be announced.

22. As regards maritime meteorology, it is desirable that each country should, if possible, collect all its meteorological observations at one place, and that the Institute for Maritime Meteorology should be established as near as possible to the sea, and that this institute might best be placed under the general management of the chief institute of the country. The convening of a maritime meteorological conference was declared to be desirable, and the preparation for this conference is entrusted to the permanent committee appointed by the Congress.
23. It is necessary that in every country, at least one but in case of necessity several central institutions should be established for the management, collection, and publication of meteorological observations.
24. The verification of all instruments supplied to meteorological stations, and the inspection of stations yearly, but at least once in the course of every five years, is necessary. With regard to instrumental errors detected on verification, or inspection, corrected results only should be published. It is intended that the Permanent Committee prepare, in conjunction with the other members of Congress, instructions for the institution and discussion of meteorological observations.
25. As regards standard barometers and thermometers, each central office is recommended to adopt a real standard barometer, *i.e.* an instrument which allows of the determination of atmospheric pressure according to its definition in absolute measure, and to prepare a standard thermometer on scientific principles.
26. The publication of observations at stations of the first order should be entirely separated from those of stations of the second order. It is handed over to the Permanent Committee to prepare, in conjunction with members of Congress, a form of publication suited for international purposes.
27. It is desirable to organise, on the model of the Smithsonian Institution at Washington and the Central Bureau at Haarlem, a similar office for the exchange of publications in every country.
28. A Permanent Committee of seven, with the right of increasing their number to nine, was appointed, with Dr. Buys Ballot as president. The duty of this committee is to care for the carrying out of the decision of the Congress, and arrange for convening a future Congress; and it shall place the delegates of the Congress in cognisance with its action and proceedings.

For the extension of meteorological knowledge it was recommended that stations provided with self-registering instruments be established on high mountain-tops; that experiments on the possibility of continuous meteorological observations with captive balloons be instituted; that stations be established in the North Polar regions, and also in the high southern latitudes;

on the north coast of Africa; that the organisation of the stations ill Turkey be made more complete, especially the Central Observatory at Constantinople, and that the meteorological station at Athens be maintained.

29. The establishment of an International Institution for the Advancement of Meteorology was declared to be really useful and desirable, and it was remitted to the Permanent Committee to prepare a detailed scheme for this purpose for the consideration of a future Meteorological Congress.

Source Endnotes

1 "Report of the Proceedings of the Meteorological Congress at Vienna." Protocols and Appendices. Translated from the Official Report. Published by the authority of the Meteorological Committee. London, 1874.

2 Nature, vol. viii. p. 468.

46

APPENDIX 1. TO THE PROTOCOL OF THE FOURTH MEETING

Hugo Von Schoder

Source: Hugo von Schoder, 'Appendix 1. to the Protocol of the Fourth Meeting', *Report of the Proceedings of the Meteorological Congress at Vienna. Protocols and Appendices* (London: G.E. Eyre and W. Spottiswoode for H.M. Stationery Office, 1874), pp. 48–49.

APP., IV. 1.

Symbols, pp. 19, 29, and 44.

REPORT of the SUB-COMMITTEE chosen at LEIPZIG on Question 15, discussed at the Meeting of September 7th; 10 a.m. till 1 p.m.

Present: the Sub-Committee, consisting of –

Ebermayer, President.
Schoder, Reporter.
Sohncke,

and in addition the co-opted members,

Wild, Lorenz, Plantamour.

1. For the designation of the hydrometeors and other phenomena, the following symbols were proposed:

   1. The following Symbols are proposed for the indication of Hydrometeors and other phenomena: –

Rain	●	Hail	▲
Snow	✱	"Graupel"	△
Thunderstorm		Mist, Fog	≡
Lightning without Thunder, or Sheet Lightning.		Hoar-frost	⊔
		Dew	

DOI: 10.4324/9781003650294-71

[1] Saver-thaw ("Rauh-frost," "Duft")	V	Solar Halo	O
Glazed Frost ("Glatteis")	~	Lunar Corona	[symbol]
Snow-drift	[symbol]	Lunar Halo	[symbol]
Ice crystals	←	Rainbow	⌒
Strong wind	[symbol]	Aurora	[symbol]
Solar Corona	⊕	Dust haze ("Höhen-rauch")	∞

APP., IV. 1.

Symbols, pp. 29, 44.

In respect of their intensity, the individual phenomena are to be distinguished by the figures 0 and 2, which should be used as exponents of the symbols, in such a way that 0 should indicate very slight, 2 strong.

e.g. $●^0$ slight Rain.
$●^2$ heavy Rain.[2]

It is further to be remarked that Fog (mist) is only to be entered when the observer is quite enveloped in it. Höhenrauch (dust haze), moreover, is not only to be indicated by the proper symbol, but at the same time the concomitant obscuration of the atmosphere is to be noticed under the head of Amount of Cloud.

CLOUDS.

Clouds, p. 15.

1. Entries as to the extension of clouds on the visible sky according to the scale 0–10 are to be made without reference to the thickness of the cloud. The latter is to be indicated by an exponent applied to the figure for the amount of Cloud (0 slight, 2 great).
2. As to the Forms of Cloud, or rather, classes of cloud, the matters most urgently desired are more thorough observations and more accurate drawings, which will correspond to the variety of clouds which are in reality observed; as all the systems which have as yet been proposed, *e.g.* that of Poëy, are not of such a nature that they can at once be recommended for general adoption.
3. For the present it is recommended to add to Howard's designations and their combinations, epithets as characteristic as possible, according to the choice of the individual observer, in order to express the actual appearance as clearly as possible.

4. In order to facilitate a correct understanding and designation of the clouds, it is recommended,

   (*a*.) That Central Offices should prepare as completely as possible lists and characteristics of the forms of cloud which occur in the district.

   (*b*.) That on the part of the Congress the preparation of sketches of the principal forms of cloud be set on foot, which should be added to the Instructions for the observers.

   (*c*.) That the study of the connexion between the form, constitution, and origin of the clouds be undertaken and supported, special attention being paid to the circumstance that one and the same mass of cloud presents a different appearance when seen from different sides or under different angles.[3]

Reporter, SCHODER.

Source Endnotes

1 "Rauh-frost" or "Duft-anhang" is the phenomenon of the deposition of a large quantity of frozen moisture on trees when the weather suddenly becomes warm after great cold, sometimes called, as above, "Silver-thaw."

2 Adopted at the Seventh Meeting (p. 29).

3 Adopted at the Fourth Meeting (p. 20).

47

HISTORY OF THE WEATHER MAP

Mark Harrington

Source: Mark Harrington, 'History of the Weather Map', in Oliver L. Fassig (ed.), *Report of the International Meteorological Congress, Chicago, Ill., August 21–24, 1893* (Washington D.C.: Weather Bureau, Part 2, 1895), pp. 327–335.

The history of meteorology falls into three distinct periods: 1. That preceding the discovery of the barometer in 1643, during which time very little real advance was made. It was practically a folklore stage. 2. The period from the discovery of the barometer to that of the general recognition of the cyclonic character of storms. The latter was the work of many men, among whom stand pre-eminent Dove, Redfield, and Espy, and the date is not definite, but may be placed at 1830. 3. The development of the cyclonic theory, from 1830 to the present time, 1893.

One of the most important elements in the development of the third and last stage was the use of the synchronous weather map. Indeed, it was by the employment of such maps (made long after the occurrence of the phenomena which it was desired to investigate) that Redfield, Espy, and others made their principal discoveries. It was not, however, until the current weather map was made – the weather map which is made at the time of the phenomena themselves – that the full harvest of facts due to it was reaped. The harvest obtained from the current weather map was so great that it has laid the basis for the next, and impending, great advance in the science, viz, its full dynamical treatment. Its importance will therefore justify a discussion of its history, and it is to this that this paper is devoted.

For the current weather map two elements were needed, simultaneous observations and immediate collection of the data. Meteorological observations were usually taken at the same local hours, but differing in absolute time. Those of eastern Russia were thus (if at the same local hours) three or four hours earlier in absolute time than those of western Europe, and those of the Atlantic coast of the United States about as much earlier than those on the Pacific coast. The need of absolutely simultaneous observations was recognized early. From 1772 to 1777 Thomas Jefferson (afterwards President of the United States) and James Madison (afterwards a bishop of the Episcopalian Church) took simultaneous observations at Monticello and Williamsburg, both in Virginia, and about 120 miles apart.[1] This appears to have been at the instigation of Jefferson, who was often far in

 DOI: 10.4324/9781003650294-72

advance of the world on physical as well as political matters. Jefferson drew some interesting conclusions from the observations. About the same time it appears that the idea had occurred to Lavoisier who proposed (probably before 1784) that instruments for such observations should be scattered over France, Europe, and the world generally, and refers to an earlier proposal of this sort by Borda.[2]

Lavoisier declared, with wonderful foresight, the possibility of forecasting the weather by such proceedings as these, for at least a day ahead, and wrote, "One would think that it would not be impossible to publish each morning a journal of predictions which would be of great use to society."[3] This was, however, impracticable without some rapid means of communication. The matter seems to have appealed so strongly to the public as the case of a want which was decidedly felt that several years after (1793) M. Romme, a deputy, was charged to present to the French Constituent Assembly a scheme for an aerial telegraph, invented by Chappe, approved by the united committees on public instruction and on war. Lavoisier himself became a victim of the French Revolution, and in the political cyclone of the time the plan dropped out of view. The first renewal of a plan of rapid collection of meteorological data which I have been able to find was that of Kreil in 1842.[4] Kreil maintains that the method of aerial telegraph is not sufficient, and proposes the use of an electromagnetic telegraph. This was after the publication of the invention of the telegraph by Morse (1837), but about coincident with its first experimental test between Washington and Baltimore (1842). A similar suggestion is said to have been made by Piddington in the same year, but I have not been able to verify it.[5] In 1846 the same suggestion was made by Redfield[6]; in 1847 by Loomis[7] (by which time the telegraph was in common use), and in 1848 by John Ball in a paper to the British Association for the Advancement of Science.[8] From 1847 on for several years Prof. Henry reported from year to year recommendations for such use, or tentative arrangements which had been made with the telegraph companies. Finally, the proposals took the form of a regular telegraphic service, and this plan was preached by Maury with great persistence from 1851 on for several years.[9]

With the idea of simultaneous observations and the completion of the telegraphic systems the road was clear for the making of the weather map. As a matter of fact, however, this had been invented and made use of long before the last requirement was fulfilled. The earliest proposal of this sort which I have seen was one by Brandes, in a letter from Breslau, dated December 1, 1816, printed in Gilbert's *Annalen der Physik*[10] in 1817. Brandes there states that he had tried making daily weather maps from newspaper reports and had found them unsatisfactory. He then proposes to get together a service of 40 or 50 stations, scattered from the Pyrenees to the Urals, and make from their reports (received, of course, by mail) a chart for each day in the year. This plan seems never to have been carried out, as his *Beiträge zur Witterungs-Kunde* (1820), which was a study of the weather of 1783, was carried out on the plan of diagrammatic, not geographic representation.

The ideas of the weather map are very similar to those involved in the special studies of storms by Redfield, Piddington, and others, though the latter are special, being the study of individual storms rather than that of the combined elements involved in weather. Still more like the weather maps of to-day are those prepared by Espy. Indeed, Prof. Henry was accustomed to say that Espy was the father of the present weather map, and Mrs. Moorhead quotes him in the following words[11]: "Charts now used in the service (Signal Service) were identical (with some slight modifications) with those the old Storm King (Prof. Espy) constructed for use in the Meteorological Bureau of the War Department when he was at its head."

It was not, however, to the best of the information available, until some time in 1856 that current charts were made of the weather, and this was done by the Smithsonian Institution in Washington. Prof. Henry's own words[12] on this subject are: "The first practical application of the principle we have mentioned was made by this Institution in 1856; the information conveyed by telegraphic dispatches in regard to the weather was daily exhibited by means of different colored tokens, on a map of the United States, so as to show at one view the meteorological condition of the atmosphere over the whole country. * * * The system, however, was necessarily discontinued at the beginning of the war, and has not yet been resumed." These maps were apparently not drawn and naturally were not preserved.

On February 26, 1858, Le Verrier began charting the telegraphic data received by him.[13] These were in manuscript and apparently were never published. Le Verrier had studied by graphic methods the storm of November 14, 1854, and on February 16, 1855, he submitted to the Emperor a plan for a meteorological telegraphic network over France. In 1856 he began such a system with thirteen stations, reporting by telegraph and eleven by post. He began publishing an international bulletin in 1857 which he made daily on the beginning of 1858. His first predictions for ports were made in 1860. The synoptic charts or weather maps were not published by him until 1860. On September 11, of that year, he printed and published the isobars and winds for September 7 and September 10, one and three days after their respective dates. On September 16, 1863, he printed the weather map for that day and this has been continued, I believe, without interruption from that day to this.

This was the first, current weather map published. It gives daily the isobars and winds for central and western Europe, from St. Petersburg to Palermo and westward. It is in black and white, 170 millimeters broad by 140 high. The winds are indicated by arrows which fly with them, the velocity being indicated by feathers on a scale of 6. The isotherms and clouds and rains were not entered until April 24, 1878, and were then given by means of an additional map. From that time to this the *Bulletin International* has contained two maps.

In 1869 Prof. Cleveland Abbe, with the assistance of the Western Union Telegraph Company, began the collection and use of telegraphic weather reports at Cincinnati, Ohio, and on February 2, 1870, Mr. Armstrong, local manager of the telegraph company, undertook the making of weather maps and their multiplication to a limited number by a manifold process. These maps were continued until October 10, 1870, when they were relinquished on account of the daily maps then undertaken by the Signal Office. These maps gave about 30 stations from the Atlantic seaboard westward to Houston, Tex., Omaha, Nebr., and Cheyenne, Wyo. They extended northward to Halifax, N. S., and southward to Habana, Cuba. Their size was 386 by 315 millimeters. They gave the temperatures, cloudiness, rain, snow, and direction of the wind, but not the pressures. No météorologic lines were entered on the maps. These were the first current weather maps in the United States, and copies are now very rare. For the opportunity to examine the set, I am indebted to Prof. Abbe.

The official American system of weather maps began with the tridaily maps, November 1, 1870. They were in manuscript, and were made both at Washington, D. C., and Chicago, 111. They were multiplied so as to be properly published only at Washington from January 14, 1871, and the printing of the morning maps began on May 2, 1871. The first maps contained reading of barometer and thermometer, wind direction, clouds, and precipitation, but no lines. The isobars began to be inserted on October 30, 1871, and the isotherms April 15, 1872.

The next series of current weather maps was that of the British Meteorological Office. They first appeared printed in the bulletin for March 23, 1872, though from a report of the Meteorological Committee[14] of the Royal Society it appears that they began on March 11. Probably the earliest maps were in manuscript. The maps of these bulletins were daily, and consisted of four at each issue; the first for isobars; the second, isotherms; the third, wind and sea; and the fourth, clouds, rain, etc.

All the current weather maps issued, with dates at which they were begun, are given in Table I.

Copies of these maps for the dates January 1–5, 1892, were kindly sent, on request, to the Weather Bureau library where they have been compared. The only exception to the above date has been for the Spanish maps, which were begun in 1893. The copy of this map examined is for April 2, 1893.

In the case of the Belgian, British, French, and Saxon services the weather map is in two equal parts, one for isobars and winds, the other for isotherms and (usually) precipitation. In the Bavarian issue the isothermal map is smaller than the other. In the German issue, besides the two larger maps for the day of issue, there are two smaller ones for the day before, making four maps with each issue. In the Italian issue there is an additional smaller map (three in all) of the isobars placed with the forecasts.

All are in black lines on blue water and white land, except the German which is black and white, the Swedish which is blue on white, and the Australian which is red and purple on white sea and yellow land.

In size the maps vary from 565 by 400 millimeters (American) to 98 by 134 millimeters (British). The Australian is next in size to the American (550 by 387 millimeters).

In Japan the map is yet tri-daily. In America it was formerly tridaily, but is now bi-daily. The Russian maps are bi-daily. The remainder (15 in number) are daily, and all, except the Australian, are issued on Sunday.

The hours of observation, outside of Europe are:

For the United States, 8 a. m. and 8 p. m., seventy-fifth meridian time.
For Australia, 1 p. m. and 4 p. m.
For Japan, 6 a. m., 2 and 10 p. m., one hundred and thirty-fifth meridian time.
For India, 8 a. m.
For Algeria, 7 a. m. (Same as France.)
For European countries, the hours are:

7 a. m., Austria, France, Russia (and 9 p. m.), and Switzerland.
8 a. m., Germany, Italy, Bavaria, Sweden, Great Britain, Belgium.
9 a. m., Spain.

For Europe, therefore, the maps that cover more than one country are not made with sychronous observations. If the above times are local (proper), then the observations would be in Greenwich time from 2 a. m. (in western Asia) or 5 a. m. (in western Russia) to 8 a. m. (in western Europe), a difference of three to six hours. If they are the times of the respective capitals, the difference in absolute time would be about three hours. If all are in Greenwich time, the difference is about two hours.

The territory covered by the maps is as follows:

Algerian. – Tripoli to Finland, westward.
American. – United States and southern Canada.
Australian. – Australia and New Zealand.
Austrian. – Greece to Russia, westward, except Spain.
Bavarian. – Italy to Finland, westward, except Spain.
Belgian. – Bavaria to Scandinavia, westward.
British. – Bavaria to Gulf of Bothnia, westward.
French. – All Europe and Algeria.
German. – Sardinia to North Cape; Ireland to Kiev.
Indian. – India.
Italian. – France to central Austria, southward.
Japanese. – Japan.
Russian. – France to North Cape, eastward to Obi River.
Saxon. – Austria to Finland, westward.
Spanish. – Spain.
Swedish. – Finland to Saxony, westward.
Swiss. – Greece to Finland, westward, except Spain.

The forecasts or probabilities are given on all the maps but the Algerian. In the American, Australian, Bavarian, Belgian, German, Japanese, Russian, and Swiss, they are definitely for "to-morrow." For the British, they end at noon "to-morrow." In the Austrian, French. Indian, Italian, Saxon, and Spanish, they are indefinite as to time covered. The floods are forecasted in France and the United States.

A synopsis is published on all the maps and comprises a statement of weather conditions prevailing with reference generally to the changes in the last twenty-four hours. Often these latter statements are tabulated. The tabular statement, with or without change since last map, is given in each case for a number of stations varying from 5 for Belgium and 12 for Saxony to 123 on the French map, 128 on the Indian, 130 on the American, and 138 on the Russian.

The wind scale employed is in 3 grades on the German, Saxon, and Swiss maps; 4 on the Australian, British, Italian, Spanish, and Russian; 6 on the Austrian, Bavarian, Belgian, French, Japanese, and Swedish; and 7 on the Indian maps. It is not given on the American maps. The international symbols are usually employed on the European maps but not elsewhere. The type for the Australian and Japanese maps are the American symbols. The Australian map makes an excellent distinction between the rain which has fallen and that which is falling. The American symbols include a shading for a rain area and an inclosing line for the region where the temperature has fallen or risen 20° F., or more, in twenty-four hours.

There are some special features of individual maps which deserve special note. The French map is the most truly international. It even contains the general conditions in the United States for the day before. The British gives its "probable changes" and forecasts separately. The Russian map has its explanations in Russian and French. On the Australian map the forecasts for different colonies are made by different persons. The Swiss map gives a tabular statement of sunshine at 6 stations. The Belgian and Algerian give the automatic trace of barometer, thermometer, and hygrometer for the preceding twenty-four hours. The Saxon map has 9 types of forecasts which it gives for the four quarters of the day at each of 9 stations. The British map is reprinted in the London "Times."

In addition to the bi-daily weather map issued at Washington, 73 stations of the United States issue maps independently, each for a considerable area about the station, as a center. There were thus issued 8,830 maps, daily, on June 1, 1893, of which 6,257 were issued in the morning and the remainder in the evening. For details of this service see Table II.

These maps are of uniform size, 750 by 325 millimeters. They are in black lines on white and blue, like the Washington map. The isobars, isotherms, wind, state of sky, and rain are given, with a brief tabular list of rainfall, weather, or river notes, on the right-hand margin, and a forecast for a limited area for the next day in the lower left-hand corner. The rain areas are given on many, and on some are marked out areas of decided rise or fall of temperature. These are the maps which are most broadly distributed to the public in the United States. They are multiplied by the milliographic process, and in some cases are reduced and printed in daily papers of large circulation.

In addition to the current weather maps already noted, there are others, the issue of which is delayed, though they are in all proper senses, daily weather maps. Among these the most important are:

1. *The Standard Weather Chart of Australasia.* – This is issued by the Queensland Post and Telegraph Department at Brisbane, Australia. It is made

by Mr. Clement L. Wragge. A rough map is made on time and posted in Brisbane. The completed map is held for ships' news. It is the largest weather map issued, being 23 by 18.5 inches. It covers the whole of Australasia and surrounding region from E. 40° to W. 140°, and from S. 60° to N. 40°. The issue was begun in 1887. The map examined was that for March 28, 1892.

2. *Hoffmeyer's Cartes Synoptiques Journalières.* – These charts were made by Dr. N. Hoffmeyer and issued by the Institut Météorologique Dansois, Copenhagen, Denmark. They covered the North Atlantic and the most of Europe, and are daily maps for the period from September 1, 1873, to November 30, 1876. They were published from three to six years after their date. Continuations of this series were made with the aid of the Hamburg *Deutsche Seewarte* from December 1, 1880, to November 30, 1884.
3. *Atlas des Mouvements Généraux de l'Atmosphère.* – These maps were issued by the Observatoire Impériale de France. They covered the North Atlantic Ocean and western Europe. They were daily maps from June 1, 1864, to December 31, 1865, and were printed at Paris, 1868-'69.
4. *Synchronous Weather Charts.* – These were issued by the Meteorological Council, London, England, covering daily the North Atlantic and adjacent land surfaces for the thirteen months, August 1, 1882, to August 31, 1883. They were printed in London in 1886.
5. *Bulletin of International Meteorological Observations taken Simultaneously.* – These were taken daily and published by the Signal Service from six months to one year after date. In 1882 the title was changed to "Bulletin of International Meteorology." Quarto charts were published from October 1, 1877, to December, 31, 1884, and from October 1, 1886, to December 31, 1887. The maps from January 1, 1885, to September 30, 1886, remained in manuscript. A folio edition was printed separately from July 1, 1884, to December 31, 1884, and from October 1, 1886, to December 31, 1887. The territory covered was the entire Northern Hemisphere.

In conclusion, it is interesting to see what parts of the earth's surface have been covered by weather maps and for how many years they are available. They are as follows:

Europe, since September 16, 1863.
North Atlantic, 1864 (from June 1), 1865, 1873 (from September 1), 1874, 1875, 1876 (to November 30), and from October 1, 1877, to December 31, 1887.
United States and southern Canada, since January 1, 1871.
Japan, since March 1, 1883.
Western Siberia, on the Russian map, since May 12, 1889.
Australia and New Zealand, since January 1, 1877.
Australasia, since 1887.
India, since October 13, 1887.
Indian Ocean, February, 1861, and since 1887.
Algeria, since April 8, 1877.
Northern Hemisphere, from October 1, 1877, to December 31, 1887.

TABLE I.—*Current weather maps.*

Minneapolis, Minn	a. m.	....	130	....	130	Oct. 25, 1890	....	....	
Mobile, Ala	a. m.	....	36	....	36	Apr. 10, 1888	July 19, 1888	July 20, 1891	
Montgomery, Ala	a. m.	....	26	....	26	July 1, 1887	Jan. 24, 1888	May 16, 1892	
Nashville, Tenn	a. m.	....	56	....	56	Oct. 1, 1887	....	....	
New Haven, Conn	a. m.	....	169	....	169	Sept. 1, 1887	Oct. 19, 1888	Mar. 2, 1891	
New Orleans, La	a. m.	....	121	....	121	July 1, 1887	....	....	
New York, N. Y	a. m.	p. m.	313	349	662	Aug. 2, 1886	....	....	
Norfolk, Va	a. m.	....	90	....	90	July 1, 1887	....	....	
Omaha, Nebr	a. m.	....	119	....	119	Nov. 7, 1891	....	....	
Parkersburg, W. Va	a. m.	....	32	....	32	Aug. 13, 1891	....	....	
Pensacola, Fla	a. m.	....	25	....	25	July 15, 1889	....	....	From June 15 to September 15, annually.
Philadelphia, Pa	a. m.	....	193	....	193	Sept. 1, 1886	Oct. 19, 1888	Nov. 21, 1889	
Pierre, S. Dak	a. m.	....	62	....	62	Jan. 5, 1893	....	....	
Pittsburg, Pa	a. m.	....	116	....	116	July 1, 1887	....	....	
Portland, Me	a. m.	....	122	....	122	Oct. 1, 1887	....	....	
Portland, Oreg	a. m.	....	81	....	81	Oct. 15, 1892	....	....	
Raleigh, N. C	a. m.	....	75	....	75	Mar. 29, 1892	....	....	
Rochester, N. Y	a. m.	....	97	....	97	July 1, 1887	Sept. 15, 1888	July 8, 1892	
St. Louis, Mo	a. m.	p. m.	205	122	327	July 1, 1887	....	....	
St. Paul, Minn	a. m.	....	113	....	113	Dec. 13, 1889	....	....	
Sandusky, Ohio	a. m.	....	38	....	38	July 1, 1887	....	....	
San Francisco, Cal	a. m.	p. m.	259	95	354	Apr. 17, 1891	....	....	
Sault Ste. Marie, Mich	a. m.	....	60	....	60	May 25, 1892	....	....	From December 1 to April 1, annually.
Savannah, Ga	a. m.	....	36	....	36	July 1, 1887	....	....	
Shreveport, La	a. m.	....	59	....	59	Apr. 1, 1891	....	....	
Sioux City, Iowa	a. m.	....	169	....	169	May 1, 1890	....	....	
Springfield, Ill	a. m.	....	39	....	39	Feb. 10, 1887	....	....	
Springfield, Mo	a. m.	....	27	....	22	Oct. 13, 1892	....	....	
Tampa, Fla	a. m.	....	32	....	32	Nov. 22, 1890	....	....	From April 1 to October 31, annually.
Toledo, Ohio	a. m.	....	106	....	106	July 1, 1887	....	....	
Vicksburg, Miss	a. m.	....	180	....	180	May 26, 1890	....	....	
Wilmington, N. C	a. m.	....	74	....	74	July 1, 1887	....	....	
Total	....	....	6,257	2,573	8,830				

NOTE.—Oswego, N. Y., published a. m. maps from July 1, 1887, to September 21, 1887; Baltimore, Md., from October 15, 1886, to October 19, 1888, and Port Huron, Mich., from June 1, 1888, to December 10, 1888. Number of stations issuing a. m. maps, 58; p. m. maps, 4; both a. m. and p. m. maps, 11; total, 73.

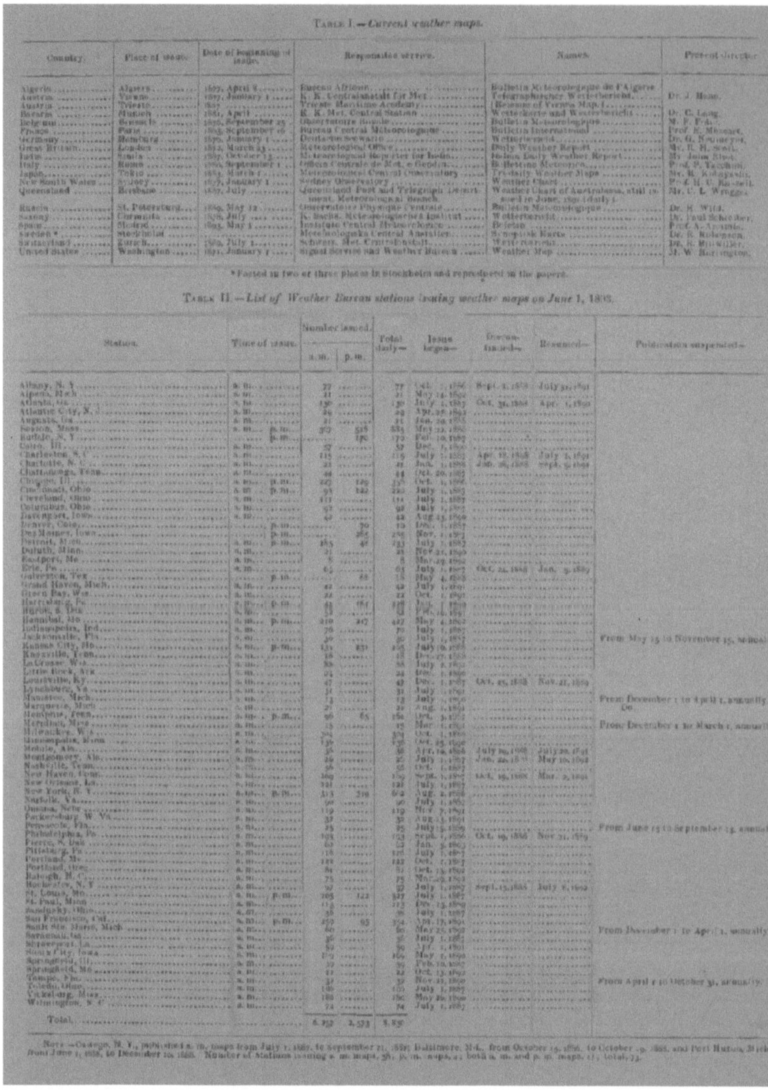

TABLE I.—*Current weather maps.*

Country.	Place of issue.	Date of beginning of issue.	Responsible service.	Names.	Present director.
Algeria	Algiers	1877, April 8	Bureau Africain	Bulletin Météorologique de l'Algérie	
Austria	Vienna	1877, January 1	K. K. Centralanstalt für Met.	Telegraphischer Wetterbericht	Dr. J. Hann.
Austria	Trieste	18[illegible]	Trieste Maritime Academy	(Reissue of Vienna Map.)	
Bavaria	Munich	1881, April	K. K. Met. Central Station	Wetterkarte und Wetterbericht	Dr. C. Lang.
Belgium	Brussels	1876, September 25	Observatoire Royale	Bulletin Météorologique	M. F. Folie.
France	Paris	1863, September 16	Bureau Central Météorologique	Bulletin International	Prof. E. Mascart.
Germany	Hamburg	1876, January 1	Deutsche Seewarte	Wetterbericht	Dr. G. Neumayer.
Great Britain	London	1872, March 25	Meteorological Office	Daily Weather Report	Mr. R. H. Scott.
India	Simla	1887, October 13	Meteorological Reporter for India	Indian Daily Weather Report	Mr. John Eliot.
Italy	Rome	1880, September 1	Officio Centrale de Met. e Geodin.	Bollettino Meteorico	Prof. P. Tacchini.
Japan	Tokio	1883, March 1	Meteorological Central Observatory	Tri-daily Weather Maps	Mr. K. Kobayashi.
New South Wales	Sydney	1879, January 1	Sydney Observatory	Weather Chart	Prof. H. C. Russell.
Queensland	Brisbane	1887, July	Queensland Post and Telegraph Department, Meteorological Branch	Weather Chart of Australasia, still issued in June, 1891 (daily)	Mr. C. L. Wragge.
Russia	St. Petersburg	1869, May 12	Observatoire Physique Centrale	Bulletin Météorologique	Dr. H. Wild.
Saxony	Chemnitz	1878, July	K. Sachs. Meteorologisches Institut	Wetterbericht	Dr. Paul Schreiber.
Spain	Madrid	1893, May 1	Instituto Central Meteorológico	Boletín	Prof. A. Arcimis.
Sweden *	Stockholm		Meteorologiska Central Anstalten	Synoptisk Karta	Dr. R. Rubenson.
Switzerland	Zurich	1880, July 1	Schweiz. Met. Centralanstalt	Wetterbericht	Dr. R. Billwiller.
United States	Washington	1871, January 1	Signal Service and Weather Bureau	Weather Map	M. W. Harrington.

* Posted in two or three places in Stockholm and reproduced in the papers.

TABLE II.—*List of Weather Bureau stations issuing weather maps on June 1, 1893.*

Station.	Time of issue.	Number issued. a.m.	Number issued. p.m.	Total daily—
Albany, N. Y.	a. m.	77		77
Alpena, Mich.	a. m.	21		21
Atlanta, Ga.	a. m.	130		130
Atlantic City, N. J.	a. m.	24		24
Augusta, Ga.	a. m.	21		21
Boston, Mass.	a. m. p. m.	313	518	831
Buffalo, N. Y.	p. m.		170	170
Cairo, Ill.	a. m.	57		57
Charleston, S. C.	a. m.	115		115
Charlotte, N. C.	a. m.	21		21
Chattanooga, Tenn.	a. m.	44		44
Chicago, Ill.	a. m. p. m.	[illegible]	129	[illegible]
Cincinnati, Ohio	a. m. p. m.	93	142	[illegible]
Cleveland, Ohio	a. m.	111		111
Columbus, Ohio	a. m.	98		98
Davenport, Iowa	a. m.	42		42
Denver, Colo.	p. m.		70	[illegible]
Des Moines, Iowa	p. m.		256	256
Detroit, Mich.	a. m. p. m.	185	48	[illegible]
Duluth, Minn.	a. m.	21		21
Eastport, Me.	a. m.	8		8
Erie, Pa.	a. m.	65		65
Galveston, Tex.	p. m.		[illegible]	[illegible]
Grand Haven, Mich.	a. m.	42		42
Green Bay, Wis.	a. m.	22		22
Harrisburg, Pa.	a. m. p. m.	[illegible]	[illegible]	[illegible]
Huron, S. Dak.	a. m.	[illegible]		[illegible]
Hannibal, Mo.	a. m. p. m.	210	217	427
Indianapolis, Ind.	a. m.	70		70
Jacksonville, Fla.	a. m.	30		30
Kansas City, Mo.	a. m. p. m.	[illegible]	[illegible]	[illegible]
Knoxville, Tenn.	a. m.	18		18
La Crosse, Wis.	a. m.	[illegible]		[illegible]
Little Rock, Ark.	a. m.	24		24
Louisville, Ky.	a. m.	47		47
Lynchburg, Va.	a. m.	31		31
Manistee, Mich.	a. m.	13		13
Marquette, Mich.	a. m.	21		21
Memphis, Tenn.	a. m. p. m.	96	65	[illegible]
Meridian, Miss.	a. m.	15		15
Milwaukee, Wis.	a. m.	[illegible]		[illegible]
Minneapolis, Minn.	a. m.	[illegible]		[illegible]
Mobile, Ala.	a. m.	[illegible]		[illegible]
Montgomery, Ala.	a. m.	[illegible]		[illegible]
Nashville, Tenn.	a. m.	[illegible]		[illegible]
New Haven, Conn.	a. m.	[illegible]		[illegible]
New Orleans, La.	a. m.	121		121
New York, N. Y.	a. m. p. m.	[illegible]	310	[illegible]
Norfolk, Va.	a. m.	[illegible]		[illegible]
Omaha, Nebr.	a. m.	119		119
Parkersburg, W. Va.	a. m.	32		32
Pensacola, Fla.	a. m.	[illegible]		[illegible]
Philadelphia, Pa.	a. m.	[illegible]		[illegible]
Pierre, S. Dak.	a. m.	[illegible]		[illegible]
Pittsburg, Pa.	a. m.	[illegible]		[illegible]
Portland, Me.	a. m.	122		122
Portland, Oreg.	a. m.	[illegible]		[illegible]
Raleigh, N. C.	a. m.	75		75
Rochester, N. Y.	a. m.	[illegible]		[illegible]
St. Louis, Mo.	a. m. p. m.	205	122	327
St. Paul, Minn.	a. m.	[illegible]		[illegible]
Sandusky, Ohio	a. m.	[illegible]		[illegible]
San Francisco, Cal.	a. m. p. m.	[illegible]	95	[illegible]
Sault Ste. Marie, Mich.	a. m.	[illegible]		[illegible]
Savannah, Ga.	a. m.	[illegible]		[illegible]
Shreveport, La.	a. m.	[illegible]		[illegible]
Sioux City, Iowa	a. m.	[illegible]		[illegible]
Springfield, Ill.	a. m.	[illegible]		[illegible]
Springfield, Mo.	a. m.	22		22
Tampa, Fla.	a. m.	[illegible]		[illegible]
Toledo, Ohio	a. m.	[illegible]		[illegible]
Vicksburg, Miss.	a. m.	[illegible]		[illegible]
Wilmington, N. C.	a. m.	74		74
Total		[illegible]	[illegible]	[illegible]

NOTE.—Oswego, N. Y., published a. m. maps from July 1, 1887, to September [illegible] … [illegible]

Source Endnotes

1 Jefferson, Notes on the State of Virginia (1825), pp. 147–149.
2 *Marié-Davy, Météorologie générale* (Paris, 1877), pp. 27–28.
3 *Ibid*, p. 28.
4 *Jelinek, Ueber die Priorität der Anwendung des el. Telegraphen zu den Sturmwarnungen; Zeitsch. der Oest. Ges. f. Met.* (1867), pp. 193–200, especially pp. 197–199.
5 *Ibid*, p. 196.
6 Amer. Jour. Sci., 1846, ii (2), p. 334.
7 Smithsonian Annual Report, 1848, p. 203.

8 British A. A. S. Rep. (1848), Abstract, p. 12–13.
9 Mrs. Corbin, Life of Maury, pp. 79–83. The telegraph was used for advanced information about the weather in 1850 by Mr. Joseph Brooks, manager of line of steamers between Boston, Mass., and Portland, Me. See Prescott, History of the Electric Telegraph, 4th ed. (1866), p. 254.
10 Vol. LV, pp. 112–114.
11 Moorhead, A few Incidents in the Life of Prof. James P. Espy (1888), p. 12.
12 Writings of Joseph Henry, Smith. Misc. Coll. Vol. xxx (2), p. 458. See also in this volume the "Meteorological work of the Smithsonian Institution", where these dates are somewhat different. My dates are those obtained from the published statements of Prof. Henry.
13 *Jelinek, l. c.*, p. 196.
14 Rep. Met. Com. Royal Soc., 1871, pp. 43–44, with specimens of chart for May 4, 1872. From this it would appear the maps began March 11, 1871, but it was probably 1872.

5.5

GLOBAL SYNOPTIC CHARTS

Editorial Headnote

5.5 Global Synoptic Charts

While many nineteenth-century weather scientists produced forecast-oriented weather charts and were immersed in bureaucratisation and mass production of maps, there were a very few for whom weather maps were a tool for synoptic analysis and research. Alexander Buchan (1829–1907) epitomises the focus on synoptic research. Unlike the maps of his colleagues at the Paris Observatory and the Weather Bureau in Washington, Buchan's synoptic maps were intended to track the movement of weather systems across Europe for the scientific purpose of ascertaining the general laws governing atmospheric changes – not for immediate weather prediction. Buchan's texts in this section illustrate the use and value of synoptic maps as analytical tools, and some of his major achievements, such as the first global mean barometric pressure charts and global wind charts.[1]

Buchan was secretary of the Scottish Meteorological Society for many years before becoming inspector of Scottish stations for the British Meteorological Office from 1877 to 1906. This outsider position, not directly linked to any national weather service, gave him access to data from a wide variety of locations and station networks, as evidenced by the fact that as early as 1865 he was able to publish a series of eighteen synoptic tables constructed from data from 135 stations scattered across Europe.[2] His mapping technique consisted of plotting isobars and isotherms as well as wind, cloud and precipitation data for a twenty-four-hour period on a single graph (see 3.2 'Scottish Analytical Method'). In these maps, the pressure observations have been reduced to sea level and the isotherms reflect temperature departures from the mean rather than absolute values. This was not his only foray into the isoline technique; in what was to become the standard English textbook in meteorology, the second edition of his influential *Handy book of Meteorology* of 1868, Buchan published isotherm maps for the year, and in his *Introductory Text-Book of Meteorology* of 1871, printed in Edinburgh and London, he published isotherms for January and July.

'On the Storms Which Passed over the United States between the 13th and 22nd March 1859 . . .' is the first of several seminal memoirs he published following his *Handy Book of Meteorology.* It was published in the *Journal of the Scottish Meteorological Society*, the official journal of the Society of which he was secretary. He drew on numerous observations made in different countries and over the sea, from which he constructed several maps, although he only published three – one of which is reproduced here.[3] He studied the case of the storms that passed over the United States and the Atlantic in 1859 to determine the influence of the baric field on wind speed and direction. This case study proved attractive to European and American meteorologists in the late nineteenth century because it provided an empirical and synoptic basis for a view of the relationship between pressure and wind that went beyond Buys Ballot's well-known baric wind law. Buchan argued that winds always blow counterclockwise around a centre of barometric depression in the northern hemisphere (as predicted by this law), but at the same time air is transported towards the centre (which contradicts Buy's vote). In fact, he added, the closer the isobarometric lines, the greater the wind speed. The actual movement of the air in storms could not be explained by the sense of rotation of winds alone, but also by the influence of the Earth's rotation. As for the cause of the west-to-east movement of the depressions, Buchan points to the hypothesis of the convective nature and origin of storms: it is the condensation of water vapour in the atmosphere that causes the storm to form and move.[4]

Buchan's second memoir, excerpted in part here, is considered an epoch-making work in climatology as it describes the world's first maps of mean barometric pressure at sea level. It also provides maps of wind distribution over the entire surface of the Earth.[5] Preliminary papers were read to the Royal Society of Edinburgh on 16 March 1868 and published in the *Proceedings* of that institution in 1869; later in 1868 he succeeded in constructing isobar maps for each month and each year, published as Part II in the *Transactions* of the Society, also in 1869. Considered by himself to be a first approximation, the isobar charts were improved and in 1889 he published what became known as 'the most accurate isobar charts available in English measurements'.[6] The extracts selected from this memoir show Buchan reflecting on the relationship between pressure and wind. He expresses great concern about two different elements represented in the charts, showing his conviction that the lines indicating mean atmospheric pressure and the arrows indicating prevailing winds are two independent but interrelated elements. What is the relationship between the two? he asks. Barometric pressure is both symbol and link, for isobars provide information about winds on the earth's surface. Thus, for example, the direction of winds is from high-pressure to low-pressure areas. 'The upper currents of the atmosphere may be inferred from the isobaric lines taken reversely, together with the isothermal lines taken directly' (Part II, pp. 589–590). Finding this sense of balance and interdependence in the atmosphere was fundamental to a global climatological perspective.

Notes

1 Aitor Anduaga, *Politics, Statistics, and Weather Forecasting, 1840–1910. Taming the Weather* (London and New York: Routledge, 2020), pp. 217–24.

2 W. Napier Shaw, 'Dr. Alexander Buchan, F.R.S.', *Nature*, 76 (1960) (1907), pp. 83–4; Arthur Mitchell et al., 'Contributions towards a Memorial Notice of Alexander Buchan, M.A., L.L.D., F.R.S.', *Journal of the Scottish Meteorological Society,* 14 (1908), pp. 101–18; Amy Ackerberg-Hastings, 'Buchan, Alexander (1829–1907)', in Bernard Lightman (ed.), *The Dictionary of Nineteenth-Century British Scientists* (Bristol: Thoemes Continuum, 2004), vol. 1, pp. 322–323.

3 Hugo H. Hildebrandsson and Léon Philippe Teisserenc de Bort, *Les bases de la météorologie dynamique* (Paris: Gauthier-Villars, 1898), pp. 98–106.

4 Gisela Kutzbach, *The Thermal Theory of Cyclones: A History of Meteorological Thought in the Nineteenth Century* (Boston, MA: American Meteorological Society, 1979), pp. 71–5.

5 Karl Schneider-Carius, *Weather Science, Weather Research: History of Their Problems and Findings from Documents during Three Thousand Years* (New Delhi: Indian National Scientific Documentation Centre, 1975); originally published as *Wetterkunde, Wetterforschung: Geschichte ihrer Probleme und Erkenntnisse in Dokumenten aus drei Jahrtausenden* (Freiburg: Verlag Karl Albert, 1955), pp. 287–89.

6 Hugo H. Hildebrandsson and Léon Philippe Teisserenc de Bort, *Les bases de la météorologie dynamique* (Paris: Gauthier-Villars, 1898), p. 211.

48

ON THE STORMS WHICH PASSED OVER THE UNITED STATES BETWEEN THE 13TH AND 22ND MARCH 1859, WITH REMARKS ON STORMS WHICH OCCURRED AT THE SAME TIME IN THE NORTH ATLANTIC, EUROPE AND WESTERN ASIA

Alexander Buchan

Source: Alexander Buchan, 'On the Storms Which Passed over the United States between the 13th and 22nd March 1859, with Remarks on Storms which Occurred at the Same Time in the North Atlantic, Europe and Western Asia', *Journal of the Scottish Meteorological Society*, 2, 1869, 198–201, 212–213.

THE relation of the weather of the British Islands to that of the Atlantic is evident, when it is considered that the storms which reach our coasts come more frequently from the westward than from any other quarter. A close connection has been stated, or surmised, to exist between our weather and that of North America. If this be so, an inquiry into this relation cannot fail to lead to a better understanding of extensive atmospherical disturbances, and to a more certain and earlier knowledge of their approach.

In the Society's Proceedings for the quarter ending September 1861, Mr. Milne Home wrote a short paper on a '*Comparison of the Fluctuations of the Barometer in Nova Scotia and in Scotland,*' concluding with the following remarks: – 'The great movements of the atmosphere which occasion the fluctuations of the barometer and changes of wind appear to be nearly simultaneous over the Atlantic Ocean. Such a result was to be expected, if the laws which produce these movements are not local, but form part of the general machinery of the globe. It seems desirable, therefore, to obtain evidence of the extent of the atmospheric movements, and to discover how far they agree, and in what respects they differ, on different parts of the earth's surface. I do not assert that

DOI: 10.4324/9781003650294-74

important inferences can clearly be deduced from the comparison I have been instituting between the barometric fluctuations in Nova Scotia and Scotland. My object is only to suggest the desirableness of such comparisons, and to indicate in what manner they may be carried out.' The present inquiry is an attempt to carry out this suggestion.

The periods selected for this purpose are from the 13th to the end of March 1859, and from the 19th September to the end of October 1859, and for the following reasons: – The Smithsonian Institution, Washington, has published copious observations of several great storms which passed over North America in March and September of that year. The observations have been prepared by Professor J. H. Coffin, Lafayette College, Pennsylvania, and they include the periods from the 13th to the 20th March, and from the 14th to the 25th September. From a discussion of these voluminous observations, including returns from 287 places, it may be hoped that some definite knowledge can be acquired respecting the general characteristics of storms. The time has been extended to the end of October, so as to include the 'Royal Charter' storm, of which copious materials were collected by the late Admiral Fitzroy. The time brought under review in the present paper extends only from the 13th to the 22d March.

MATERIALS USED IN THE INQUIRY.

Land Observations. – The Smithsonian observations just referred to; barometric observations made at 37 places in Europe, published by Dr. Buys Ballot in the *Jaarbœh* for 1859; observations made in different parts of the Russian Empire, published under the direction of Kuppfer; observations made by the Royal Engineers in the British Colonies and other places, under the direction of Sir Henry James, procured from the Ordnance Survey Office; in Scotland and adjoining islands, by the Society's observers, – together with observations made at isolated stations, and published in the transactions of learned societies: the chief of which are, 'The Fox' in Port Kennedy, in the Arctic regions; Toronto and St. Martin's, Canada; New Providence, Rhode Island; Montpellier and Paris, France; Brussels; Copenhagen; Christiania; Upsal; Vienna; Prague; and Munich.

In addition to these, the following manuscript observations have been received, for which the Council desire to return special thanks: – Mr. Thorlacius, Stykkisholm, Iceland; Dr. Bertherand, Algiers; Dr. Carl Jelinek, from ten stations in Austria; Dr. Buys Ballot, Utrecht, observations of the wind from 21 places in different parts of Europe; Mr. John Hartnup, Observatory, Liverpool; Mr. Matthew P. Moyle, M.R.C.S., Helston, Cornwall; and Dr. Richard Caulfield, Royal Institution, Cork.

Sea Observations. – Extracts from the logs of 36 ships in the North Atlantic, and 3 in the North Pacific, procured from the Meteorological Committee of the Royal Society, London; logs of 2 ships in the Atlantic, from Messrs. M'Iver and Bums, Liverpool; logs of 12 vessels from the Royal Mail Company of Southampton.

[. . .]

Charts Constructed. – 1. Seven large charts of the United States, on which the height of the barometer at 32° and sea-level, and the direction and force of the wind, were entered at each of the 287 places as observed each morning from the 13th to the 19th inclusive.

2. Ten charts, representing a large portion of the northern hemisphere, polar projection, on which were put down the height of the barometer and the winds at upwards of 400 places over the wide region which extends from the N. Pacific, through the United States, the N. Atlantic, Europe, and Western Asia as far as Tobolsk and Barnaul. Isolated observations were also entered frem Nertchinsk in Eastern Siberia, Japan, China, Hindostan, and in the Atlantic south of lat. 20°. The A.M. observations local time were adopted in constructing the charts. In America, the hour was almost uniformly 7 A.M.; and in the Atlantic, Europe, and Asia, 8 A.M.
3. Several smaller charts, for the representation of the midday or the P.M. observations, when it was desirable to trace a storm more particularly than could be done with charts separated by an interval of 24 hours.
4. Four charts, showing cloud, rain, and snow in America on the mornings of the 16th, 17th, 18th, and 19th March.

When the heights of places were not known or doubtful, they were calculated by the method of the barometric measurement of heights from places whose heights were known, on those days when the barometer remained equable and steady over a considerable extent. This method gives an approximate height sufficiently near the truth, to serve the purposes of this inquiry. The following readings were underlined – viz., doubtful readings of the barometer, and daily means, or the means of two or more observations made on that day; and little value was given to these barometric readings in drawing the isobaric lines.

Three of these charts, on a reduced scale, are presented with this number of the *Journal;* all the other charts are deposited in the Society's library.

The isobaric lines, or lines passing through all places where atmospheric pressure (reduced to 32° and sea-level) is the same, are drawn for every two-tenths of an inch at 30·4, 30·2, 30·0, 29·8, etc. The lines which show a pressure of 30 inches and upwards are the heavy lines on the charts, and the lines indicating a pressure less than 30 inches are the thin lines. Thus the position of storms, in so far as characterized by a lower pressure than the average, is marked out by thin lines; and the varying position of these systems of lines from day to day shows the course the storm is pursuing. When the number of trustworthy barometric observations is sufficient to enable the isobaric lines to be drawn with approximate exactness, the lines are solid; but when the observations are so few that the lines are necessarily to some extent hypothetical, the lines are dotted.

The arrows are represented flying with the wind. A plain arrow represents a very light wind; and as regards the feathered arrows, each feather indicates the force on the scale 0 to 6; two feathers a moderate breeze, four a strong wind or gale, six a hurricane, etc. When the force of the wind is not known, the shaft of the arrow is terminated by a transverse line. Two transverse lines mean that the wind is not the A.M. direction, but the average direction during the day. This kind of arrow is wholly confined to Russia. Calms are shown by a small circle with a dot in the centre, and variable winds by a large asterisk (*).

In each of the regions where thin lines are grouped together, indicating low pressures and a disturbed state of the atmosphere, there is a comparatively small space where pressures are lowest. This space will be assumed in this paper as the centre of the storm. With reference to the several storms indicated on the charts, it was determined from day to day, and the different positions connected by lines. These are the hatched lines on the charts, and are intended to show the probable track of the storm in its progressive course from day to day. Each storm is designated on the charts by a capital letter – as A, B, C, D, E, F – printed near its centre. These storms may therefore be referred to as Storm A, Storm B, etc. Each storm track will be seen to have a capital letter at its extremities, with a date attached. Thus Storm A 13th March, A 23d March, mean that this storm was first observed at that part of the United States on the 13th March 1859, and was last observed in Russia on the 23d.

[. . .]

The three charts exhibiting the weather at one time over a large portion of the northern hemisphere present some instructive features. Of these, the most remarkable, perhaps, is the course of the isobaric 30·0 inches on Plate I., from Vancouver Island, in the Pacific, eastward through America, the Atlantic, and Europe, into Asia. It will be observed to rise in higher and fall in lower latitudes, presenting in its course a succession of loops, which enclose systems of lower pressures marked off by the concentrical isobaric curves characteristic of the storms just described. At this date, the morning of the 16th, there were seven distinct areas of low pressure, viz. in north-west of America, the States west of the Mississippi, the New England States, Mid-Atlantic, Iceland, Denmark, and the east of Russia.

Leaving out of view the first of these, since the observations are too scanty to trace it, the other depressions may be regarded as moving forward in some easterly direction. On comparing the three charts, the progressive motion is found to proceed with different velocities and directions in different places; and all of the spaces defined by the isobarics of high and of low pressure underwent constant change as regards outlines and dimensions.

It is to the progressive movement of these systems of high and low pressures that the rising and falling of the barometer at particular places is due. These risings and fallings, when represented on a diagram, exhibit what appears to be a succession of waves, in the examination of which, under the name of '*barometric waves*,' much labour has been bestowed by Mr. Birt and others. The November wave is best known of these, a short account of which appeared in the Society's Proceedings for the quarter ending March 1861.

In the inquiry into the temperature of Europe for the year from 1st April 1865 to 31st March 1866, the results of which were given in Nos. IX. and XI. of this

GENERAL CIRCULATION AND THE CYCLONE

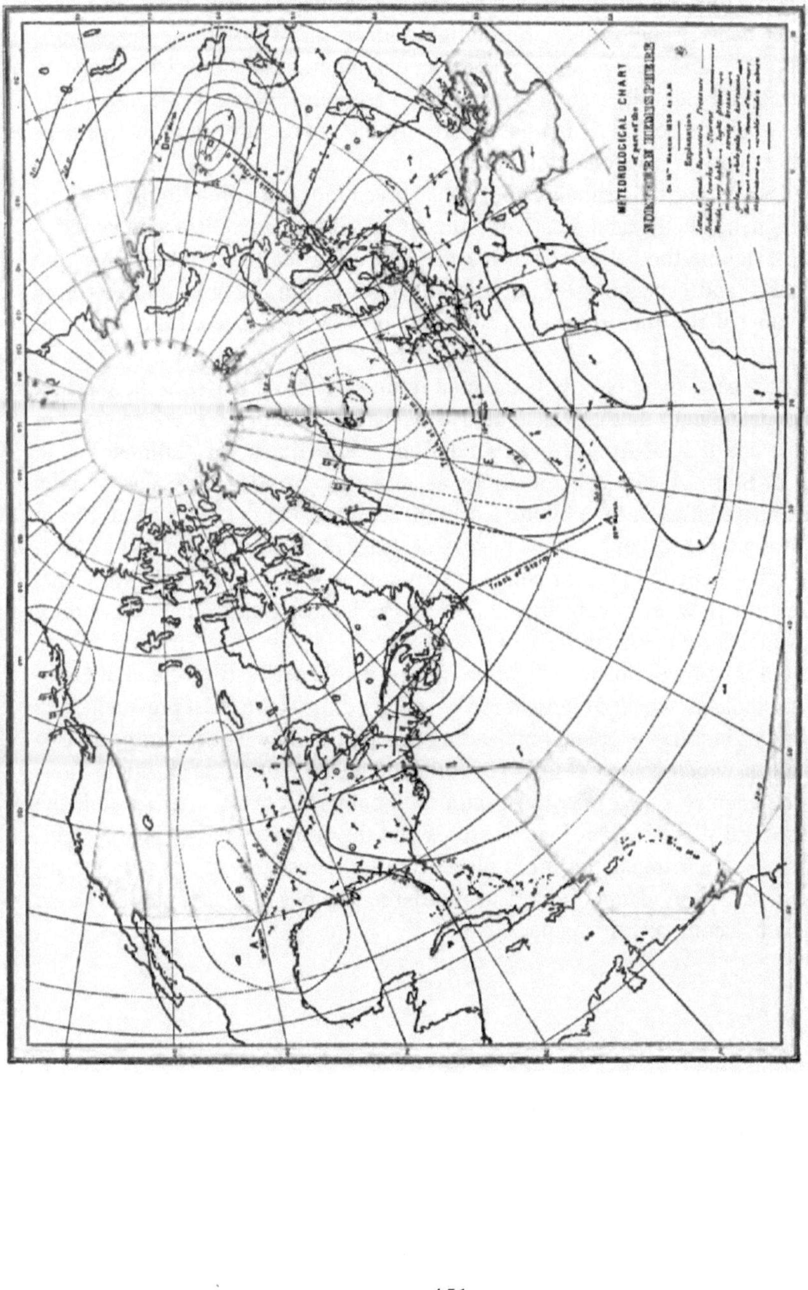

Fig. 121 Reproduction on a reduced scale of one of the charts prepared by Alexander Buchan to show the travel of depressions over the Atlantic.

(From the *Journal of the Scottish Meteorological Society*, October, 1868.)

Journal, the relations of these all-important conditions of the weather, heat and cold, to each other, were likened to a continual contest, to decide which should have the predominance; as the one gathered strength, it pressed on the other, first checking its advance, and then driving it back; these advances and retreats were generally made over the whole of Europe, and not unfrequently the whole of Europe could represent no more of these vast waves of heat or cold than what was only a fragment of one of them. Some light will be thrown on these fluctuations of temperature, and their great extent, by an examination of the isobaric lines on the charts and the distribution of high and low pressures they indicate, especially when the relations between high temperature and low pressure, and low temperature and high pressure, are kept in view.

Father Secchi, of the Roman College, has given much attention to the motions of the magnetic needle as indicative of the approach of storms. It would be interesting to examine the behaviour of the needle during the time embraced by the three charts, and especially on the morning of the 16th, in order to determine whether any of the four storms on the chart of that day affected the needle at Rome.

It thus appears, that not only is there an intimate relation between the weather of Europe and that of the Atlantic, but also, as was surmised, between the weather of Europe and that of America. This relation is sometimes very intimate, as in the case of Storm A, which began in Texas, and after crossing the States and the Atlantic, passed into Europe by the North of Scotland. On the other hand, Storm E does not appear to have passed over America, but to have originated in Mid-Atlantic, and thence pursued a course nearly due N.E. towards the North Cape. Again, Storm B, after having travelled over the United States, died out shortly after advancing on the Atlantic.

The continuation of this investigation, for the whole period till the end of October, might indicate which of these three types of storms is the most prevalent; and thus a line of inquiry be opened up leading to a better knowledge of atmospheric disturbances, which could not fail to be turned to the benefit of the maritime interests of the country. One of the most important points yet to be determined is the course generally pursued by low pressures and the stormy winds which accompany them, in their translation from place to place. These points can only be satisfactorily settled by data collected from a large number of charts similar to the three which accompany this paper.

49

THE MEAN PRESSURE OF THE ATMOSPHERE AND THE PREVAILING WINDS OF THE GLOBE FOR THE MONTHS AND FOR THE YEAR

Alexander Buchan

Source: Alexander Buchan, 1869. 'The Mean Pressure of the Atmosphere and the Prevailing Winds of the Globe for the Months and for the Year', *Proceedings of the Royal Society of Edinburgh*, 6, 1869, 303–07, Part II, *Transactions of the Royal Society of Edinburgh*, 25, 1869, 575–577, 579, 581–583, 589–590, 592.

Monday, 16th March 1868.
Dr CHRISTISON, Vice-President, in the Chair.

The following Communications were read: –

1. The Mean Pressure of the Atmosphere over the Globe for the Months and for the Year. Part I. – January, July, and the Year. By Alexander Buchan, M.A., Secretary to the Scottish Meteorological Society.

The three charts which were exhibited, showing, by isobarometric lines, the mean atmospheric pressure over the globe, during January, July, and the year, were constructed from observations made at 358 places thus distributed over the earth, – 167 in Europe; 51 in Asia; 22 in Africa and adjoining islands; 35 in South America, West Indian Islands, and Atlantic; 63 in North America; and 20 in Australasia and Antarctic Ocean. Of the European stations, 12 are in Scotland, 14 in England, 27 in Austria, 12 in Italy, 10 in France, 10 in the Netherlands, 9 in Norway, and 57 in the Russian empire, &c. The list might have been largely increased; thus a larger number might have been given from the 80 Scottish stations; but the 12 given were judged sufficient to represent the mean atmospheric pressure of this country.

In the British islands the means were uniformly taken for the ten years from 1857 to 1866, in order that they might be strictly comparable with each other;

and the means of several European stations were calculated for the same year. In the United States of America the means were uniform for the six years from 1854 to 1859. In selecting the stations, respect was had to the obtaining of a good mean, – that is, observations for a sufficient number of years to show, as nearly as possible, the true mean. For example, at Bombay, from 1847 to 1860, the lowest mean for July was 29·598 inches in 1851, and the highest 29·673 in 1853. This regularity in the pressure of the same month, from year to year, is a feature common to all tropical countries; and hence, in such places, a few years were accepted as a good mean. On the other hand, since the mean pressure at Reykjavik, Iceland, during January 1867, was 29·913 inches; during February, 29·359 inches; and during March, 30·037 inches, it is evident that a good many years are required to represent the mean of the month at Reykjavik. Hence, a very subordinate position, if any at all, has been given to observations from such places, unless they embraced a considerable number of years. The following are a few of the places, with the number of years, for which the means are given: – Sitka, 15; Algiers, 10; Hobart Town, 25; St Louis, Mauritius, 13; Bogoslovsk, 26; Nigni-Tagilsk, 21; Barnaul, 19; Nertchinsk, 18; Pekin, 14; Calcutta, 11; Tiflis, 14; Baku, 17; Alagir, 15; Jakobshavn, 10; Reykjavik, 13; Hammerfest, 13; St Petersburg, 19; Archangel, 18; Zlatoust, 28; Lugan, 22; Christiania, Cracow, and Kursk, 27; Brussels, 33; Gand, 26: Geneva, 25; Ahun, 34; Verona, 73; Bologna, 45; Milan, 25; Turin, 74; most of the Austrian stations, 14 to 18, &c.

Tables were prepared ruled with columns for the insertion of the following information: – 1. The place and country; 2. The source or authority from whence the observations were obtained; 3. The number of years of the means; 4. These years specified; 5. The hours of observation; 6. The latitude; 7. The longitude; 8. The height above the sea in English feet; 9 to 21. The simple arithmetic means for each month and for the year.

For reducing to sea-level a table was prepared from the formula and Table XVI. given in Guyot's Meteorological and Physical Tables, D, p. 89. This table, calculated for each 5° F. of the temperature of the air – from 40° to 90° – was used in all cases where the height did not exceed 800 feet. For higher situations, the reduction was made by means of Dippe's method, as detailed in Guyot's tables, D, p. 60. With the correction for height was included the correction for daily range.

For many means I have been indebted to the labours and writings of Buys Ballot, Carl Jelinek, Dove, Quetelet, James, and Kuppfer, and particularly to Secchi's admirable abstracts, which have appeared from time to time in the "Bulletino Metereologico."

The means, so reduced, were then entered on charts of the globe drawn on Mercator's Projection, from which isobarometric lines were drawn for every tenth of an English inch in the difference of the pressure. The lines of 30 inches and upwards were coloured red, and the lines representing pressures less than 30 inches were coloured blue. Thus, the portions of the earth's surface where the pressure was above or below the average (30 inches) could be seen at once.

Mean Atmospheric Pressure for July. – The lowest pressures occur over the land and the highest over the ocean; or, the lowest pressures occur in the north hemisphere and the highest in the south hemisphere. The greatest extent of high pressure extends quite round the globe from a little south of the equator to 40° lat. south. In the South Atlantic, in lat. 20°, this rises to 30·348 inches; in the North Atlantic there is a corresponding area of high pressure rising at lat. 35° to full 30·348 inches; and this region of high pressure extends over the south-west of Europe and the southeast of the United States. There is also a region of high pressure in the North Pacific, but it only amounts to about 30·1 inches.

The greatest extent of low pressure occurs in Asia, amounting in the central regions of the Continent only to 29·5 inches. Pressures are also low in the interior of North America, and round the north and the south poles.

Mean Atmospheric Pressure for January. – In January the highest pressures are over the land and the lowest over the ocean; or, the highest are in the north hemisphere and the lowest in the south hemisphere. Thus, taken in a general sense, the mean pressures of July and January are reversed.

The region of highest pressure occurs in the interior of Asia, where it amounts to 30·4 inches, being thus nearly one inch greater in winter than in summer. The area of high barometer (above 30 inches) is continued westward through Europe south of the North Sea and the Baltic; the north of Africa; the North Atlantic, between 15° and 45° lat.; North America, except the north and northwest; the West India Islands; and the North Pacific, as far west probably as 150° long. W. The effect of the Mediterranean and adjoining seas, which are at this season warmer than the land, in lowering the mean winter pressure, and thus breaking the continuity of the isobarometric 30·2 inches, and preventing its extension from the Pacific to the north of the Lena, in Siberia, is very striking. All the charts show similar disturbances of the lines of equal pressure over the same region. There are two other regions, though of comparatively small extent, where the mean pressure exceeds 30 inches, – one in South Atlantic, and the other between South America and Australia.

There occurs in the North Atlantic an extensive diminution of pressure, which deepens northwards till the greatest depression (29·5 inches) is reached in Iceland.[1] It is this low pressure over the North Atlantic, together with the high pressures in North America, in the Atlantic south of lat. 40°, and in Asia, which furnishes the key to the winter climates of North America and Europe. Another and equally remarkable depression occurs in the North Pacific, having its course of greatest depression (29·6 inches) in the ocean between Kamschatka and Sitka in the north-west of America. The pressure is also under the average in the south of Africa and in South America.

The equatorial depression stretches quite across the globe in an irregular belt, which attains its greatest breadth in Africa and its least in the Pacific. In crossing the Indian Ocean it does not lie parallel to the equator, but slants from Tamatave in Madagascar 18° lat. S. to the coast of Sumatra, in 5° lat. S. It is in this trough that nearly all the tropical storms of the Indian Ocean have their origin.

Mean Atmospheric Pressure for the Year. – There are two broad belts of high pressure passing completely round the globe, – the one north, the other south of the equator, – enclosing between them the low pressure of the tropics, through the centre of which runs a narrow belt of still lower pressure, towards which the trade winds blow. The southern belt of high pressure lies nearly parallel to the equator, and is generally of uniform breadth throughout. But the belt north of the equator has a very irregular outline and great differences in its breadth and in its inclination to the equator, – these irregularities being due to the unequal distribution of land and water in the northern hemisphere.

Considered in a broad sense, there are only two regions of low pressure, one round each pole, bounded by or contained within the belts of high pressure just referred to. The most remarkable of these is the region of low pressure round the south pole, which probably is subject to little variation throughout the year. The depression in the neighbourhood of the north pole is divided into distinct centres, at each of which a diminution of pressure still further below the average pressure prevails. These are the north part of the Atlantic and the north part of the Pacific oceans. There is also a smaller area of low pressure in Hindustan, caused altogether by the low summer pressure of that region during the south-west monsoon. The centre of Asia is also a little below the average, owing to the very low summer pressure.

These differences in the pressure arise from the unequal distribution either of the temperature or of the moisture of the atmosphere. Of these, considered as disturbing agents, by far the most important is the moisture, – thus giving to this element a paramount claim on our regard in studying winds, storms, and other atmospheric changes.

The relations of the barometric pressure to the prevailing winds, the varying temperature, and the rainfall in different parts of the world, were pointed out. Isobarometric and isothermal charts, – (1.) Of Europe and Western Asia, from the 18th to the 26th of December 1866; and (2.) Of Europe, from the 19th July to the 1st of August 1867, were exhibited; and it was shown that the remarkable deviations from the mean pressure of these months which then prevailed were accompanied with equally remarkable deviations from the mean temperature of the same months.

[. . .]

XVII. – *The Mean Pressure of the Atmosphere and the Prevailing Winds over the Globe, for the Months and for the Year.* Part II. By Alexander Buchan, M. A., Secretary of the Scottish Meteorological Society.

(Plates XXV. to XXVII.)

(Read 19th April 1869.)

Charts, showing by *Isobaric Lines* the mean pressure of the atmosphere over the globe during the months of the year, may be justly regarded as furnishing the key to all questions of meteorological inquiry; for without the information conveyed by such charts it is impossible to discuss satisfactorily those questions which relate to prevailing winds, the varying temperature, and the rainfall throughout the year in the different countries of the world. It is to meet this desideratum that the Charts of Mean Atmospheric Pressure of the globe which are given with this paper are offered as the first approximate solution of this great physical problem.

Since Part I.[2] was read in March 1868, valuable additional information has been obtained from Australia, New Zealand, Tasmania, Africa, South America, the west coast of North America, Iceland, Norway, and Sweden, and from several isolated stations in different parts of Europe and Asia. The period for the British Islands and a large portion of Europe has been extended so as to include the eleven years from 1857 to 1867.

In this Part the complete set of Charts for the twelve months and for the year are given, together with the data from which the Charts have been constructed.

As regards PRESSURE, the stations were selected with the view of representing as well as possible the geographical distribution of the pressure. The first place was assigned to those stations at which the barometric observations were known to be, or presumably were, of the best quality; and in drawing the isobaric curves, the greatest weight was given to means deduced from those observations. Since it is the mean pressure at sea-level which is here inquired into, and since the manner of the geographical distribution of the pressure doubtless varies at different heights, stations, at low elevations were preferred to those at greater heights. The pressures at a few elevated places, such as Great St Bernard and Dodabetta, are given in the Tables; but they were not made use of in drawing the curves, – their value consisting in the light they tend to throw on the movements of the upper currents of the atmosphere.

In an inquiry into the comparative distribution of atmospheric pressure, it is evident that the first requisite, as regards time, is, that the means be deduced from observations made in the same years. In the tropics, where there is great regularity in the mean pressure of the same month from year to year, observations in the same years are of less importance; but in extra-tropical regions, where the mean pressure of the same month varies considerably from year to year, it is of the utmost importance to obtain observations for the same years. This has been the guiding principle in selecting the years from which the means of the different places in Table I. have been calculated. Thus, in the British Islands, the means are uniformly given for the eleven years from 1857 to 1867, and the means of many European stations are given for the same years; and in the United States of America the means are for the six years from 1854 to 1859.

[. . .]

The next step was to apply to the figures in Table I. corrections (1) for daily range and (2) for height. So far as possible, that hour, or those hours, of observation were selected when the pressure of the atmosphere is nearly the mean of the day. For places for which this could not be done, a collection of *Mean Hourly Variations of the Barometer* was made from a considerable number of stations in different parts of the world. From these, approximate corrections for daily range were deduced, and applied to the monthly means of the stations.

For reducing to sea-level, a table was prepared from the Formula and Table XVI., given in GUYOT's Meteorological and Physical Tables, D, p. 89. This table, calculated for each 5° Fahr. of the temperature of the air, from −40° to 90°, was used in all cases where the height did not exceed 800 feet. For higher situations, the reduction was made by means of DIPPE's method, as detailed in GUYOT's Tables, D, p. 60.

The means,[3] so corrected, were then entered on large polar projections of the northern hemisphere, from which the *isobars* were drawn for every tenth of an English inch of pressure. The isobars for the southern hemisphere were drawn on charts of MERCATOR's Projection. The whole was ultimately transferred to charts of the projection on Plates XXV. to XXVII. The isobars, indicating a pressure of 30 inches, which is nearly the average pressure, and upwards, are represented on the charts by heavy lines, and lower pressures are represented by light lines.

For many of the means I have been indebted to the labours and writings of DOVE, BUYS BALLOT, SECCHI, CARL JELINCK, MOHN, JAMES, QUETELET, and KUPPFER. For the means of single stations and groups of stations, I have received most valuable assistance from Meteorologists in all parts of the world, for which I beg to return them my grateful thanks.

DISTRIBUTION OF ATMOSPHERIC PRESSURE, *in December, January, and February.* – In these months, the highest pressures are grouped over the land portions of the northern hemisphere, and the larger the extent of the land the greater is the pressure. The area of high barometer (30 inches and upwards) embraces nearly all Asia; all Europe, south of the North and Baltic Seas; the North Atlantic, between 15° and 45° lat.; the West Indies; North America, except the north and north-west; and the North Pacific, between 8° and 24°lat. There are also two regions of high pressure of comparatively small extent – the one in the South Atlantic, and the other in the South Pacific.

[. . .]

PREVAILING WINDS. – It will be seen that every one of the charts shows considerable disturbance of the equilibrium of the atmosphere at the surface of the earth. If the pressure was equal in all parts of the globe, we should have the physical conditions of a stagnant atmosphere. But such is not the case. From the different pressures which the charts show in different regions, it might be expected, from the laws of aërial fluids, that movements of the atmosphere would set in, giving rise to the prevailing winds of these regions.

[. . .]

Thus two distinct sets of facts are exhibited on the charts, viz., lines showing the mean pressure of the atmosphere, and arrows showing the prevailing winds at the earth's surface, each being independently arrived at by the summing and averaging of observed facts. What relation is there between these two classes of facts?

I. WINDS *within, or near, a space of Low Pressure.* – Of this class, the best example is the low pressure which prevails in the north of the Atlantic and adjoining regions in the winter months. This region of low pressure is bounded to the S.W. by the high pressure of North America; to the S. by the high pressure in the Atlantic, about 30° lat. N.; to the S.E. by the high pressure in the interior of Asia. In January, the difference between the average pressure of Iceland and the interior of Asia is fully an inch.

It is seen from the charts, that in Baffin's Bay and east of the Rocky Mountains, as far south as 40° lat., the winds are N.N.W., N.W., and W.N.W. Crossing the Atlantic, winds in the British Islands, in France, and the north of Germany, are from W.S.W. to S.W.; in Denmark, S.S.W.; near Bergen, in Norway, S.; and at Christiansund and Hammerfest, S.S.E. The relation of these winds to the isobaric lines is the same as that which is illustrated by the winds in storms, in their relation to the isobaric lines of these storms. This has been already stated in a Paper by the author, published in the "Transactions of the Society," vol. xxiv. Part i. p. 201, in the following words: – "The wind in storms neither blows round the centre of least pressure in circles, or as tangents to the concentric isobaric curves, nor does it blow directly towards that centre. It takes a direction intermediate, approaching, however, more nearly to the direction and course of the circular curves than of the radii to the centre." Or, according to Dr Buys Ballot, the angle is not a right angle, but from about 60° to 80°. This relation is usually called "Buys Ballot's Law of the Winds."

Another well-marked depression is the low summer pressure in the interior of Asia; with reference to which, it is seen from the charts that the winds of Eastern Europe and Western Asia are from N.W. to W.N.W. and W.; at Ceylon, S.W.; at Shanghai, S.E.; and on the Sea of Okotsk, N.E.; whilst in the interior, calms generally prevail.

The behaviour of the winds, as regards the low pressure of North America, is exactly similar to that of the winds in Asia at this season. In all these cases the wind appears to flow round and in upon the space where pressures are low. Even in those instances where the depression over a limited space is comparatively small, such as in Australia during the summer months, the winds observe the same course with respect to it.

A well-known and remarkable diminution of pressure is that of the Antarctic regions; and though, except in Tasmania and the south of New Zealand, observations are wanting at particular points for a sufficiently long time to give good averages, yet the concurrent testimony of sailors and the inhabitants of these regions all go to show that, at least on the outskirts of the region, winds are chiefly N.W. or W.N.W. – that is, they appear to flow in upon the space of low pressure. The low pressure in the equatorial regions, towards which the trades blow, is an illustration of the same principle.

Winds *within, or near, a space of High Pressure.* – The most prominent illustration of this is the high pressure in the interior of Asia in winter. It is seen from a single glance at the charts that the winds flow *out of* this space in every direction. The same *outflow* is seen with respect to the less strongly marked, but still very distinct space of high pressure in North America; owing to the large number of stations available here, this principle is amply illustrated.

The next most noteworthy area of high pressure occurs in summer between Africa and North America, out of which also the charts show the winds blowing in all directions towards and round upon the surrounding low pressures.

[. . .]

This behaviour of the winds with respect to spaces of high pressure differs in no respect from what occurs on particular days on which the isobaric lines present the same conditions of pressure. Mr FRANCIS GALTON first drew attention to this peculiarity, under the name of *Anticyclones,* by which name he intended to convey the idea that in cases of high pressure occurring over a limited area, the course of the winds is exactly the reverse of what is seen to prevail in cyclones in which the winds blow round and in upon a space of low pressure.

The *outflow* of the air from a region of high pressure, and the *inflow* upon a region of low pressure, appears to be reducible to a single principle, viz., the principle of gravitation. Given as observed facts the differences of pressure, it might almost be predicted, before calculating the averages, what the prevailing winds are. Indeed, so predominating is the influence of gravitation that it may be regarded as the sole force immediately concerned in determining the movements of the atmosphere. If there be any other force or forces which set the winds in motion, their influence must be altogether insignificant as compared with gravitation.

The effect of a mountain range interposed in the course of one of these great atmospheric currents is interesting. Of this, the best example is furnished by the mountain range of the Scandinavian peninsula, in its effect on the prevailing winds in winter. It will be observed that this mountain range lies between the low pressure about Iceland, and the high pressure in the interior of Asia.

[. . .]

From these considerations, it may be concluded that the winds on the surface of the earth are approximately known from the isobaric lines, – the direction being from regions of high towards those of low pressure, subject to the changes in the direction of the currents produced by the earth's rotation; and that the upper currents of the atmosphere may be inferred from the isobaric lines taken reversely, together with the isothermal lines taken directly. In other words, the regions of lowest pressure, by giving the ascending currents, point out the sources or fountains whence the upper currents flow; and the isothermals, by showing where, on account of the low temperature, the greater portion of the air is condensed in the lower beds, and so diminishing the pressure in the upper beds, point out the regions towards and over which the upper currents diffuse themselves.

1. To travellers in the interior of continents and regions at a great distance from places where Meteorological Observations are made, the Charts will be of use in showing the approximate sea-level pressures for each month of the year. They show at a glance the *zero points* from which the heights of places may be calculated, at which observations of the pressure of the atmosphere have been made.[4]

2. To sailors, the Charts will be useful as showing the prevailing winds at many places in each of the twelve months, and still further as suggesting, from the connection which is here pointed out between mean atmospheric pressure and prevailing winds, the winds which are most likely to be met with in regions where little is known of the general course of the winds from actual observations. The charts of least prevailing winds in January and July (Plate XXVII.), will also be useful in this respect.

   [. . .]

4. An inquiry of still greater interest is suggested by the isobaric lines of the Charts. Their position appears to be altogether determined by the geographical distribution of land and water on the surface of the earth; and since the isobaric lines determine the prevailing winds, and these in their turn the peculiar distribution of temperature and rainfall, – in other words, the climates of the globe, – it is evident that we have here a principle applicable not merely to the present state of the earth, but also *to different distributions of land and water in past time.* In other words, there is here a principle which the geologist will require to apply in attempting to account for glacial and warm epochs, through which the climates of great Britain and other countries have passed. In this way it is possible to arrive at an approximate numerical statement, as regards temperature and rainfall, of Sir CHARLES LYELL's idea of the changes of climate brought about through the displacements of continents.

Source Endnotes

1 The diminution of pressure in this region is even greater than this in February.
2 Proceedings of the Society, vol. vi. p. 303.
3 The original observations arc given in Table 1. in preference to the corrected means deduced from them.
4 Note of the Determination of Heights, chiefly in the interior of Continents, from Observations of Atmospheric Pressure. – "Proceedings of the Society," vol. vi. p. 465.

VOLUME I REFERENCES

Achbari, Azadeh and Lunteren, Frans van, 'Dutch Skies, Global Laws: The British Creation of "Buys Ballot's Law"', *Historical Studies in the Natural Sciences*, 46:1 (2016), pp. 1–43.

Adas, Michael, *Machines as the Measure of Men: Science, Technology, and Ideologies of Western Dominance* (Ithaca: Cornell University Press, 1989).

Agnew, Duncan Carr, 'Robert FitzRoy and the Myth of the 'Marsden Square': Transatlantic Rivalries in Early Marine Meteorology', *Notes and Records of the Royal Society of London*, 58:1 (2004), pp. 21–46.

Ahrens, C. Donald, *Meteorology Today: An Introduction to Weather, Climate, and the Environment*, 6th edition (Pacific Grove: Brooks and Cole, 2000).

Altamore, Aldo and Maffeo, Sabino (eds.), *Angelo Secchi. L'avventura scientifica del Collegio Romano* (Vaticano: Specola Vaticana, 2012).

Anderson, Katharine, 'The Weather Prophets: Science and Reputation in Victorian Meteorology', *History of Science*, 37 (1999), pp. 179–216.

Anderson, Katharine, *Predicting the Weather: Victorians and the Science of Meteorology* (Chicago: The University of Chicago Press, 2005).

Anderson, Katharine, 'Mapping Meteorology', in James Rodger Fleming, Vladimir Jankovic, and Deborah R. Coen (eds), *Intimate Universality. Local and Global Themes in the History of Weather and Climate* (Sagamore Beach: Watson Publishing International, 2006), pp. 69–91.

Anduaga, Aitor, *Cyclones and Earthquakes. The Jesuits, Prediction, Trade, and Spanish Dominion in Cuba and the Philippines, 1850–1898* (Quezon City, Ateneo de Manila University Press, 2017).

Anduaga, Aitor, 'Los jesuitas y el estudio de los tifones en Filipinas, 1865–1899', *Illes i Imperis*, 22 (2020), pp. 63–83.

Anduaga, Aitor, *Politics, Statistics, and Weather Forecasting, 1840–1910. Taming the Weather* (London and New York: Routledge, 2020).

Anduaga, Aitor, 'Epistemic Network: The Jesuits and Tropical Cyclone Prediction, 1860–1900', *Isis,* 113:3 (2022), pp. 513–36.

Anduaga, Aitor, 'The Jesuits and Meteorological Observatories'. *Quaderni di Storia della Fisica*, 27 (2023), 85–100.

Anduaga, Aitor, 'The Jesuit Culture of Correlation in Observatory Sciences', *Isis*, 116:1 (2025), pp. 3–22.

Aubin, David, 'Observatory Mathematics in the Nineteenth Century', in Eleanor Robson and Jacqueline Stedall (eds.), *The Oxford Handbook of the History of Mathematics* (Oxford: Oxford University Press, 2009), pp. 273–98.

Aubin, David, Bigg, Charlotte and Sibum, H. Otto, 'Introduction: Observatory Techniques in Nineteenth-Century Science and Society', in David Aubin, Charlotte Bigg and H. Otto Sibum (eds.), *The Heavens on Earth: Observatories and Astronomy in Nineteenth-Century Science* (Durham, NC: Duke University Press, 2010), pp. 1–32.

Baneke, David, 'Who Predicts? Scientific Authority and User Expertise in Dutch Storm Warnings 1860–1920', *Science, Technology, & Human Values,* 50:1 (2024), pp. 1–29.

Bankoff, Greg, 'Aeolian Empires: The Influence of Winds and Currents on European Maritime Expansion in the Days of Sail', *Environment and History,* 23 (2017), pp. 163–96.

Barboza, Christina Helena da Motta, 'Nice Weather, Meteors at the End of the Day', in Stefan Emeis and Cornelia Lüdecke (eds), *From Beaufort to Bjerknes and Beyond: Critical Perspectives on Observing, Analyzing, and Predicting Weather and Climate* (Augsburg: Rauner, 2005), pp. 157–68.

Barboza, Christina Helena da Motta, *As viagens do tempo. Uma história da meteorologia em meados do século XIX* (Rio de Janeiro: E-papers Serviços Editoriais Ltda., 2012).

Barboza, Christina and Janković, Vladimir, 'The Many Lives of Weather', in Vladimir Janković and Christina Barboza (eds), *Weather, Local Knowledge and Everyday Life: Integrated Climate Studies* (Rio de Janeiro: MAST, 2009), pp. 11–17.

Beattie, James, O'Gorman, Emily and Henry, Matthew, *Climate, Science, and Colonization: Histories from Australia and New Zealand* (Basingstoke: Palgrave Macmillan, 2014).

Bennett, Jim A., *Church, State and Astronomy in Ireland – 200 years of Armagh Observatory* (Armagh: Armagh Observatory, 1990).

Bergeron, Tor, 'Synoptic Meteorology: An Historical Review', *Pure and Applied Geophysics,* 119 (1981), pp. 443–73.

Boumans, Marcel, *Science Outside the Laboratory: Measurement in Field Science and Economics* (Oxford: Oxford University Press, 2015).

Bowen, Margarita J., *Empiricism and Geographical Thought: from Francis Bacon to Alexander von Humboldt* (London: Cambridge University Press, 1981).

Brooke, John H., *Science and Religion* (Cambridge: Cambridge University Press, 1991).

Burton, James, 'Robert FitzRoy and the Early History of the Meteorological Office', *British Journal for the History of Science*, 19 (1986), pp. 147–76.

Burton, James, 'The History of the British Meteorological Office to 1905', (The Open University, PhD diss., 1988).

Caglioti, Angelo Matteo, *Meteorological Imperialism. Climate Science, Environment, and Empire in Liberal and Fascist Italy (1870–1940)*, PhD diss. (University of California, Berkeley, 2017).

Cannegieter, Hendrik Gerrit, 'The History of the International Meteorological Organization, 1872–1951', *Annalen der Meteorologie*, 1 (1963), pp. 7–280.

Cantor, Geoffrey, 'Quakers and Science', in Stephen W. Angell and Pink Dandelion (eds.), *The Oxford Handbook of Quaker Studies* (Oxford: Oxford University Press, 2013), pp. 520–34.

Cantor, Geoffrey, *Quakers, Jews and Science* (Oxford: Oxford University Press, 2013).

Carter, Christopher, 'Herschel, Humboldt, and Imperial Science', in Raymond Erickson, Mauricio A. Font, and Brian Schwartz (eds.), *Alexander von Humboldt. From the Americas to the Cosmos: An International Interdisciplinary Conference, October 14–16, 2004* (New York: The City University of New York, Bildner Center for Western Hemisphere Studies, 2004), pp. 509–18.

Carter, Christopher, 'Magnetic Fever: Global Imperialism and Empiricism in the Nineteenth Century', *Transactions of the American Philosophical Society*, New Series, 99:4 (2009), pp. 1–163.

Cawood, John, 'The Magnetic Crusade: Science and Politics in Early Victorian Britain', *Isis*, 70 (1979), pp. 492–518.

Chinnici, Ileana, *Decoding the Stars: A Biography of Angelo Secchi, Jesuit and Scientist* (Leiden, Boston: Brill, 2019).

Chinnici, Ileana and Consolmagno, Guy (eds.), *Angelo Secchi and Nineteenth Century Science. The Multidisciplinary Contributions of A Pioneer and Innovator* (Cham, Switzerland: Springer, 2021).

Chinnici, Ileana and Gramatowski, Wiktor, 'Le carte di Angelo Secchi S.J. (1818–1878) conservate presso la Pontificia Università Gregoriana. Un inventario inedito rivisitato', *Nuncius*, 16:2 (2001), pp. 571–627.

Coen, Deborah R., 'Scaling Down: Mapping the Austrian Climate between Empire and Republic', in James Rodger Fleming, Vladimir Jankovic, and Deborah R. Coen (eds.), *Intimate Universality. Local and Global Themes in the History of Weather and Climate* (Sagamore Beach, MA: Watson Publishing International, 2006), pp. 115–40.

Coen, Deborah R., *Climate in Motion: Science, Empire, and the Problem of Scale* (Chicago: University of Chicago Press, 2018).

Cox, John D., *Storm Watchers: The Turbulent History of Weather Prediction from Franklin's Kite to El Niño* (Hoboken, New Jersey: John Wiley & Sons, Inc., 2002).

Curry, Patrick, *A Confusion of Prophets: Victorian and Edwardian Astrology* (London: Collins & Brown, 1992).

Cushman, Gregory, 'The Imperial Politics of Hurricane Prediction: From Calcutta and Havana to Manila and Galveston, 1839–1900', in Erika Marie Bsumek, David Kinkela, and Mark Atwood Lawrence (eds.), *New Approaches to International Environmental History* (Nueva York: Oxford University Press, 2013), pp. 137–62.

Darrigol, Olivier, *Worlds of Flow. A History of Hydrodynamics from the Bernoullis to Prandtl* (Oxford: Oxford University Press, 2008).

Davis, John L., 'Weather Forecasting and the Development of Meteorological Theory at the Paris Observatory, 1853–1878', *Annals of Science*, 41:4 (1984), pp. 359–82.

Dettelbach, Michael, 'Humboldtian Science', in Nicholas Jardine, James A. Secord, and Emma C. Spary (eds.), *Cultures of Natural History* (Cambridge: Cambridge University Press, 1996), pp. 287–304.

Dry, Sarah, *Fishermen and Forecasts: How Barometers Helped Make the Meteorological Department Safer in Victorian Britain* (London: London School of Economics and Political Science, 2007), discussion paper nº 46.

Dry, Sarah, 'Safety Networks: Fishery Barometers and the Outsourcing of Judgement at the Early Meteorological Department', *The British Journal for the History of Science*, 42:1 (2009), pp. 35–56.

Fine, Gary Alen. *Authors of the Storm: Meteorologists and the Culture of Prediction* (Chicago: University of Chicago Press, 2007).

Fleming, James Rodger, *Meteorology in America, 1800–1870* (Baltimore: Johns Hopkins University Press, 1990).

Fleming, James Rodger and Janković, Vladimir, 'Revisiting Klima', *Osiris*, 26 (2011), pp. 1–15.

Forrest, Derek, *Francis Galton: The Life of a Victorian Genius* (London: Paul Elek, 1974).

Fox, Robert. *The Savant and the State: Science and Cultural Politics in Nineteenth-Century France* (Baltimore: Maryland: The John Hopkins University Press, 2012).

Friedman, Robert Marc, *Appropriating the Weather: Vilhelm Bjerknes and the Construction of a Modern Meteorology* (Ithaca, NY: Cornell University Press, 1989).

Friendly, Michael, 'The Golden Age of Statistical Graphics', *Statistical Science*, 29:4 (2008), pp. 502–35.

Gibbons, Carey, 'Grasping the Elusive: Victorian Weather Forecasting and Arthur Hughes's Illustrations for George Macdonald's *At the Back of the North Wind'*, in Nancy Rose Marshall (ed.), *Victorian Science and Imagery: Representation and Knowledge in Nineteenth-Century Visual Culture* (Pittsburgh, Pa.: University of Pittsburgh Press, 2021), pp. 79–110.

Gillham, Nicholas Wright, *A Life of Sir Francis Galton: From African Exploration to the Birth of Eugenics* (Oxford: Oxford University Press, 2001).

Golinski, Jan, *British Weather and the Climate of Enlightenment* (Chicago: University of Chicago Press, 2007).

Good, Gregory A., 'A Shift of View: Meteorology in John Herschel's Terrestrial Physics', in James Rodger Fleming, Vladimir Jankovic, and Deborah R. Coen (eds.), *Intimate Universality. Local and Global Themes in the History of Weather and Climate* (Sagamore Beach, MA: Watson Publishing International, 2006), pp. 35–67.

Goodman, Matthew, 'Proving Instruments Credible in the Early Nineteenth Century: The British Magnetic Survey and Site-Specific Experimentation', *Notes and Records: The Royal Society Journal of the History of Science*, 70 (2016), pp. 251–68.

Goss, Andrew, 'Introduction: An Imperial Turn in the History of Science', in Andrew Goss (ed.), *The Routledge Handbook of Science and Empire* (London and New York: Routledge, 2021), pp. 1–9.

Halford, Pauline, *Storm Warning. The Origins of the Weather Forecast* (Stroud: Sutton Publishing, 2004).

Harley, John Brian, 'Maps, Knowledge and Power', in Denis Cosgrove and Stephen Daniels (eds), *The Iconography of Landscape. Essays on the Symbolic Representation, Design and Use of Past Environments* (Cambridge: Cambridge University Press, 1988), pp. 277–312.

Hearn, Chester G., *Tracks in the Sea: Matthew Fontaine Maury and the Mapping of the Oceans* (Camden, ME: International Marine/McGraw-Hill, 2002).

Herschel, John Frederick William. *Instructions for Making and Registering Meteorological Observations in Southern Africa and other Countries in the South Seas, as Also at Sea* (London: Bradbury and Evans, 1835).

Hersey, Mark D. and Vetter, Jeremy, 'Shared ground: Between environmental history and the history of science', *History of Science,* 57:4 (2019), pp. 403–40.

Howe, Joshua P. and Sutter, Paul S., *Making Climate Change History: Documents from Global Warming's Past* (Seattle: University of Washington Press, 2017).

Iafrate, Luigi, 'Padre Angelo Secchi: l'astronomo italiano che inventò le previsioni del tempo', *Nova Historica: rivista internazionale di storia*, 36 (2011), pp. 110–30.

Jankovic, Vladimir, 'Ideological Crests versus Empirical Troughs: John Herschel's and William Radcliffe Birt's Research on Atmospheric Waves, 1843–50', *British Journal for the History of Science*, 31 (1998), pp. 21–40.

Jankovic, Vladimir, *Reading the Skies: A Cultural History of English Weather, 1650–1820* (Chicago: University of Chicago Press, 2001).

Jankovic, Vladimir, *Confronting the Climate: British Airs and the Making of Environmental Medicine* (New York: Palgrave Macmillan Ltd., 2010).

Javelle, Jean-Pierre, Rochas, Michel, Pastre, Claude, Beaurepaire Michel and Jacomy, Bruno, *La météorologie, du baromètre au satellite. Mesurer l'atmosphère et prévoir le temps* (Lausanne: Éditions Delachaux et Niestlé, 2000).

Khrgian, Aleksandr Khristoforovich, *Meteorology: A Historical Survey* (Jerusalem: Israel Program for Scientific Translations, 1970), 2nd ed.

Kutzbach, Gisela, *The Thermal Theory of Cyclones: A History of Meteorological Thought in the Nineteenth Century* (Boston: American Meteorological Society, 1979).

Lawrence-Mathers, Anne, *Medieval Meteorology. Forecasting the Weather from Aristotle to the Almanac* (Cambridge: Cambridge University Press, 2020).

Lequeux, James, *Le Verrier–Magnificent and Detestable Astronomer* (New York: Springer, 2013).

Livingstone, David N., *Putting Science in its Place: Geographies of Scientific Knowledge* (Chicago: University of Chicago Press, 2003).

Locher, Fabien, 'L'empire de l'astronome: Urbain Le Verrier, l'Ordre et le Pouvoir', *Cahiers d'histoire. Revue d'histoire critique*, 102 (2007), pp. 33–48.

Locher, Fabien, *Le savant et la tempête. Étudier l'atmosphère et prévoir le temps au XIX siècle* (Rennes: Presses Universitaires de Rennes, 2008).

Locher, Fabien, 'Atmosphere of Globalisation. Depressions, the Astronomer and the Telegraph (1850–1914)', *Revue d'histoire moderne et contemporaine*, 56:4 (2009), pp. 77–103.

Ludlam, Frank Henry, *The Cyclone Problem: A History of Models of the Cyclonic Storm* (London: Imperial College of Science and Technology, 1967).

Mahony, Martin and Caglioti, Angelo Matteo, 'Relocating Meteorology', *History of Meteorology*, 8 (2017), pp. 1–14.

Mahony, Martin, 'The 'Genie of the Storm': Cyclonic Reasoning and the Spaces of Weather Observation in the Southern Indian Ocean, 1851–1925', *British Journal for the History of Science*, 51:4 (2018), pp. 607–33.

Mahony, Martin, 'Meteorology and Empire', in Andrew Goss (ed.), *The Routledge Handbook of Science and Empire* (London and New York: Routledge, 2021), pp. 47–58.

Mahony, Martin and Enfield, Georgina H., 'Climate and Colonialism', *Wiley Interdisciplinary Reviews: Climate Change*, 9:2 (2018), e510 https://doi.org/10.1002/wcc.510

Mercer, Harriett and Simpson, Thomas, 'Imperialism, Colonialism, and Climate Change Science', *WIREs Climate Change*, 14(2023), e851. https://doi.org/10.1002/wcc.851

Monmonier, Mark, *Air Apparent. How Meteorologists Learned to Map, Predict, and Dramatize Weather* (Chicago: The University of Chicago Press, 1999).

Morgan, Ruth, 'Prophecy and Prediction: Forecasting Drought and Famine in British India and the Australian Colonies', *Global Environment,* 13:1 (2020), pp. 95–132.

Naylor, Simon, 'Log Books and the Law of Storms: Maritime Meteorology and the British Admiralty in the Nineteenth Century', *Isis*, 106:4 (2015), pp. 771–97.

Naylor, Simon, 'Weather Instruments all at Sea: Meteorology and the Royal Navy in the Nineteenth Century', in Fraser MacDonald, Charles W.J. Withers (eds.), *Geography, Technology and Instruments of Exploration* (London: Routledge, 2015), pp. 77–96.

Naylor, Simon, *The Observatory Experiment. Meteorology in Britain and Its Empire* (Cambridge: Cambridge University Press, 2024).

Naylor, Simon, and Goodman, Matthew, 'Atmospheric Empire: Historical Geographies of Meteorology at the Colonial Observatories', in Martin Mahony and Samuel Randalls (eds.), *Weather, Climate, and the Geographical Imagination* (Pittsburgh, PA: University of Pittsburgh Press, 2020), pp. 25–42.

Nebeker, Frederik. *Calculating the Weather. Meteorology in the 20th-Century* (San Diego: Academic Press, 1995).

Nicolson, Malcolm, 'Alexander von Humboldt and the Geography of Vegetation,' in Nicholas Jardine and Andrew Cunningham (eds.) *Romanticism and the Sciences* (Cambridge: Cambridge University Press, 1996), pp. 169–88.

Pearson, Karl, *Life, Letters and Labours of Francis Galton* (Cambridge: Cambridge University Press, 1940–1930), 3 vols.

Porter, Theodore M., *The Rise of Statistical Thinking 1820–1900* (Princeton: Princeton University Press, 1986).

Prévost, Jean-Guy and Beaud, Jean-Pierre, *Statistics, Public Debate and the State, 1800–1945: A Social, Political and Intellectual History of Numbers* (London: Pickering & Chatto, 2012).

Pyenson, Lewis, *Cultural Imperialism and Exact Sciences: German Expansion Overseas, 1900–1930* (New York: Lang, 1985).

Pyenson, Lewis, 'Functionaries and Seekers in Latin America: Missionary Diffusion of the Exact Sciences, 1850–1930', *Quipu*, 2:3 (1985), pp 387–420.

Pyenson, Lewis, *Empire of Reason: Exact Sciences in Indonesia, 1840–1940* (Leiden: Brill, 1989).

Pyenson, Lewis, 'Pure Research, Jesuit Institution, and Metropolitan Ambitions: The Evolution of French Policies Overseas, 1880–1940', in *Chine et Europe: Évolution et particularités des rapports est-ouest du XVIe au XXe siècle. Actes du IVe Colloque International de Sinologie de Chantilly* (Paris: Institut Ricci, 1991), pp. 249–71.

Pyenson, Lewis, *Civilizing Mission. Exact Sciences and French Overseas Expansion, 1830–1940* (Baltimore: The Johns Hopkins University Press, 1993).

Raj, Kapil, *Relocating Modern Science: Circulation and the Construction of Knowledge in South Asia and Europe, 1650–1900* (Basingstoke: Palgrave Macmillan, 2007).

Reardon, Bernard M. G., *Religious Thought in the Victorian Age*. 2nd ed. (London: Longman, 1995).

Reidy, Michael S., 'From the Oceans to the Mountains: Spatial Science in an Age of Empire', in Jeremy Vetter (ed.), *Knowing Global Environments* (New Brunswick, N.J.: Rutgers University Press, 2011), pp. 17–38.

Reidy, Michael S., *Tides of History: Ocean Science and Her Majesty's Navy* (Chicago: University of Chicago Press, 2008).

Robinson, Arthur H., *Early Thematic Mapping in the History of Cartography* (Chicago, IL: The University of Chicago Press, 1982).

Robinson, Arthur H. and Pentchenik, Barbara Bartz, *The Nature of Maps: Essays towards Understanding Maps and Mapping* (Chicago: Chicago University Press, 1976).

Robinson, Arthur H. and Wallis, Helen, 'Humboldt's Map of Isothermal Lines: A Milestone in the History of Thematic Cartography', *Cartographic Journal*, 2 (1967), pp. 119–23.

Roy, Tirthankar, 'The Law of Storms': European and Indigenous Responses to Natural Disasters in Colonial India, *c.* 1800–1850', *Australian Economic History Review*, 50:1 (2010), pp. 6–22.

Schneider-Carius, Karl, *Weather Science, Weather Research: History of Their Problems and Findings from Documents during Three Thousand Years* (New Delhi: Indian National Scientific Documentation Centre, 1975); originally published as *Wetterkunde,*

Wetterforschung: Geschichte ihrer Probleme und Erkenntnisse in Dokumenten aus drei Jahrtausenden (Freiburg: Verlag Karl Albert, 1955).

Sheynin, Oscar Boris, 'On the History of the Statistical Method in Meteorology', *Archive for History of Exact Sciences*, 31:1 (1984), pp. 53–95.

Stigler, Stephen M., *The History of Statistics. The Measurement of Uncertainty before 1900* (Cambridge, Mass.: The Belknap Press of Harvard University Press, 1986).

Symons, George J., 'History of the English Meteorological Societies', *Quarterly Journal of the Royal Meteorological Society,* 7 (1881), pp. 66–107.

Topham, Jonathan, 'Science and Popular Education in the 1830s: The Role of the Bridgewater Treatises', *British Journal for the History of Science,* 15 (1992), pp. 397–430.

Turner, Frank, *Between Science and Religion: The Reaction to Scientific Naturalisn in Later Victorian England* (New Haven, CT: Yale University Press, 1974).

Turner, Frank, 'The Victorian Conflict between Science and Religion: A Professional Development', *Isis,* 69 (1978), pp. 356–76.

Udías, Agustín, *Searching the Heavens and the Earth: The History of Jesuit Observatories* (Boston: Kluwer, 2003).

Udías, Agustín, *Jesuit Contribution to Science: A History* (New York: Springer, 2015).

Walker, J. Malcolm, *History of the Meteorological Office* (Cambridge: Cambridge University Press, 2012).

Waller, John C., 'Gentlemanly Men of Science: Sir Francis Galton and the Professionalization of the British Life-Sciences', *Journal of the History of Biology*, 34 (2001), pp. 83–114.

Wheeler, Dennis and Wilkinson, Clive, 'The Determination of Logbook Wind Force and Weather Terms: The English Case', *Climatic Change*, 73 (2005), pp. 57–77.

Whitnah, Donald R., *A History of the United States Weather Bureau* (Urbana: University of the Illinois Press, 1961).

Williamson, Fiona, 'Uncertain Skies: 'Forecasting' Typhoons in Hong Kong c. 1874–1906', *Quaderni Storici*, 52:3 (2017), 777–802.

Williamson, Fiona, 'Just Doing their Job: The Hidden Meteorologists of Colonial Hong Kong c.1883–1914', *British Journal for the History of Science*, 54:3 (2021), pp. 341–359.

Winter, Alison, 'Compasses All Awry': The Iron Ship and the Ambiguities of Cultural Authority in Victorian Britain', *Victorian Studies*, 38 (1994), pp. 69–98.

Wright, Sarah and Tofa, Matalena, 'Weather Geographies: Talking About the Weather, Considering Diverse Sovereignties', *Progress in Human Geography,* 45:5 (2021), https://doi.org/10.1177/0309132520982

Young, Donald F., *A Brief Introduction to Fluid Mechanics*, 5th ed. (Hoboken, NJ: Wiley, 2011).

INDEX

Note: Page numbers in **bold** indicate tables and page numbers in *italics* indicate figures.